Ashraf M.T. Elewa
Predation in Organisms
A Distinct Phenomenon

Ashraf M.T. Elewa
Editor

Predation in Organisms
A Distinct Phenomenon

with 48 Figures and 9 Tables

Springer

Professor Ashraf M.T. Elewa
Minia University
Faculty of Science
Department of Geology
61519 Minia
Egypt

Email: aelewa@link.net

Library of Congress Control Number: 2006932817

ISBN-10   3-540-46044-6 Springer Berlin Heidelberg New York
ISBN-13   978-3-540-46044-2 Springer Berlin Heidelberg New York

This work is subject to copyright. All rights are reserved, whether the whole or part of the material is concerned, specifically the rights of translation, reprinting, reuse of illustrations, recitation, broadcasting, reproduction on microfilm or in any other way, and storage in data banks. Duplication of this publication or parts thereof is permitted only under the provisions of the German Copyright Law of September 9, 1965, in its current version, and permission for use must always be obtained from Springer-Verlag. Violations are liable to prosecution under the German Copyright Law.

Springer is a part of Springer Science+Business Media
springer.com
© Springer-Verlag Berlin Heidelberg 2007

The use of general descriptive names, registered names, trademarks, etc. in this publication does not imply, even in the absence of a specific statement, that such names are exempt from the relevant protective laws and regulations and therefore free for general use.

Cover design: E. Kirchner, Heidelberg
Production: Almas Schimmel
Typesetting: camera-ready by the editor

Printed on acid-free paper   30/3141/as 5 4 3 2 1 0

Dedication

This book is dedicated to my advisor:

Prof. Dr. Richard Arthur Reyment of the Swedish Natural History Museum

Actually, he is one of many Christians who could understand Islam in its right way as a great religion inviting all peoples to **PEACE**

Foreword

André J. Veldmeijer

PalArch Foundation Amsterdam; Natural History Museum Rotterdam, The Netherlands, veldmeijer@palarch.nl

When Professor Ashraf Elewa asked me to start this volume on predator prey interactions, I felt privileged. As a palaeontologist, I came in contact with Professor Ashraf Elewa when corresponding on his previous book on morphometry. Currently, Professor Ashraf Elewa works at Minia University in Egypt, as president of the Palaeontology Group of the Geology Department of the Faculty of Science. Egypt, including Amarna at the opposite side of the river Nile of Minya, also happens to be the focus of my archaeological research on ancient Egyptian leatherworks. It's a small world…..

The science of palaeontology has changed considerably over the last few decades. The use of powerful techniques and high tech tools such as X-rays and CT-scanning enables the palaeontologist access to information previously not within reach. Furthermore, scientists look more and more at other sciences, borrowing whatever may give new impulse to their research. These developments have, for instance, resulted in the extensive use of cladistics, but also provoked a shift in palaeontology from the more descriptive way of the early pioneers towards a more 'experimental' approach nowadays. More and more, palaeontologists focus on the animals itself, trying to reconstruct their way of life: behaviour, reproduction, food gathering etc. rather than regard the taxonomy as the ultimate goal. Ideally, a holistic approach could follow, combining data from other disciplines such as palaeobotany and (palaeo)geology but also biological sciences of the present-day fauna. To get insight in a complex system as predator-prey interactions, this is an absolute necessity or, as Richard K. Bambach put it in his foreword to Kelley's et al. 'Predator-Prey Interactions in the Fossil Record'(2003): "It will only be by compiling and evaluating data on predator-prey relations as they are recorded in the fossil record that we can hope to tease apart their role in the tangled web of evolutionary interaction over time."

The present volume is just such a work in its totality but within the various chapters as well. The myriad of topics discusses predation in both invertebrates and vertebrates, in a variety of ways and on various levels.

Examples of studies that combine fossil and modern-day animals are the study on feeding strategies of fossil Ostracoda compared to modern analogues, and a paper in which modern and fossil shells as indicator of biotic interactions are compared. A paper on biological control of mosquito populations shows that the focus is not entirely on the fossil record. A more theoretical paper deals with the evolutionary consequences of predation. Due to a biased fossil record, which favours fossilization of invertebrates, as remarked by Carpenter et al. (2005: 325): "Unequivocal evidence of predator-prey relationships in the vertebrate fossil record is rare owing to the vagaries of preservation and the difficulties of interpretation." This comes not entirely as a surprise given the better fossilization changes of invertebrates such as shells and the larger number of individuals. Nevertheless, the present volume includes various chapters in which feeding and predation in vertebrates are being discussed with a remarkable variety in topics, ranging from a study on predation in fossil eggs, predation tactics in flightless birds and non-avian dinosaurs to an overview of predator-prey interaction in pterosaurs. Not only will this book be of great value to invertebrate palaeontologist, it will also provide a challenge for those working in the field of vertebrate palaeontology.

References

Carpenter K, Sanders F, McWhinney LA, Wood L (2005) Evidence for Predator-Prey Relationships. Examples for Allosaurus and Stegosaurus. In Carpenter K (ed) The Carnivorous Dinosaurs. Bloomington/Indianapolis: Indiana University Press 325-350

Kelley PH, Kowalewski M, Hansen TA (eds) (2003) Predator-Prey Interactions in the Fossil Record. New York, Kluwer Academic/Plenum Publishers, series Topics in Geobiology 20

Table of Contents

1 An introduction to predation in organisms 1
   Ashraf M. T. Elewa 1
      References 5

2 Predation due to changes in environment: Ostracod provinciality at the Paleocene-Eocene thermal maximum in North and West Africa and the Middle East 7
   Ashraf M. T. Elewa 7
      2.1 Abstract 7
      2.2 Introduction 7
      2.3 Methodology 8
      2.4 Results 10
      2.5 Paleoenvironments 16
      2.6 Predation as a strong factor affecting ostracod abundances in the studied regions 20
      2.7 Acknowledgement 21
      References 21

3 Predation on Miocene ostracods of Wadi Um Ashtan, Mersa Matruh, Western Desert, Egypt 27
   Ashraf M. T. Elewa 27
      3.1 Abstract 27

3.2 Introduction 27
3.3 Predation and survival 28
3.4 Material for the present study 29
3.5 Echinoid drillholes 31
3.6 Holes by marine fungi 32
3.7 Discussion 33
3.8 Acknowledgement 36
References 36

4 Ostracod carnivory through time 39
I. Wilkinson[1], P. Wilby[1], P. Williams[2], D. Siveter[3] and J. Vannier[4] 39
4.1 Abstract 39
4.2 Introduction 39
4.3 Carnivory in modern Ostracods 40
4.3.1 Predation 40
4.3.2 Scavenging 42
4.3.3 Parasitism 43
4.4 Carnivory in the fossil record 44
4.4.1 Early Cretaceous Pattersoncypris micopapillosa from Brazil 45
4.4.2 Jurassic Juraleberis jubata from Russia 46
4.4.3 Triassic Triadocypris spitzbergensis from Spitzbergen 46
4.4.4 Carboniferous Eocypridina carsingtonensis from Central England 48
4.4.5 Early Silurian Colymbosathon ecplecticos from Herefordshire, England 49
4.4.6 Late Ordovician Myodoprimigenia fistuca from South Africa 50
4.4.7 Cambrian bradoriids and the origin of carnivory 50
4.5 Conclusions 53
4.6 Acknowledgements 53
References 53

5 Trophic relationships in crustacean decapods of a river with a floodplain 59
P. Collins[1,2,3], V. Williner[1,2] and F. Giri,[1,3] 59
5.1 Abstract 59
5.2 Introduction 60
5.2.1 Physical Environment: A river with a floodplain 61
5.2.2 Potential predators 62
5.2.3 Potential preys 63
5.3 Feeding ecology 66

    5.3.1 Selectivity 71
    5.3.2 Circadian rhythms 72
    5.3.3 Annual rhythms 76
  5.4 How does Crustacea Decapoda obtain its food? 76
  References 79
  Appendix 1 85

6 The role of predation in shaping biological communities, with particular emphasis to insects 87
  Panos V. Petrakis[1] and Anastasios Legakis[2] 87
  6.1 Abstract 87
  6.2 Predation and its types in insects 89
  6.3 Prey-predator interaction seen through models 90
  6.4 Predation in relation to competition, parasitism, cannibalism and size 95
    6.4.1 Competition 95
    6.4.2 Parasitism 97
    6.4.3 Size and predation 98
    6.4.4 Cannibalism 101
  6.5 The control of prey population by predators 103
  6.6 The relation of predation to biodiversity 106
  6.7 The chemical ecology of the prey-predator systems 110
  6.8 Predator Confusion Hypothesis 114
  6.9 Search Image Behaviour 114
  6.10 Sensory Exploitation Hypothesis 114
  6.11 Predator Interference Hypothesis 115
  6.12 Pest Release Hypothesis 115
  6.13 Optimal foraging theory 115
  6.14 Concluding remarks 116
  References 116

7 Biological control of mosquito populations: An applied aspect of pest control by means of natural enemies 123
  Anna Samanidou–Voyadjoglou[1], Vassilios Roussis[2] and Panos V. Petrakis[3] 123
  7.1 Abstract 123
  7.2 Introduction 124
  7.3 The basic suppression agents of mosquitoes in natural and anthropogenic ecosystems 127
  7.4 The problem posed by synthetic chemical treatments and some toxins from biological preparations 133
  7.5 The chemical basis of predation on mosquitoes 137

7.6 Towards an integrated system of mosquito control 139
7.7 Acknowledgements 142
References 143

8 A case for cannibalism: Confamilial and conspecific predation by naticid gastropods, Cretaceous through Pleistocene of the United States Coastal Plain 151
Patricia H. Kelley[1] and Thor A. Hansen[2] 151
8.1 Abstract 151
8.2 Introduction 152
8.3 Materials and methods 154
8.4 Results 157
8.5 Discussion 161
8.6 Conclusions 166
8.7 Acknowledgements 167
References 167

9 On models for the dynamics of predator-prey interaction 171
Richard A. Reyment 171
9.1 Abstract 171
References 175

10 Evolutionary consequences of predation: avoidance, escape, reproduction, and diversification 177
R. Brian Langerhans 177
10.1 Abstract 177
10.2 Introduction 178
10.3 Solving the problem of being eaten: avoidance and escape 180
10.4 Predator avoidance: winning without a fight 182
   10.4.1 Steering clear of a predator's realm: avoiding a predator's sensory field 183
   10.4.2 Hiding in plain sight: avoiding detection within a predator's sensory field 184
10.5 Predator escape: prey fight back 185
   10.5.1 Don't even think about it: attack deterrence 185
   10.5.2 Catch me if you can: capture deterrence 187
   10.5.3 Go ahead, try and eat me: consumption deterrence 189
   10.5.4 Multitasking prey: all-purpose antipredator traits 190
10.6 Reproductive strategies: transcending predators through life history traits 192
   10.6.1 Know when to hold 'em, know when to fold 'em: reproductive timing 193

10.6.2 Putting all your eggs in one basket and flooding the market: reproductive effort 193
10.7 Predators spawn phenotypic diversity of prey: plasticity, divergence, and speciation... 195
    10.7.1 To induce or not to induce: tradeoffs can drive predator-induced plasticity... 196
    10.7.2 Divergent selection between predator regimes: evolutionary divergence among prey 197
    10.7.3 Divergent selection within predator regimes: the search for enemy-free space... 202
    10.7.4 Divergent selection involving other selective agent(s): predation as a context shift.. 203
    10.7.5 Predation as a driver of speciation: eating individuals, spitting out species 204
10.8 Conclusions and future directions 205
10.9 Acknowledgements 206
References 206

11 Predation impacts and management strategies for wildlife protection.. 221
Michael J. Bodenchuk[1] and David J. Hayes[2] 221
  11.1 Abstract 221
  11.2 Introduction 221
  11.3 Predator-prey relationships... 223
  11.4 Habitat v. predators 227
  11.5 Predation and management effects... 227
    11.5.1 Deer... 229
    11.5.2 Pronghorn antelope 230
    11.5.3 Sage grouse 232
    11.5.4 Ring-necked pheasants and turkeys 234
    11.5.5 Waterfowl.. 235
  11.6 Factors affecting predation rates 236
    11.6.1 Habitat factors 236
    11.6.2 Prey factors 238
    11.6.3 Predator factors 239
  11.7 Specific strategies... 242
    11.7.1 Non-lethal strategies 242
    11.7.2 Mule deer protection strategies 243
    11.7.3 Pronghorn antelope protection strategies 245
    11.7.4 Bighorn sheep protection strategies 245
    11.7.5 Sage grouse protection strategies 246
    11.7.6 Ring-necked pheasant protection strategies 247

    11.7.7 Waterfowl and shore bird protection strategies 247
    11.7.8 Endangered and threatened species protection strategies 248
    11.8 Conclusion..250
    11.9 Acknowledgement..252
    References 252

12 Invasive Predators: a synthesis of the past, present, and future ..265
    William C. Pitt[1] and Gary W. Witmer[2] 265
    12.1 Abstract 265
    12.2 Introduction 266
    12.3 Species profiles..268
        12.3.1 Mammals..269
        12.3.2 Birds..275
        12.3.3 Reptiles 277
        12.3.4 Amphibians...279
        12.3.5 Fish..281
    12.4 Regulation of invasive species 282
    12.5 Priorities of invasive species 283
    References 285

13 Predator-prey interaction of Brazilian Cretaceous toothed pterosaurs: a case example..295
    André J. Veldmeijer[1], Marco Signore[2] and Enrico Bucci[3]..................295
    13.1 Abstract 295
    13.2 Introduction 295
    13.3 Pterosaurs as prey..296
    13.4 Pterosaurs as predator..297
    13.5 Fishing technique: the model...300
    13.6 Conclusions 305
    13.7 Acknowledgements 306
    References 307

Index 309

1 An introduction to predation in organisms

Ashraf M. T. Elewa

Geology Department, Faculty of Science, Minia University, Egypt,
aelewa@link.net

Predation is considered as one of the distinct phenomena related to relations between species to each other on the Earth. According to the Wikipedia, the free encyclopedia, predation is an interaction between organisms (animals) in which one organism captures and feeds upon another called the prey. Some others consider predation as an interaction between two species in which one of them gains and the other loses. As to me, I define predation as a phenomenon of "Antagonism".

There are several predators living on the Earth, ranging in size from micro-creatures, like ostracods, to big mammals like lions and tigers. Of course, we, humans, think of these big cats as well as reptiles, like crocodiles and snakes, as typical predators. However, spiders, centipedes, most lizards and turtles, and frogs are also voracious predators, some of them are dangerous to humans (see the Wikipedia: http://en.wikipedia.org/wiki/Predation). In general, predation is widespread not only in wildlife but also in marine environments where big fishes eat small fishes and other small organisms of the sea.

Anyhow, some important questions arise to mind when discussing this subject: what is behind predation? Why some predators do not benefit from their prey after killing them? Are there genetic origins of this antagonism between organisms? Why some female organisms kill their males after completion of sex? How can we avoid predation? We, editor and contributors, tried to answer these questions through the study of many aspects of predation as well as some relations between species to each other.

In the following I am presenting a summary of the most important books on predation in the last forty years.

Since 1969, when James Frederick Clarke published his book "Man is the prey", predation has taken a substantial consideration by many scientists on different groups of organisms. Chimpanzees (Teleki 1973), arthropods (Hassell 1978), fishes (Noakes 1983), red foxes and breeding ducks (Sargeant et al. 1984), coyote (Leydet 1988), wolves and black-tailed deers (Atkinson and Janz 1994), reptiles (Cloudsley-Thompson

1994), fish-eating birds (Russell 1996), chimpanzee and red colobus (Stanford 1998), ladybird beetles (Dixon 2000), barn owls (Taylor 2004). However, the book by Michael Bright (2002) on the "Man-eaters" is the most interesting to read. This book describes horrifying true stories of savage, flesh-eating predators and their human prey. Nonetheless, this book is mentioning just stories about predators of human without discussing scientific hypotheses related to this phenomenon.

Geza Teleki has spent two years observing wild chimpanzees at very close quarters in the Gombe National Park of Tanzania. He has compiled his report on predatory behavior, based in part upon a decade of observations by a research team living in the park, but primarily upon numerous episodes he observed since early 1968. For details on the predatory behavior of wild chimpanzees see Teleki (1973).

Sargeant et al. (1984) stated that the average annual take of adult ducks by foxes in the midcontinent area was estimated to be about 900,000. This estimate included both scavenged and fox-killed ducks, as well as ducks taken after the denning season. Fox impact on midcontinent ducks was greatest in eastern North Dakota where both fox and duck densities were relatively high. Predation in that area was likely increased by environmental factors, especially intensive agriculture that concentrated nesting and reduced prey abundance. Predation by red foxes and other predators severely reduces duck production in the midcontinent area. Effective management to increase waterfowl production will necessitate coping with or reducing high levels of predation.

Atkinson and Janz (1994) concluded that black-tailed deer provided most of the wolf diet on Vancouver Island, while elk and beaver were secondary prey sources. They added, furthermore, that mortality factors, other than wolf predation (cougar, black bear, hunter harvest, winter kill), were not responsible for initiating deer declines in their study area.

Stanford (1998) argued that the chimpanzees are familiar enough--bright and ornery and promiscuous. But they also kill and eat their kin, in this case the red colobus monkey, which may say something about primate, even hominid, evolution. This book is based on a first long-term field study of predator-prey relationship, involving two wild primates, documents a six-year investigation into how the risk of predation molds primate society.

Dixon (2000) clued that ladybird beetles have long been used in the biological control of insect pests, but as with many biological control agents, they have not always been succesful. Dixon's book explores the biology and interactions of these predators and their prey to develop a better understanding of what makes a successful predator for biological control.

Reyment and Elewa, in Kelley et al. (2002), concluded that identifiable evidence for predation on fossil ostracods is limited to the drilling gastropods of the families Naticidae and Muricidae. It is worth to mention that this edited book, which is titled "predator-prey interactions in the fossil record", covers several aspects of predation that occurred during the geological times, nevertheless, most of its chapters represent predation in the past with little knowledge on recent organisms.

Taylor (2004), in his book titled "Barn owls predator-prey relationships and conservation", discusses the relationship between barn owls, their prey and prospects for conservation.

To my surprise, the prey could be a predator not only on smaller and weaker organisms but also on big predators. This is the case in Ostracoda, where some ostracod species (especially *Cypretta kawatai* Sohn and Kornicker) can be effective predators in laboratory experiments on 1- to 3-day-old *Planorbis glabratus* Say (= *Biomphalaria glabrata*), a vector snail of the blood fluke that causes the tropical and subtropical disease schistosomiasis (see Sohn and Kornicker 1972, 1975). The question arises to mind is "how these tiny creatures can do that?" It seems that ostracods so irritated adult snails that the snails left the water, then weakened, and either died or returned to the water and were killed by ostracods (Kawata 1971).

Dr. Richard Reyment of the Swedish Natural History Museum declared that Dr. Emma Johnston, Biology faculty of Sydney University, is leading a group dealing with environmental degeneration in Sydney Harbour. Her group has identified an important predator on barnacles, namely a group of flatworms. This is of potential interest for the ostracod world. Just how the worms manage to obtain access to the soft parts of the barnacles is currently under investigation.

Dr. Anne Cohen, of the Bodega Marine Laboratory (University of California at Davis), informed me that she once saw some pet freshwater ostracods that she kept in a jar on her desk kill snails that she put in the jar. They would nibble away at the snail foot making the snail withdraw its foot and eventually the snail died - possibly from being unable to move and feed. The snails were of course much larger than the little ostracods.

At any case, as it is usual, most of the published books on predation just focus on limited groups of organisms and could not answer several questions concerning predation philosophy and predator-prey interactions.

Conversely, the present book introduces diverse organisms ranging from small invertebrates to mammals and includes the most popular subjects discussing predation in insects, mammals, fishes, ostracods and molluscs. Furthermore, our book discusses predation as a phenomenon predominant not only in recent life but also in the fossil record. All over all, I selected

an excellent group of experts working on this phenomenon to discuss the following main topics that are not collected together, I think, in any published material in the past:
1. What is behind predation in organisms?
2. Factors affecting predation in organisms.
3. Predator-prey interaction.
4. The distinct role of predation in keeping the environmental equilibrium.
5. Examples of predation in the fossil record.
6. Examples of predation in marine and non-marine organisms.
7. Avoiding and preventing predation.
8. Interference between predators.
9. Herbivory, carnivory, cannibalism, parasitoidism, and parasitism.
10. Scavenging compared to predation.
11. Future trends in this subject.

On the whole, predation in organisms is one of the most popular topics for students, at all levels, and professionals alike. Moreover, this effort represents an up to date project ideas as well as a valuable synopsis of the advancement of this branch of learning.

I would like to convey my deep gratefulness to all people who participated in the achievement of this book. I especially express my thanks to Dr. André Veldmeijer (Natuurmuseum Rotterdam; the founder of the PalArch Foundation Amsterdam, The Netherlands) for writing the foreword, as well as contributing one chapter and reviewing one chapter for this book. The reviewers, an outstanding group of professionals, are also appreciated for their reviews of chapters in this book (Vamosi from Canada; Ulla Schudack from Germany; Colin from France; Giokas, Savopoulou, Sfenthourakis and Tzakou from Greece; Gherardi from Italy; Meijer and de Vos from The Netherlands; Marçal from Portugal; Reyment from Sweden; Purnel from UK; Kelley, Mason, Polly, Roopnarine and West from USA; arranged alphabetically according to their countries). The contributors are sincerely acknowledged for offering their time in preparing their chapters for this book. As usual, exceptional appreciations are due to the publishers of Springer-Verlag for their incessant facilities they offered during editing this book. Finally, I am much indebted to the people of Minia University of Egypt.

References

Atkinson, KT, Janz D (1994) Effect of Wolf Control on Black-Tailed Deer in the Nimpkish Valley on Vancouver Island, Ministry of Environment, Lands & Parks, Wildlife Branch, Bulletin No. B-73

Bright M (2002) Man-Eaters. St Martins Pr

Clarke JF (1969) Man is the prey. Stein and Day, New York

Cloudsley-Thompson JL (1994) Predation and Defense Amongst Reptiles. Serpents Tale Publishers

Craig BS (1998) Chimpanzee and Red Colobus: The Ecology of Predator and Prey, Harvard Univ Pr

Dixon AFG (2000) Insect Predator-Prey Dynamics: Ladybird Beetles and Biological Control, Cambridge Univ Pr

Hassell MP (1978) The Dynamics of Arthropod Predator-Prey Systems, Princeton Univ Pr

Kawata K (1971) Survival studies of *Biomphalaria glabratus* in polluted waters. Rockefeller Foundation Grant GA MNS 6846, Technical Report

Kelley PH, Kowalewski M, Hansen TA (eds) (2002) Predator-Prey Interactions In The Fossil Record. Kluwer Academic/Plenum Publishers, New York

Leydet F (1988) The Coyote: Defiant Songdog of the West, Univ Oklahoma Pr

Noakes DLG (1983) Predators and Prey in Fishes: Proceedings of the 3rd Biennial Conference on the Ethology and Behavioral Ecology of Fishes, Held at Normal, Illinois, U.S.A., May 19-22, 1981, W Junk

Reyment RA, Elewa AMT (2002) Predation by drills on Ostracoda. In Kelley PH, Kowalewski M, Hansen TA (eds) (2002) Predator-Prey Interactions In The Fossil Record. Kluwer Academic/Plenum Publishers, 4: 93-111, New York

Russell I (1996) Assessment of the Problem of Fish-Eating Birds in Inland Fisheries in England and Wales: Summary Report: Report to the Ministry of Agriculture, Fisheries and Food (MAFF Project VC0104), Directorate of Fisheries Research, Fisheries Laboratory

Sargeant AB, Allen SH, Eberhardt RT (1984) Red fox predation on breeding ducks in midcontinent North America. Wildlife Monographs 89:1-41. Jamestown, ND: Northern Prairie Wildlife Research Center Online. http://www.npwrc.usgs.gov/resource/mammals/redfox/redfox.htm (Version 02JUN99)

Sohn IG, Kornicker LS (1972) Predation of schistosomiasis vector snails by Ostracoda (Crustacea). Science 175: 1258-1259

Sohn IG, Kornicker LS (1975) Variation in predation behavior of ostracode species on schistosomiasis vector snails. Bull Amer Paleont 65 (282): 217-223

Taylor I (2004) Barn Owls: Predator-Prey Relationships and Conservation Cambridge Univ Pr

Teleki, G (1973) The Predatory Behavior of Wild Chimpanzees. Lewisburg, Bucknell Univ Pr

2 Predation due to changes in environment: Ostracod provinciality at the Paleocene-Eocene thermal maximum in North and West Africa and the Middle East

Ashraf M. T. Elewa

Geology Department, Faculty of Science, Minia University, Egypt, aelewa@link.net

2.1 Abstract

This study successfully interprets the ostracod provinciality associated with the Paleocene-Eocene thermal maximum (PETM) in North and West Africa and the Middle East. The results indicate that the West African Province (after Elewa 2002b) demonstrates a distinct local turnover in ostracod assemblages, due to predation activity associated with the high temperature at the Paleocene-Eocene thermal maximum. Simultaneously, the South Tethyan Province (after Elewa 2002b) exemplifies a minor faunal change in ostracods as a result of oxygen depletion during the late Paleocene-early Eocene interval. While, Egyptian ostracod assemblages show evidence of faunal adaptability at the Paleocene-Eocene boundary. On the whole, a reduced ostracod migration rate through the Trans-Saharan Seaway has been detected between the WAP, from one side, and the STP and Egypt, from the other side, just at the end of the late Paleocene and the earliest Eocene times. Notably, *Cytherella* showed distinct reaction against high predation activity that associated the oxygen depletion occurred during the PETM in the South Tethyan Province.

Keywords: Predation, ostracods, PETM, Africa, Middle East.

2.2 Introduction

The ostracod faunas of North and West Africa and the Middle East were the subject of numerous studies, starting from the sixties of the Twentieth

Century, when Reyment presented to science his published works on the Upper Cretaceous/Lower Tertiary ostracods of Nigeria and Niger Delta (e.g. Reyment and Reyment 1959; Reyment 1960, 1963, 1966a,b). Since then, several publications have dealt with ostracods of this wide area of the world (Apostolescu 1961; Barsotti 1963; Salahi 1966; Esker 1968; Grekoff 1969; Siddiqui 1971; Bassiouni et al. 1977; Haggag 1979; Yassini 1979; Al Furaih 1980; Reyment and Reyment 1980; Reyment 1981; Shamah 1981; Al Sheikhly 1981; Aref 1982; Boukhary et al. 1982; Donze et al. 1982; Foster et al. 1983; Al Sheikhly 1985; Carbonnel 1986; Peypouquet et al. 1986; Damotte and Fleury 1987; Reyment et al. 1987; Carbonnel and Johnson 1989; Helal 1990; Bassiouni and Luger 1990; Carbonnel et al. 1990; El Sogher 1991; Honigstein et al. 1991; Guernet et al. 1991; El Waer 1992; Ismail 1992; Whatley and Arias 1993; Keen et al. 1994; Elewa 1994; Elewa and ishizaki 1994; Aref 1995; Damotte 1995; Honigstein and Rosenfeld 1995; Gammudi and Keen 1996; Ismail 1996; Colin et al. 1998; Elewa 1998; Sarr 1998, 1999; Bassiouni and Elewa 1999; Elewa et al. 1999; Bassiouni and Morsi 2000; Shahin 2000; Honigstein et al. 2002; Elewa 2002a).

Consequently, Elewa (2002b), in his study of 103 ostracod species and subspecies from Maastrichtian to lower Eocene localities in North and West Africa and the Middle East, arrived at distinguishing two main provinces (WAP or the West African province and STP or South Tethyan Province) and Egypt, or GAT/ATT, in-between with its characteristic ostracod faunas that are more similar to STP. Furthermore, he concluded that the P/E boundary shows no turnover but faunal changes resulting from lateral migration of certain ostracod genera.

In fact, Elewa's study did not yield itself insight into the control mechanisms for ostracod provinciality and for any changes across the P/E boundary, and hence there is a need to the present study. I herein try to spot the light on the reasons led to late Paleocene-early Eocene ostracod provinciality, with reference to the most affecting environmental factors on the distribution and migration of ostracod assemblages in the studied regions during that time.

2. 3 Methodology

The regions covered in the present study includes countries from West and North Africa and the Middle East, and extends from Senegal in the west to Pakistan in the East with special allusion to Egypt (Fig. 1).

The 103 ostracod species and subspecies of Elewa (2002b) have been re-examined and more data have been investigated to accomplish the goal of the present study. Elewa (2002b) selected 46 species for faunal analysis, after excluding both rare species and countries with ostracod species that cannot be compared with those of Egypt (e.g. Saudi Arabia, Iraq and Pakistan).

Although the Paleocene-Eocene thermal maximum is not yet exactly defined in West Africa, however I use here the term PETM for the West African Province according to the frequent occurrences of ostracod taxa that could tolerate high temperature. Thus, this term is applied here in its broad sense as Paleocene/Eocene transition. Seemingly, the predation activity of drills on ostracods of the WAP is correlated positively with the increase in temperature. Even though, further faunal studies supported with geochemical data are urgently needed to establish the PETM in this region.

Statistical analyses of important ostracod families and genera have been compiled on those 46 species, and the cluster analysis of Elewa's work is re-considered, in combination with the ostracod statistics, to find out supplementary results to the ostracod data in hand. Table 1 shows distribution of the 46 analyzed ostracod species in the significant countries of North and West Africa and the Middle East. Finally, cluster analysis based on the Jaccard coefficient of similarity (the paired-group method) was applied to find out similarities between the 13 examined countries.

The used statistical software in the present study is STATISTICA for Windows, release 4.5, StatSoft Inc. (1993). The cluster analysis program is included with the PAST statistical package, version 1.14 of September 2003.

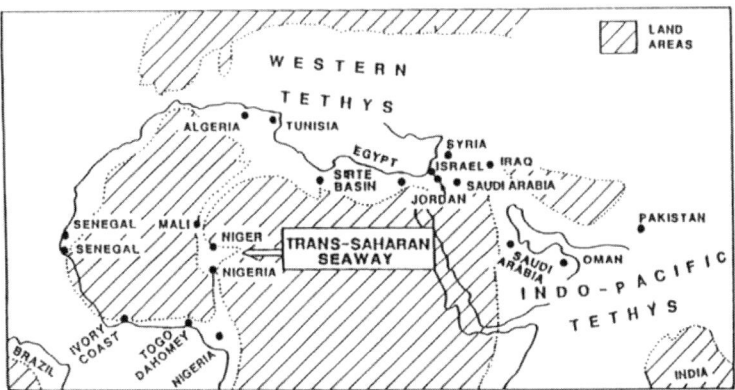

Fig. 1. Map showing the studied regions of North and West Africa and the Middle East, with a late Paleocene paleogeography (after Keen et al., 1994)

Table 1. Distribution of the analyzed 46 ostracod species in the significant countries of North and West Africa and the Middle East (after Elewa, 2002b)

Species	Senegal	Ivory Coast	Togo	SW Nigeria	NW Nigeria	Mali	Tunisia	Libya	Egypt	Jordan	Israel	Algeria	Niger
Leguminocythereis senegalensis	1	1	0	1	0	0	0	0	0	0	0	0	0
Buntonia virgulata	1	0	0	0	0	1	0	1	0	0	0	0	0
Reticulina sangalkamensis	1	0	0	0	0	0	0	1	1	0	0	0	0
Oertliella vesiculosa	1	0	0	0	1	1	0	0	1	0	0	0	0
Dahomeya alata alata	1	1	0	0	1	0	1	1	1	0	0	0	1
Soudanella laciniosa triangulata	1	0	0	1	0	0	1	1	1	1	0	0	0
Buntonia issabaensis	0	1	1	1	0	0	0	0	0	0	0	0	0
B. livida	0	1	1	1	0	0	0	0	0	0	0	0	0
Soudanella laciniosa laciniosa	0	1	1	1	0	0	0	0	0	0	0	0	0
Quadracythere lagaghiroboensis	0	1	0	1	1	0	0	0	0	0	0	0	0
Buntonia bopaensis	0	1	1	1	1	0	0	0	0	0	0	0	0
B. fortunata	0	1	1	1	1	0	0	0	0	0	0	0	0
B. sehouensis	0	1	1	0	1	1	0	1	0	0	0	0	0
Leguminocythereis lokossaensis	0	1	1	0	0	0	1	1	0	0	0	0	0
L. lagaghiroboensis	0	1	0	1	0	0	1	1	0	0	0	0	0
Trachyleberis modesta	0	1	0	0	1	1	0	1	0	0	0	0	0
Phalcocythere cultrata	0	1	0	0	1	1	0	1	1	0	0	0	0
Buntonia attitogoensis	0	0	0	0	0	0	1	1	1	0	0	0	0
Protobasslerites bopaensis	0	0	1	1	1	0	0	0	0	0	0	0	0
Limburgina sehouensis	0	0	1	1	1	0	0	0	0	0	0	0	0
Leguminocythereis bopaensis	0	0	1	0	1	1	0	1	0	0	0	0	0
Buntonia pulvinata	0	0	1	1	1	0	0	1	0	0	0	0	0
Isohabrocythere teiskotensis	0	0	1	1	1	0	0	1	0	0	0	0	1
Bairdia ilaroensis	0	0	1	0	1	1	0	0	1	0	0	0	1
Trachyleberis teiskotensis	0	0	1	1	1	0	0	1	0	0	0	0	0
Mauritsina teiskotensis teiskotensis	0	0	0	1	1	1	0	1	0	0	0	0	0
Buntonia tichittensis	0	0	0	0	1	0	1	0	1	0	0	0	0
Paracypris ? nigeriensis	0	0	0	1	1	1	0	0	1	0	0	0	0
Alocopocythere teiskotensis	0	0	0	0	1	1	0	1	0	0	0	0	0
Nucleolina tatteuliensis	0	0	0	0	1	1	0	1	1	0	0	0	1
Bythocypris eskeri	0	0	0	0	1	0	1	0	1	0	0	0	0
Paracypris jonesi	0	0	0	0	0	0	1	1	1	0	0	0	0
Reymenticosta bensoni	0	0	0	0	1	0	1	0	1	0	0	0	0
Cytheropteron lekefens	0	0	0	0	0	0	1	1	1	0	0	0	0
Mauritsina coronata	0	0	0	0	0	0	1	0	1	1	0	0	0
M. jordanica nodoreticulata	0	0	0	0	0	0	1	0	1	1	1	0	0
Protobuntonia nakkadii	0	0	0	0	0	0	1	0	1	1	0	0	0
Reticulina proteros	0	0	0	0	0	0	1	0	1	1	1	0	0
Ordoniya ordoniya	0	0	0	0	0	0	1	0	1	1	0	0	0
O. hasaensis	0	0	0	0	0	0	0	0	1	1	1	0	0
O. bulaqensis	0	0	0	0	0	0	1	0	1	0	1	0	0
Megommatocythere denticulata	0	0	0	0	0	0	1	0	1	0	1	1	0
Xestoleberis tunisiensis	0	0	0	0	0	0	1	0	1	0	0	1	0
Paragrenocythere cultrata	0	1	0	0	1	1	0	1	1	0	0	0	1
Dahomeya alata anteroglabrata	0	0	1	1	0	0	0	0	1	0	0	0	0
Uroleberis glabella	0	0	0	0	0	1	0	1	1	0	0	0	0

2. 4 Results

The dendrogram (Fig. 2), resulting from cluster analysis, led to distinguish three ostracod groups (STP, GAT/ATT, WAP). Examination of the ostracod species of these groups disclosed the following results (Table 2):

1. The West African Province (WAP) discloses species that are restricted to the late Paleocene;
2. The South Tethyan Province (STP) contains species that are mostly survived to the early Eocene, and
3. The Egyptian material (GAT/ATT) has much resemblance to the STP in having early Eocene survived species.

Distinguishing the environmental factors affected the ostracod distribution of these three groups could interpret the ostracod reactions at the Paleocene/Eocene thermal maximum (PETM) of the studied regions.

In general, the 46 species subjected to faunal analysis are belonging to 12 families in which Buntoniidae, Trachyleberididae and Campylocytheridae are the most common with more than 78% of the total number of species.

Quantitative analysis of ostracod families revealed that Trachyleberididae represents 54.3% of the total number of species in the STP (Fig. 3).

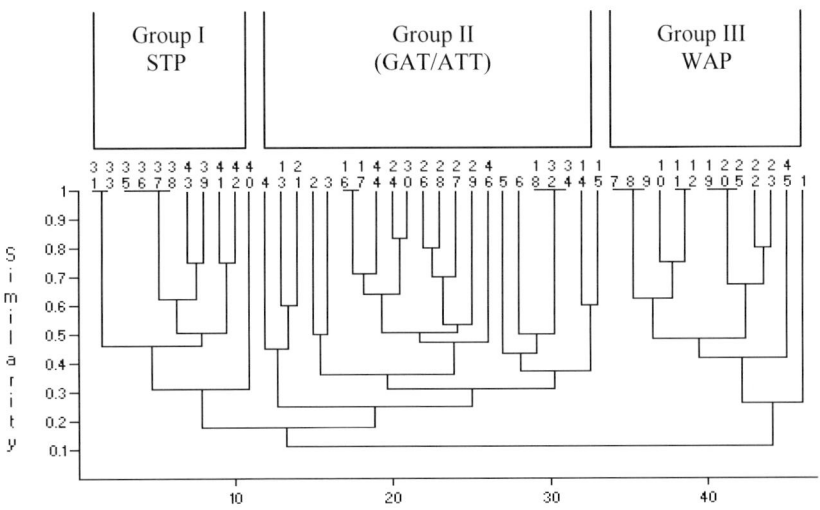

Fig. 2. . Dendrogram resulted from cluster analysis of the analyzed 46 ostracod species (after Elewa, 2002b)

The GAT/ATT (Fig. 4) is mostly occupied by Buntoniidae (22.7%), Campylocytheridae (22.7) and Trachyleberididae (18.2%). Buntoniidae dominates the WAP with 53.8% of the total number of species (Fig. 5).

Table 2. Last occurrences of the 46 analyzed ostracod species of the studied regions. P = late Paleocene; E = early Eocene

Province	Ostracod species	Last occurrence
STP	Bythocypris eskeri	E
	Reymenticosta bensoni	?E
	Mauritsina coronata	E
	M. jordanica nodoreticulata	E
	Protobuntonia nakkadii	P
	Reticulina proteros	E
	Ordoniya ordoniya	E
	O. hasaensis	E
	O. bulaqensis	E
	Megommatocythere denticulate	E
	Xestoleberis tunisiensis	E
GAT/ATT	Buntonia virgulata	E
	Reticulina sangalkamensis	E
	Oertliella vesiculosa	P
	Dahomeya alata alata	E
	Soudanella laciniosa triangulata	E
	Buntonia sehouensis	P
	Leguminocythereis lokossaensis	E
	L. lagaghiroboensis	?E
	Trachyleberis modesta	?E
	Phalcocythere cultrata	E
	Buntonia attitogoensis	E
	Leguminocythereis bopaensis	E
	Bairdia ilaroensis	E
	Mauritsina teiskotensis teiskotensis	P
	Buntonia tichittensis	E
	Paracypris ? nigeriensis	E
	Alocopocythere teiskotensis	P
	Nucleolina tatteuliensis	E
	Paracypris jonesi	E
	Cytheropteron lekefense	E
	Paragrenocythere cultrata	E
	Uroleberis glabella	E
ATT	Leguminocythereis senegalensis	E
	Buntonia issabaensis	P
	B. livida	P
	Soudanella laciniosa laciniosa	P
	Quadracythere lagaghiroboensis	P
	Buntonia bopaensis	P
	B. fortunata	P
	Protobasslerites bopaensis	P
	Limburgina sehouensis	P
	Buntonia pulvinata	P
	Isohabrocythere teiskotensis	?E
	Trachyleberis teiskotensis	?E
	Dahomeya alata anteroglabrata	E

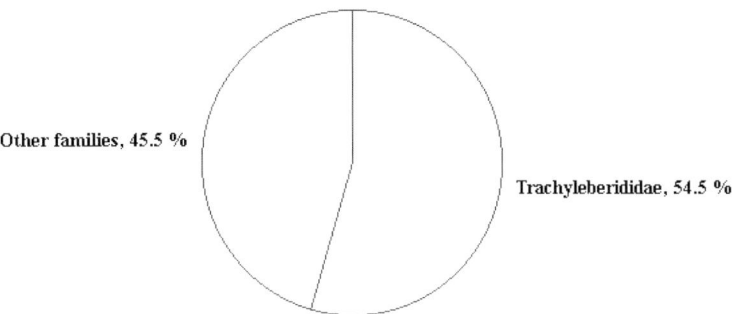

Fig. 3. Pie chart for Trachyleberididae percentage in comparison to other families in the South Tethyan Province (STP)

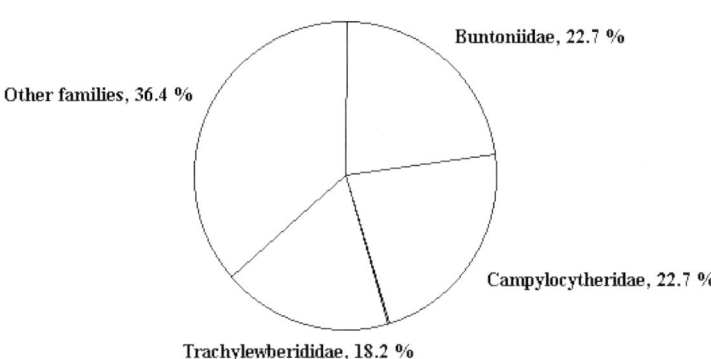

Fig. 4. Pie chart for ostracod families percentages in the Garra/Afro-Tethyan region (GAT/ATT; Egypt in the present study)

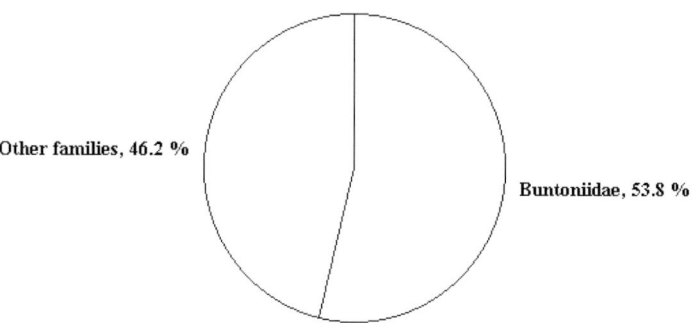

Fig. 5. Pie chart for Buntoniidae percentage in comparison to other families in the West African Province (WAP)

With regard to genera, *Buntonia* is distinguished by its abundance in the WAP (38.5%) (Fig. 6), but less common in the GAT/ATT (18.2%) (Fig. 7), while absent in the STP. Conversely, *Cytherella* is more abundant in the STP than in the WAP.

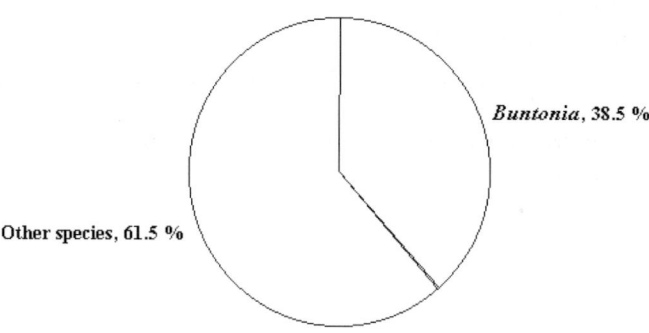

Fig. 6. Pie chart for *Buntonia* percentage in comparison to other genera in the WAP

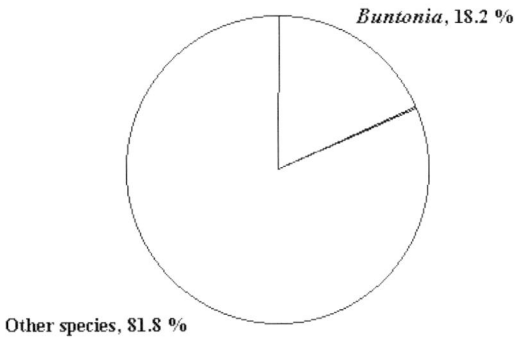

Fig. 7. Pie chart for *Buntonia* percentage in comparison to other genera in the GAT/ATT

Finally, the examined countries were subjected to cluster analysis for the aim of finding out any sub-provinces within the distinguished provinces. The data matrix of 13 countries and 46 ostracod species revealed the distinction of two major provinces (WAP, STP) (Fig. 8), and attributed Egypt to the STP. Moreover, a distinction of two sub-provinces within the WAP could be detected. These two sub-provinces geographically divided the WAP into the Northern Sub-Province (NSP), including Mali, NW Nigeria and Libya, and the Southern Sub-Province (SSP), comprising Ivory Coast, Togo and SW Nigeria (See Fig. 1 for locations of countries). However, Senegal and Niger have close similarities to both sub-provinces.

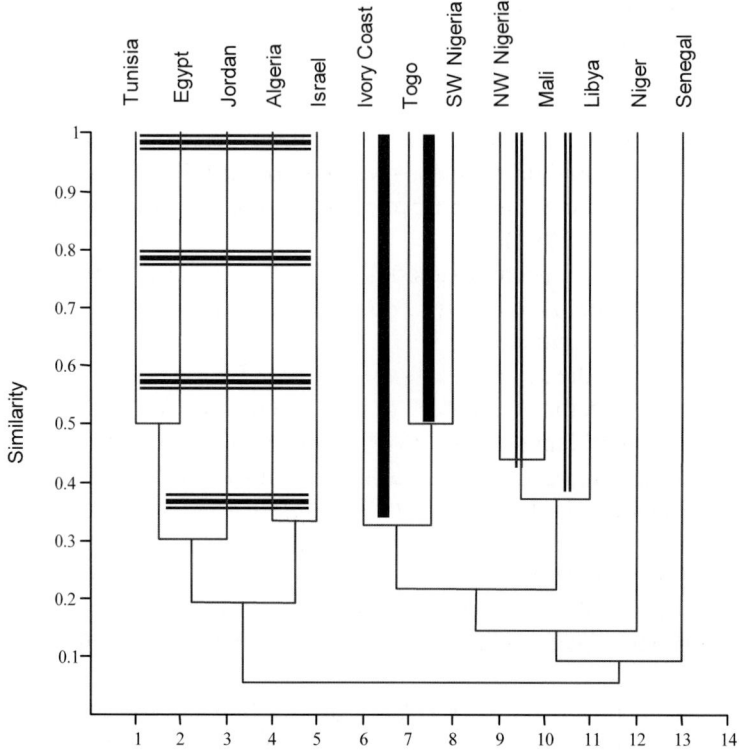

Fig. 8. Dendrogram resulting from cluster analysis (Jaccard coefficient of similarity; the paired-group method) for the 13 investigated countries

2. 5 Paleoenvironments

In their work on ostracod assemblages of two sections situated on the southwestern slope of Gebel Duwi of Egypt, Speijer and Morsi (2002) mentioned that these assemblages show a sharp turnover at the PETM. They indicated this turnover by abundance changes, local extinctions and immigrations. In contradiction, the present results indicate no turnover in ostracod faunas of Egypt during the PETM and the faunal changes are due to the effect of environmental factors in the paleogeographical distribution of these organisms rather than origination or extinction (for details see Elewa 2002b; Elewa and Morsi 2004).

The results obtained by quantitative analysis of the 46 selected ostracod species could interpret the history of their depositional environments as well as their reactions against worse environmental conditions during the late Paleocene-early Eocene times.

From the historical viewpoint, the late Paleocene shows the greatest transgression in the studied regions, thus it is believed that this was the maximum extent of the Trans-Saharan Seaway (see Fig. 1 of the present study; after Keen et al. 1994).

On the other hand, Eocene ostracods are not as widely distributed as those of the Paleocene. Accordingly, no distinctive early Eocene ostracod faunas have been distinguished.

The abundance of the family Buntoniidae (Fig. 5), especially the genus *Buntonia* (Fig. 6), together with more or less absence of *Cytherella* species in the ostracod assemblages of the WAP, designate that these assemblages were subject to normal oxygenation associated with warm water conditions at the late Paleocene-early Eocene times (See Carbonnel and Johnson 1989; Whatley 1991). Yet, the relative abundance of *Cytherella* species and the simultaneous absence of *Buntonia* species in the South Tethyan Province during the PETM may probably indicate more or less reduced oxygen conditions associated with high predation activity for this region, as it will be seen below.

With regard to ostracods of the WAP in the Paleocene/Eocene times, exhibiting almost epicontinental ostracod faunas (Fig. 9; Tabl. 2), it is concluded by Elewa (2002b; 2nd factor resulted from correspondence analysis) that temperature was the most affecting factor on the ostracod distribution. The highest degrees of temperature led to local turnover of ostracod assemblages of the WAP (Fig. 10; Tabl. 2), as indicated by the vanishing of most ostracod species at the PETM, and appearance of several new species (Afro-Tethyan Type of Bassiouni and Luger 1990) just after the PETM. These post-PETM ostracod species characterized the WAP and were absent in both Egypt and the South Tethyan Province (STP).

Whereas, the South Tethyan Province was prominently affected by change in the dissolved oxygen content of water associated with fluctuation in water depth, from outer neritic to upper bathyal (Fig. 9; Tabl. 2), during PETM. The depletion of oxygen in the water led to a minor faunal change in the STP during the late Paleocene-early Eocene times (Fig. 10; Tabl. 2).

With regard to Egypt (GAT/ATT), the ostracod assemblages were affected by water depth, which ranges from epineritic to inner neritic (Fig. 9, Tabl. 2), as well as turbulence. Consequently, a faunal adaptability to change in water depth characterized the ostracods of Egypt during PETM

(Fig. 10; Tabl. 2). It is worth noting that ostracod faunas show migration of deeper dwelling forms (e.g. *Leguminocythereis lokossaensis*, *Soudanella laciniosa triangulata*) from the WAP to Egypt and North Africa. Bassiouni and Luger (1990) stated that these deeper dwelling forms could not reproduce in the shallow marine environments of the Trans-Saharan Seaway. However, Reyment (1983), Stinnesbeck and Reyment (1988), and Reyment and Aranki (1991) strictly contradict the opinion of Bassiouni and Luger. They believe that the genus *Soudanella* has a remarkable distributional capability, as it was able to migrate to Brazil from West Africa and formed part of the impressive ostracod migration between North Africa and West Africa.

Anyway, whether *Soudanella* has a wide distributional capability or it migrated to North Africa for any other reason, its existence, together with other deeper dwelling forms of West Africa, in Egypt during the PETM, helped their survival there to the early Eocene time. Concurrently they disappeared at the end of the late Paleocene time of the WAP.

The work by Shahin (2000) supports the attained results, hence he stated that there is general agreement that the marine connection across the Sahara was broken towards the end of Paleocene, therefore the faunal migration occurred during the Paleocene and earlier. Consequently, most migrated Paleocene ostracods possessed relatively long chronostratigraphical ranges in North Africa and other southern tethyan realms and survived at least to the late Eocene.

In conclusion, the differences in environmental factors and predation activity (as it will be seen below), between the regions of North Africa, West Africa, and the Middle East, were the main controller of turnover, migration, and provinciality in the studied ostracod assemblages. This provinciality became clearer with time, where middle Eocene ostracods of Egypt (Nile Valley) as well as other areas of North Africa and the Middle East exhibit particular ostracod provinciality, simultaneously with another tendency in West Africa (Elewa 2005). All over all, the present study revealed the distinction of two sub-provinces within the West African Province (e.g. NSP, SSP). Counting the ostracod species of these sub-provinces led to conclude that *Buntonia* was rich in the SSP than the NSP indicating the effect of high temperature on the ostracod turnover in the SSP than for the NSP. Also, the strong similarity between Libya (of North Africa) with the NSP (Mali and NW Nigeria) supports the conclusion of Barsotti (1963) in suggesting a connection, up through late Paleocene time, between the southern Tethys and the Guinean Province through Mali and NW Nigeria.

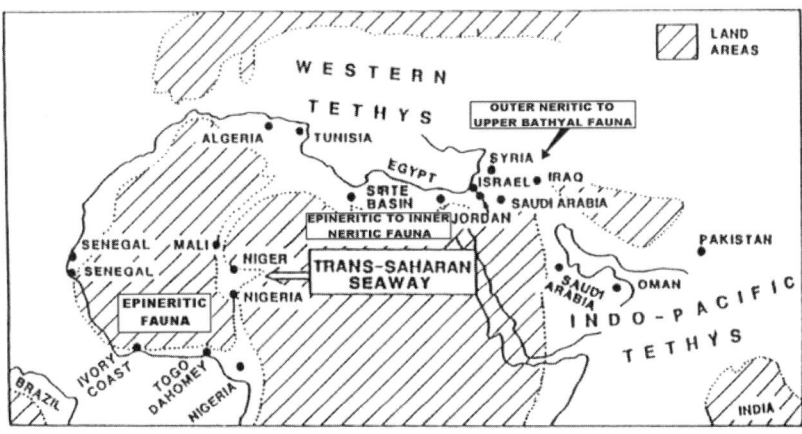

Fig. 9. Map showing environments of ostracods during the late Paleocene/early Eocene of North and West Africa and the Middle East (modification of Fig. 1)

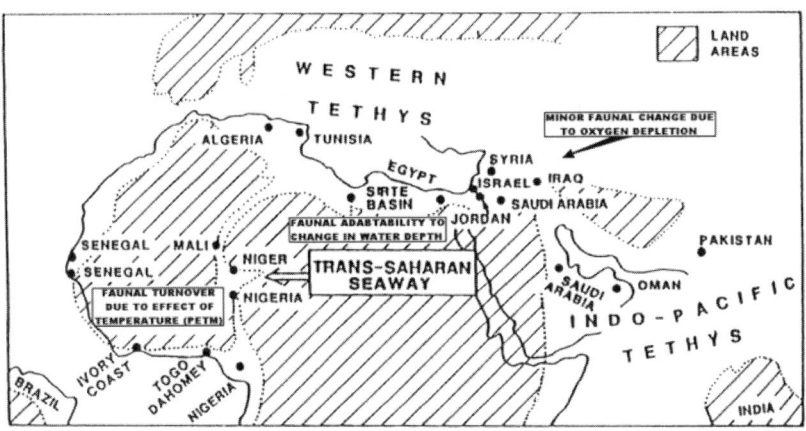

Fig. 10. Map showing most important reactions of ostracod faunas against the worse environmental conditions occurred during the late Paleocene/early Eocene of North and West Africa and the Middle East (modification of Fig. 1)

2. 6 Predation as a strong factor affecting ostracod abundances in the studied regions

In her study regarding predation on ostracods of Texas (1988), Maddocks cited that abrupt increases in naticid predation on ostracods occurred near the end of the Campanian and at the Cretaceous-Paleogene boundary. She added that high levels of naticid and other predation continued to characterize Paleocene and middle Eocene assemblages. I noticed in her material that species related to the genus *Cytherella* (Family: Cytherellidae) have been strongly attacked by drills in the Late Cretaceous and Paleocene. Species belonging to the genus *Buntonia* (Family: Buntoniidae) and the genus *Trachyleberis* (Family: Trachyleberididae) have been attacked above the Paleocene-Eocene thermal maximum (with high predation activity at Lutetian).

Reyment (1966a) concluded that the genus *Cytherella* was strongly attacked by drills in the western Niger Delta. He attributed this to the interstitial mode of life of species belonging to this genus.

Reyment et al. (1987) have shown several predation cases on ostracods from the Santonian of Israel and the Paleocene of Nigeria. They concluded that muricids and naticids prefer less strongly ornamented ostracods.

Reyment and Elewa (2002) stated that the Paleocene ostracods of Nigeria (studied by Reyment et al. 1987) indicate a reasonable model where feeding of drills on ostracods was for a short initial period in their life, and then migrated to shallower water.

Corbari et al. (2005) could give reasonable interpretation for higher proportion of platycopids in ostracod populations during hypoxic crises. They argued that during a hypoxic event in marine sediments, the redox front rises towards the surface. They suggest that platycopids (including *Cytherella*) could keep still in the deep layers of the sediment for longer, while other podocopids migrate to shallower sediment layers. When they reach the upper zones, the animals closest to the surface would then be more exposed to predation by fishes and other crustaceans (I add to them naticids), which would account for the relative increase in the proportion of platycopids. Corbari et al. concluded this scenario based on physiological and anatomical observations of the ostracod order Podocopida (see Corbari et al., 2004, for more discussion).

Following the above terminology, I could interpret the abundance of *Cytherella* in the relatively deeper water of the STP (South Tethyan Province) as a result of its reaction against the oxygen depletion occurred during the PETM. The STP exemplifies a minor faunal change in ostracods due to oxygen depletion during the late Paleocene-early Eocene interval. Podocopids (except those forms that could tolerate oxygen depletion) migrated to shallower sediment layers when oxygen decreased, while *Cytherella* could resist these hard conditions and stayed in deeper layers away from predators in the surface waters. Notice that some Trachyleberidids could tolerate the reduced oxygen conditions during this interval and stayed in deeper water, hence the apparent abundance of this family in the STP. Generally, I call this phenomenon as "Obligatory Predation" by drills on the available ostracod species during that time.

In the WAP, the situation during the PETM was different, where warm temperature and normal oxygenation permitted the high abundance of species belonging to Buntoniidae and Trachyleberididae, however predation activity was more distinct on ostracod carapaces resembling in their shapes to pelecypods (e.g. *Cytherella*, which has smooth and delicate carapace). This phenomenon is named here as "Selective Predation".

In Egypt, ostracod assemblages show evidence of faunal adaptability at the Paleocene-Eocene boundary. It seems that predation rate was low relative to the high activity occurred in the WAP at the PETM.

Over all, this paper suggests that combination of predation with changes in paleoenvironmental conditions led to the ostracod provinciality in North and West Africa and the Middle East during the Paleocene-Eocene thermal maximum.

2. 7 Acknowledgement

A special word of thanks is due to Prof. Dr. Reyment of the Swedish Natural History Museum, and Prof. Dr. Colin of France for critical reviewing of this manuscript.

References

Al Furaih AA (1980) Upper Cretaceous and Lower Tertiary Ostracoda (Superfamily: Cytheracea) from Saudi Arabia. Geol Dept Collage Sci Univ Riyadh 209 p

Al Sheikhly SAJ (1981) Maastrichtian – upper Eocene Ostracoda of the subfamily Trachyleberidinae from Iraq, Jordan and Syria. Unpublished Ph D Thesis, University of Glasgow 229 p

Al Sheikhly SAJ (1985) The new genus *Ordoniya* (Ostracoda) from the Paleogene of the Middle East. Proceedings of the Second Jordan Geological Conference, Amman 238-273

Apostolescu V (1961) Contribution à l'étude paléontologique (Ostracodes) et stratigraphique des bassins cretacés et tertiaires de l'Afrique Occidentale. Rév Inst Franc Pétrol, Paris 16 (7/8): 779-867

Aref M (1982) Micropaleontology and biostratigraphy of Eocene rocks in the area between Assiut and Beni Suef of the Nile Valley, Egypt. Ph D Thesis, Assiut University, Egypt

Aref M (1995) Early Eocene Ostracoda from the Thebes Formation along the Red Sea coastal area, Egypt. Egyptian J Geol 39 (1): 202-217

Barsotti G (1963) Paleocenic ostracods of Libya (Sirte Basin) and their wide African distribution. Rév Inst Franc Pétrol, Paris 18 (10/11): 779-867

Bassiouni MA, Elewa AMT (1999) Studies on some important ostracod groups from the Paleogene of Egypt. Geosound, Turkey 35: 13-27

Bassiouni MA, Luger P (1990) Maastrichtian to early Eocene Ostracoda from Southern Egypt (palaeontology, palaeoecology, palaeobiogeography and biostratigraphy). Berliner Geowiss Abh (A) 120 (2): 755-928

Bassiouni MA, Morsi AM (2000) Paleocene-Lower Eocene ostracods from El Quss Abu Said Plateau (Farafra Oasis), Western Desert, Egypt. Paleontographica (A), Stuttgart 257: 27-84

Bassiouni MA, Boukhary MA, Anan HS (1977) Ostracodes from Gebel Gurnah, Nile Valley. Egypt. Proc Egypt Acad Sci 30: 1-9

Boukhary MA, Guernet C, Mansour H (1982) Ostracodes du Tértiaire inférieur de l'Égypte. Cahiers Micropaléont, Paris 1982 (1): 13-20

Carbonnel G (1986). Ostracodes tertiaires (Paléogène et Néogène) du bassin Senegalo – Guineen. Documents de la Bureau de Récherche de Geologie Mineralogie 101: 34-231

Carbonnel G Johnson A (1989) Les ostracodes Paleogenes du Togo: taxonomie, biostratigraphie, apports dans l'organisation et l'evolution du basin. Géobios, Lyon 22: 409-443

Carbonnel G, Alzouma K, Dikouma M (1990). Les ostracodes Paléocènes du Niger: Taxonomie – Un témoignage de l'éxistence éventuelle de la mer transsaharienne?. Géobios, Lyon 23 (6): 671-697

Colin JP, Tambareau Y, Krasheninnikov VA (1998) Maastrichtian and Paleocene ostracod assemblages of Mali (Western Africa). Dela-Opera 34 (2): 273-345

Corbari L, Carbonel P, Massabuau J-C (2004) How a low tissue O2 strategy could be conserved in early crustaceans: The example of the podocopid ostracods. J Exp Biol 207: 4415-4425

Corbari L, Mesmer-Dudons N, Carbonel P, Massabuau J-C (2005) *Cytherella* as a tool to reconstruct deep-sea paleo-oxygen levels: The respiratory physiology of the platycopid ostracod *Cytherella* cf. *abyssorum*. Marine Biol

Damotte R (1995) The biostratigraphy and paleobiogeography of the Upper Cretaceous-basal Tertiary ostracods from North Africa, Mali and Congo. Cretaceous Research 16: 357-366

Damotte R, Fleury JJ (1987) Ostracodes Maastrichtiens et Paléocènes du Djebel Dyr, près de Tebessa (Algérie orientale). Rév Géol Méditerran 14 (2): 87-107

Donze P, Colin JP, Damotte R, Oertli H, Peypouquet JP, Said R (1982) Les ostracodes du Campanien Terminal a l'Éocène inférieur de la coupe de Kef, Tunisie Nord-Occidental. Bulletin de la Céntre Recherche Elf-Aquitaine 6: 273-355

Elewa AMT (1994) Biostratigraphical studies of Eocene rocks of the El Sheikh Fadl-Ras Gharib stretch, Eastern Desert, Egypt. Ph D Thesis, Minia University, Egypt

Elewa AMT (1998) Ostracod assemblages at the Lower/Middle Eocene boundary in the Nile Valley, Egypt. Neues Jahrbuch für Geologie und Paläontologie Abhandlungen, Stuttgart 210 (1): 1-17

Elewa AMT (2002a) Microevolution applied to ostracod biostratigraphy: the middle Cretaceous to middle Eocene ostracods of Egypt (with special reference to the genus *Paracosta* Siddiqui). Neues Jahrbuch für Geologie und Paläontologie Abhandlungen 226 (3): 289-318

Elewa AMT (2002b) Palaeobiogeography of Maastrichtian to early Eocene Ostracoda of North and West Africa and the Middle East. Micropaleontology 48 (4), 391-398

Elewa AMT (2005) Paleoecology and paleogeography of Eocene ostracod faunas from the Nile Valley between Minia and Maghagha, Upper Egypt. In Elewa AMT (ed). Migration of Organisms: Climate, Geography, Ecology. Springer, Heidelberg 25-70

Elewa AMT, Ishizaki K (1994) Ostracods from Eocene rocks of the El Sheikh Fadl-Ras Gharib stretch, the Eastern desert, Egypt- Biostratigraphy and paleoenvironment. Earth Sci (Chkyu Kagaku) 48 (2): 143-157

Elewa AMT, Morsi AM (2004) Palaeobiotope analysis and palaeoenvironmental reconstruction of the Palaeocene-early Eocene ostracodes from east-central Sinai, Egypt. In Beaudoin AB, Head MJ (eds). The Palynology and Micropalaeontology of Boundaries. Geol. Soc., London, Special Publications 230: 293-308

Elewa AMT, Bassiouni MA, Luger P (1999) Multivariate data analysis as a tool for reconstructing paleoenvironments: The Maastrichtian to early Eocene Ostracoda of southern Egypt. Bulletin of the Faculty of Science, Minia University 12 (2) 1-20

El Sogher AMS (1991) Late Cretaceous and Paleocene ostracods from the Waha and Heira formations of the Sirte Basin, Libya. Unpublished M Sc Thesis, University of Glasgow 210 p

El Waer AA (1992) Tertiary and Upper Cretaceous Ostracoda from NW offshore Libya: Their taxonomy, biostratigraphy and correlation with adjacent areas. Petrol Res Centr Publ, Tripoli 1-445

Esker GL (1968) Danian ostracods from Tunisia. Micropaleontology, 14 (3): 319-333

Foster CA, Swain FM, Petters SW (1983) Late Paleocene Ostracoda from Nigeria. Rev Español Micropaleont 15: 103-166

Gammudi AM, Keen MC (1996) The Paleocene/Eocene boundary in the Sirt Basin, Libya -recognized by ostracod faunas. 3éme Congrès Européen des Ostracodologistes, Paris-Bierville 27 (Abstract)

Grekoff N (1969) Sur la valeur stratigraphique et les relations paléogéographique de quelques ostracodes du Crétacé, du Paléocène et de l'Éocène inférieur d'Algérie orientale. Proceedings of the Third African Micropaleontology Colloquium, Cairo 227-248

Guernet C, Bourdillon De Grissac C, Roger J (1991) Ostracodes Paleogenes du Dhofar (Oman). Interet stratigraphique et paleogeographique. Rév Micropaléont 34 (4): 297-311

Haggag, MAY (1979) Biostratigraphy of the west-central part of the Western Desert, Egypt. Ph D Thesis, Ain Shams University, Egypt

Helal SA (1990) Stratigraphic and paleontologic studies on the Eocene sediments at Gebel Shabrawet area, Eastern Desert, Egypt. M.Sc Thesis, Cairo University, Egypt

Honigstein A, Rosenfeld A (1995) Paleocene ostracods from southern Israel. Rév Micropaléont 38 (1): 49-62

Honigstein A, Rosenfeld A, Benjamini C (1991) Ostracodes and foraminifera from the early-middle Eocene of Qeren Sartaba, Jordan Valley. J Micropalaeont 10 (1): 95-107

Honigstein A, Rosenfeld A, Benjamini C (2002) Eocene ostracode faunas from the Negev, southern Israel: Taxonomy, stratigraphy and paleobiogeography. Micropaleontology 48 (4) 365-389

Ismail AA (1992) Late Campanian to early Eocene Ostracoda from Esh El Mallaha area, Eastern Desert, Egypt. Rév Micropaléont 35 (1): 39-52

Ismail AA (1996) Biostratigraphy and paleoecology of Maastrichtian-early Eocene ostracods of west central Sinai, Egypt. Rév Paléobiol 15 (1): 37-54

Keen MC, Al Sheikhly SSJ, Elsogher A, Gammudi AM (1994) Tertiary ostracods of North Africa and the Middle East. In Simmons MD (ed). Micropalaeontology and hydrocarbon exploration in the Middle East, Chapman and Hall Inc., London 371-400

Maddocks RF (1988) One hundred million years of predation on ostracods: The fossil record in Texas. In Hanai T, Ikeya N, Ishizaki K (eds). Evolutionary Biology of Ostracoda, Elsevier 637-657

Peypouquet JP, Grousset F, Mourguiart P (1986) Paleoceanography of the Mesogean Sea based on ostracods of the Northern Tunisian Continental Shelf between the Late Cretaceous and Early Paleogene. Geologische Rundshau 75: 159-174

Reyment RA (1960) Studies on Nigerian Upper Cretaceous and Lower Tertiary Ostracoda: Part 1. Stockh Contr Geol 7: 238 p

Reyment RA (1963) Studies on Nigerian Upper Cretaceous and Lower Tertiary Ostracoda. Part 2, Danian, Palaeocene and Eocene Ostracoda. Stockh Contr Geol, Stockholm 10: 286 p

Reyment RA (1966a) Preliminary observations on gastropod predation in the western Niger Delta. Palaeogeogr Plalaeoclimat Palaeoecol 2: 81-102

Reyment RA (1966b) Studies on Nigerian Upper Cretaceous and Lower Tertiary Ostracoda. Part 3, Stratigraphical, palaeoecological and Biometrical conclusions. Stockh Contr Geol, Stockholm 14: 151 p

Reyment RA (1981) The Ostracoda of the Kalambaina Formation (Paleocene), north western Nigeria. Bull Geol Inst Univ Uppsala 9: 51-65

Reyment RA (1983) Phenotypic evolution in microfossils. Evol Biol 16: 209-254

Reyment RA, Aranki, JF (1991) On the Tertiary genus *Soudanella* Apostolescu (1961) (Ostracoda, Crustacea). J Micropalaeont 10 (1): 23-28

Reyment RA, Elewa AMT (2002) Predation by drills on Ostracoda. In Kelley P, Kowalewski M, Hansen T (eds). Predator-Prey Interactions in the Fossil record, Kluwer Academic/Plenum Publishers 93-112

Reyment RA, Reyment ER (1959) *Bairdia ilaroensis* sp nov aus dem Paläozän Nigeriens und die Gueltigkeit der Gattung *Bairdoppilata* (Ostr, Crust). Stockholm Contributions in Geology 3: 59-70

Reyment RA, Reyment ER (1980) The Paleocene Trans-Saharan Transgression and its Ostracod Fauna. In Salem MJ, Busrevil MI (eds). The Geology of Libya. London: Academic Press 1: 245-254

Reyment RA, Reyment ER, Honigstein A (1987) Predation by boring gastropods on Late Cretaceous and early Paleocene ostracods. Cretaceous Res 8: 189-209

Salahi D (1966) Ostracodes du Crétacé supérieur et du Tertiaire en provenanced'un Sondage de la Région de Zelten (Lybie). Rév Inst Franc. Pétrol 21 (1): 3-43

Sarr R (1998) Les ostracodes du Paléocène du Horst de Diass (Senegal): Biostratigraphie, systematique, paléoenvironment. Rév Micropaléont 41 (2): 151-174

Sarr R (1999) Le Paléogene de la region de Mbour-Joal (Senegal occidental): Biostratigraphie, etude systematique des ostracodes, paléoenvironment. Rév Paléobiol 18 (1): 1-29

Shahin A (2000) Tertiary ostracods of Gebel Withr, southwestern Sinai, Egypt: palaeontology, biostratigraphy and palaeobiogeography. J African Earth Sciences 31 (2): 285-315

Shamah K (1981) Le Paléogène de la province du Fayoum, Égypte. Mem Sci de la Terre, Paris 1: 383 p

Siddiqui QA (1971) Early Tertiary Ostracoda of the Family Trachyleberididae from West Pakistan. Bull Brit Mus Nat Hist Geol, London 9: 98 pp

Speijer R, Morsi AM (2002) Ostracode turnover and sea-level changes associated with the Paleocene-Eocene thermal maximum. Geology 30 (1): 23-26

Stinnesbeck WS, Reyment RA (1988) Note on a further occurrence of *Soudanella laciniosa* Apostolescu in northeastern Brazil. J African Earth Sciences 7 (5/6): 779-781

Whatley RC (1991) The platycopid signal: a means of detecting kenoxic events using Ostracoda. J Micropaleont 10 (2), 181-185

Whatley, RC, Arias C (1993) Paleogene Ostracoda from the Tripoli Basin, Libya. Rev Español Micropaleont 15 (2): 125-154

Yassini I (1979) The littoral system ostracods from the Bay of Bou Ismail, Algiers, Algeria. Rev Español Micropaleont 11 (3): 353-416

3 Predation on Miocene ostracods of Wadi Um Ashtan, Mersa Matruh, Western Desert, Egypt

Ashraf M. T. Elewa

Geology Department, Faculty of Science, Minia University, Egypt,
aelewa@link.net

3. 1 Abstract

Predation on Miocene ostracods of Egypt received little attention due to the lack of information on taxonomy of these ostracods. This situation continued until Bassiouni and Elewa introduced, probably, the first Egyptian Miocene ostracod taxonomic record in 2000. Two years later, Reyment and Elewa (2002) noticed post-mortem holes made by marine fungi on carapaces of these ostracods. Consequently, the present study is the first detailed description of predation along with change in environment in Wadi Um Ashtan of Mersa Matruh on the Mediterranean coast of Egypt.

Keywords: Predation, Ostracoda, echinoids, marine fungi, Miocene, Wadi Um Ashtan, Egypt.

3. 2 Introduction

Ostracods are, undoubtedly, ingested by small fish. Scott (1902) recorded several marine benthonic ostracods from the stomach contents of fish in Scottish waters. Leonard (1983) studied the diet of the butterfish, *Pholis gunnellus*, in rock pools on the Northumberland coast, finding that some immature individuals fed almost exclusively on ostracods. Indeed, it has been suggested that since ostracod eggs, and possibly even the ostracods themselves, can survive passing through the digestive tracts of freshwater fish, the same might be true of marine species, providing an effective means of dispersal (Kornicker and Sohn 1971).

Invertebrates that feed on marine and brackish-water ostracods include gastropods (Reyment 1963, 1966a, b, 1967) and echinoids (Neale 1983).

However, decomposers of ostracods are attributed to fungi (Reyment and Elewa 2002). The first detailed work of predation by drills on ostracods is that of Maddocks (1988). She recognized twenty kinds of predation scars, including drillholes of naticids and other gastropods, digestive-solution holes and holes made by unknown animals, on Cretaceous, Paleocene and Holocene ostracods of Texas. These twenty ichnophena were grouped by Maddocks into four intergradational categories according to their inferred origins:

1. Naticid drillholes: These are best expressed in smooth, robust shells.
2. Gastropod drillholes: Some may be naticid holes in thin or ornamented shells, and some others may have been drilled by other gastropods or perhaps by octopus or other predators.
3. Holes made by unknown, less patient predators: These holes appear to have been ripped, gouged, punctured, or dug out rather than drilled.
4. Solution holes: Some closely resemble the digestive solution holes illustrated by Kornicker and Sohn (1971), while others show enlarged normal pore-canals and characteristic frosting. Solution features occasionally accompany other ichnophena.

Reyment (1971) stated that ostracods are not the main source of food for predators when bivalves become rich in the same environment with ostracods. Reyment and Elewa (2002) supported this opinion of Reyment. They added that there are isolated reports of echinoids having consumed ostracods, but whether this is by design or by accident has never been resolved. Therefore this study focuses on predation on Miocene ostracods of Wadi Um Ashtan in Mersa Matruh, Western Desert of Egypt. Deposits of this area contain rare mollusk shells but several fossil irregular echinoids that lived in turbulent, shallow marine environment with slight fluctuation in water depth and turbulence at the end of the Middle Miocene (Bassiouni and Elewa 2000).

The chapter by Wilkinson et al. (this volume) presents an interesting review of ostracod carnivory through time. I recommend reading that chapter for more understanding of predation, scavenging and parasitism in ostracod community.

3. 3 Predation and survival

In the introduction of this book I asked, among some important queries, the following question: why some predators do not benefit from their prey after killing them and leave them without eating? I meant by this question to shed the light on one of the ignored sides of predation, which is environmental equilibrium and the food chain. As we all know there are two functional units or components in the biosphere, autotrophic and heterotrophic. The autotrophic organisms (plants) photosynthesize the sun's energy to produce their cells, in other words they are self-nourishing (see Boardman et al. 1987), while heterotrophic (other nourishing) organisms need autotrophic organisms as suppliers of energy. These heterotrophic organisms were divided into herbivores, or consumers of plants, and carnivores, or flesh eaters. Also, other carnivores eat these carnivores and so on. In summary, herbivores consume plants, carnivores eat herbivores and other carnivores eat these carnivores; this food chain should not be disturbed for survival of organisms. However, we should not forget the important group of organisms that decompose dead organisms into the elements required for growth (e.g. bacteria and fungi). Returning to our question, it is now understandable that some carnivores kill other heterotrophs not only to eat but also to let other decomposers to live and for life to continue. Of course, predators do not understand this equilibrium when attacking their preys, and thus I attribute their behavior either to antagonism or survival. In other words, as we all know, some creatures are physically antagonistic and would like to attack other creatures just for attack, and some others defend themselves for survival. Fortunately, these features in organisms are necessary for the environmental equilibrium.

In the present study I introduce two types of organisms that play important role in the continuity of the food chain, the first (echinoids) are a greedy predators and the second (marine fungi) are a strong decomposers. Unfortunately, these two types had little interest to researchers until the end of the Twentieth Century.

3. 4 Material for the present study

Miocene ostracod assemblages of Wadi Um Ashtan were systematically studied by Bassiouni and Elewa (2000). They, further, made detailed work on ostracod stratigraphical, paleoecological and paleobiogeographical significance.

Material for the present study was collected from the Marmarica Formation (Serravallian) of Wadi Um Ashtan (Fig. 1), which represents the ostracod *Keijella punctigibba* Zone in equivalence with the planktonic foraminifera *Orbulina universa* Zone and the benthonic foraminifera *Heterolepa dutemplei* Zone and *Amphistegina radiata* Zone (Fig. 2).

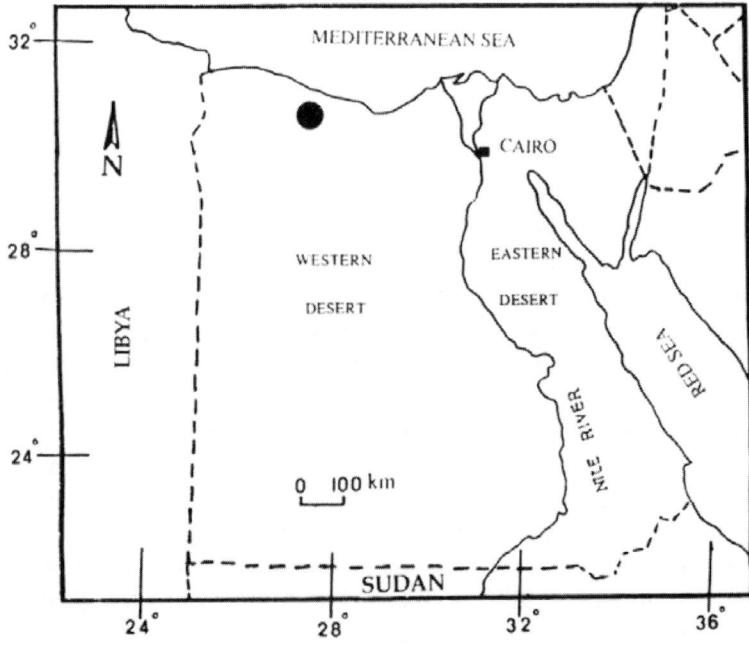

Fig. 1. Location map (after Bassiouni and Elewa 2000). Black circle refers to the study area

Age	Rock unit	Lithology	B. foraminifera Zones	P. foraminifera Zone	Ostracoda Zone
Middle Miocene (Serravallian)	Marmarica Formation	White, moderately hard limestone at the base, to yellowish white sandy limestone and greenish white marly limestone at the middle, to chalky limestone at the top	*Amphistegina radiata* / *Heterolepa dutemplei*	*Orbulina universa*	*Keijella punctigibba*

Fig. 2. Litho- and biostratigraphy of the studied Wadi Um Ashtan area

3. 5 Echinoid drillholes

Our knowledge of predation by echinoids on ostracods is very limited due to scarce publications on the subject. On the other hand, gastropod predators, especially naticids and muricids, are often studied in ostracodology as a result of the pioneer works of Reyment (1963, 1966a, b, 1967), and consequently the work of Maddocks (1988). This situation extended, with some isolated papers on echinoids as predators on ostracod prey, until I noticed during my study of the fauna of Wadi Um Ashtan that molluscs are very rare, even though ostracods seem to be attacked by less patient predators than gastropods. My investigations led to discovery of this unknown enemy; it is, no doubt, echinoids. These creatures are richly present in the deposits with several shapes and forms, from small tests that should be investigated under the microscope for identifying their features, like *Echinocyamus*, to very large tests like *Echinolampas*. Richness of these echinoids in the study area points to shallow, near shore shelf marine environments with normal salinity ranges (Boardman et al. 1987). Fortunately, the shapes of holes made by echinoids are similar to those described by Maddocks (1988) as holes of unknown, less patient predators. She mentioned that the main difference between gastropod holes and these holes is the mechanism of making these holes. Gastropods drill these holes but these predators rip, gouge, puncture and dig them out instead.

In the studied material, I noticed that echinoid holes are almost always located at the posterior parts of the ostracod carapaces (Figs. 3-1, 3-2, 3-3, 3-5) indicating preferential attack from behind. Some of these holes are crescentic in shape (Figs. 4-1, 4-2), and some others are rhomboidal (Fig. 4-3) or sub-triangular (Figs. 4-6, 4-7). However, Figure 4-4 shows an incomplete borehole of unknown origin. Loren Babcock has discussed a similar behavior in his chapter in Kelley et al. (2002). His paper discusses predators preferentially attacking the rear right side of trilobite prey and might be consulted usefully here.

Figure 5 shows distribution of the main predators and their favorite ostracod species in the standard microfacies types of the study area. From this figure it is obvious that echinoid predation has taken its maximum attack during the interval of the benthonic foraminiferal *Heterolepa dutemplei* Zone, which is equivalent to the base of the planktonic foraminiferal *Orbulina universa* Zone and the base of the ostracod *Keijella punctigibba* Zone. Deposits of this interval lie in the bioclastic wackestone facies type of Wilson (1975), which indicates open platform (shelf lagoon) environment (FZ7) according to Flügel (1982). This Facies Zone 7 (FZ7) represents open shallow platform with open circulation at or just below wave base. These results coincide with that of Boardman et al. (1987) for occurrence of echinoids.

3.6 Holes by marine fungi

Amphistegina radiata, the nominate species of the zone representing the upper part of the Marmarica Formation in the study area, is one of the foraminiferal species commonly found with coral reefs in shelf tropical regions (Renema and Troelestra 2001; Renema 2003). Renema (2003) added that *Amphistegina radiata* prefers solid substrates, and high densities occur on well-structured coral rubbles. This species was attributed to neritic environment (0-150 m); however Renema (2003) found it in samples >10 m depth. This indicates the possibility of reef communities (including fungi) to occur in the Marmarica Formation of the area under consideration.

Clipson et al. (2001) argued that fungi from coastal and marine ecosystems are a neglected but significant part of marine biodiversity. Fungi in general are able to degrade a wide range of recalcitrant biological molecules and particularly in coastal ecosystems fungal activity may be critical in the early stages of biodegradative pathways. They added, furthermore, that marine fungi have shown high decadal indices (% increases in species number over a 10 year period). Hawksworth (1991) calculated that marine fungi had the highest decadal index (49%) for any fungal group.

In the studied material, post-mortem holes made by marine fungi are located at the adductor scar region as well as the posterior part of the ostracod carapaces (Fig. 3-8). Reyment and Elewa (2002) mentioned that these minute holes, although largely represented by holes through the shell, could also occur as a kind of scoring of the surface (see Figs. 4-5, 4-8 of the present study).

Meanwhile, figure 5 shows that holes made by marine fungi were abundant during the interval of the benthonic foraminiferal *Amphstegina radiata* Zone, which is equivalent to the top of the planktonic foraminiferal *Orbulina universa* Zone and the top of the ostracod *Keijella punctigibba* Zone. Deposits of this interval lie in the packstone-wackestone facies type of Wilson (1975).

3. 7 Discussion

Ibrahim and Mansour (2002) stated that the deposition of the fossiliferous carbonate rocks of the Marmarica Formation probably took place in a warm water, inner shelf environment (0–20 m palaeodepth) subjected to some current activity, and with salinity ranging from normal to slightly hypersaline (35–50‰). Omar (1992) gave more definite environmental ranges for the Miocene succession of Mersa Matruh, where he concluded that this succession was deposited under shallow marine conditions with temperatures ranging from 9 to 13 degrees and salinity ranging from 35 to 38‰.

On the other hand, my investigations indicate that temperature increases from the base of the Marmarica Formation to its top with the increase of reef communities, salinity ranges and relatively deeper water conditions, within the open platform environment, in the same direction. In contrast, turbulence decreases from base to top.

These conclusions are based on the following:
1. The interpreted environment of the bioclastic wackestone is shallow water with open circulation close to wave-base. The packstone-wackestone represents textural inversion (particles from high-energy environment have moved down local slopes to low-energy settings).
2. Echinoids, which are common in the basal deposits of the formation, indicate shallow, near shore shelf marine environments with normal salinity ranges (Boardman et al. 1987).
3. Bryzoa, which are abundant in the top of the succession, are sensitive to wave action, where they are liable to damage. Thus, relatively few bryozoans are inter-tidal since the high environmental energy and the problems of desiccation between tides are too great for these delicate organisms. The sub-littoral zone, however, has a wealth of bryozoans encrusting the rocks and seaweeds, and they are common at depths between 20 and 80 m (for more details see Clarkson 1987).

In general, this study aimed to shed light on echinoids as an important component of predation on ostracods when usual predators, gastropods, become rare or absent, and on marine fungi as a distinct decomposer of ostracods in marine environments dominated by reef communities.

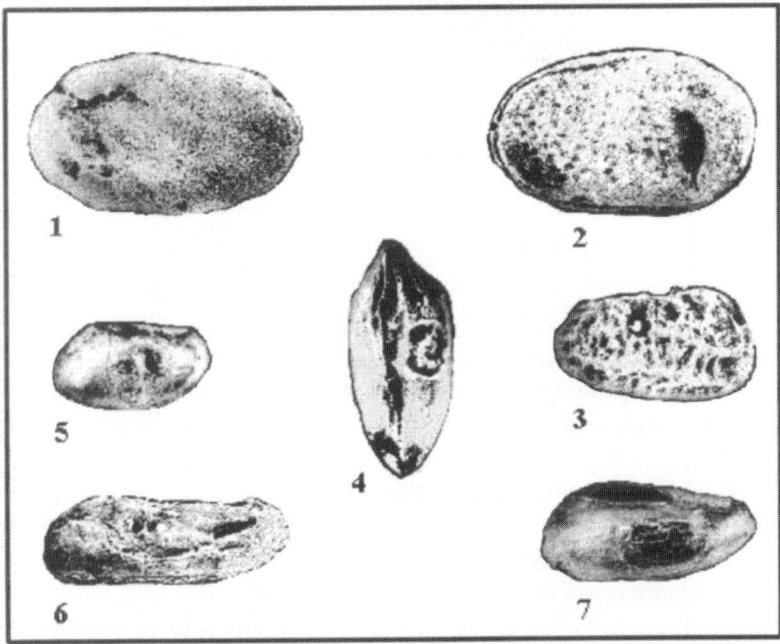

Fig. 3. Holes made by echinoids and marine fungi in Miocene ostracods of Wadi Um Ashtan. All specimens have x60 magnification except specimens 1, 2 (x75). 1, 2. *Cytherella inaequalis* Moyes; 1. right side view of carapace (incomplete predation by echinoids on the posterior part); 2. left side view of carapace (complete predation by echinoids on the posterior part). 3. *Chrysocythere cataphracta* Ruggieri; right side view of carapace (complete predation by echinoids on the posterior part). 4, 5, 7. *Keijella punctigibba* (Capeder); 4. dorsal view of carapace (complete predation by echinoids on the dorsal part of the right valve); 5. left side view of carapace (complete predation by echinoids on the posterior part); 7. left side view of carapace (incomplete predation of unknown origin). 6. *Ruggieria tetraptera tetraptera* (Seguenza); left side view of carapace (post-mortem holes made by marine fungi).

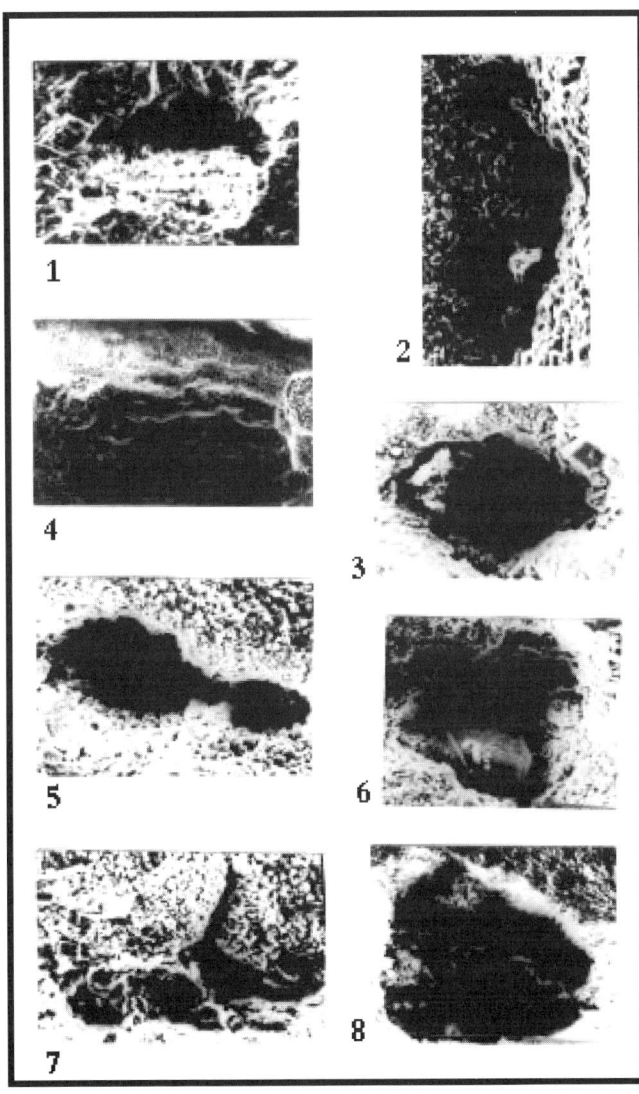

Fig. 4. Photomicrographs of boreholes made by echinoids and marine fungi in the ostracods of the study area. All photos have x375 magnification except photo 4 (x175) and photos 1, 8 (x500)

Age	Rock unit	Standard Microfacies Types (SMT)	Environment (Facies Zone)	Predator type	Prey species (Ostracoda) arranged from most favorite to less for each SMT
Middle Miocene (Serravallian)	Marmarica Formation	Bioclastic wackestone at the base	Open platform (shelf lagoon) (FZ7)	Echinoids	*Keijella punctigibba* *Cytherella inaequalis* *Chrysocythere cataphracta* *Ruggieria tetraptera tetraptera*
		Packstone-wackestone at the top		Marine fungi	*Keijella punctigibba* *Ruggieria tetraptera tetraptera*

Fig. 5. Distribution of the main predators and their favorite ostracod species in the standard microfacies types of the study area

3. 8 Acknowledgement

I would like to thank Prof. Dr. Reyment of the Swedish Natural history Museum and Prof. Dr. Patricia Kelley of the University of North Carolina at Wilmington (USA) for valuable comments as well as critical reviewing of this manuscript.

References

Babcock LE (2002): Trilobites in Paleozoic Predator-Prey Systems, and Their Role in Reorganization of Early Paleozoic Ecosystems. In Kelley P, Kowalewski M, Hansen T (eds). Predator-Prey Interactions in the Fossil record, Kluwer Academic/Plenum Publishers

Bassiouni MA, Elewa AMT (2000) Miocene ostracods of the Mediterranean: A first record from south Wadi Um Ashtan, Mersa Matruh, Western Desert, Egypt. N J Geol Paläont Abh, Stuttgart 2000: 449-466

Boardman RS, Cheetham AH, Rowell AJ (1987) Fossil Invertebrates. Blackwell Scientific Publications 713 p

Clarkson ENK (1987) Invertebrate paleontology and evolution. Allen and Unwin publishers, London 382 p

Clipson N, Landy E, Otte M (2001) Fungi. In Costello MJ et al (ed) European register of marine species: a checklist of the marine species in Europe and a bibliography of guides to their identification. Collection Patrimoines Naturels 50: 15-19

Flügel E (1982) Microfacies analysis of limestones. Springer, Berlin 633 p

Hawksworth DL (1991) The fungal dimension of biodiversity: magnitude, significance, and conservation. Mycological Research 95: 641-655

Ibrahim MIA, Mansour AMS (2002) Biostratigraphy and paleogeographical interpretation of the Miocene-Pleistocene sequence at El Dabaa northwestern Egypt. Marine Micropalaeont 21 (1): 51-65

Kornicker LS, Sohn IG (1971) Viability of ostracod eggs egested by fish and effect of digestive fluids on ostracod shells – Ecologic and paleoecologic implications. In Oertli HJ (ed) Paléoécologie ostracodes. Bull Cent Rech Pau 5: 125-135

Leonard SR (1983) Some aspects of the biology of three species of littoral teleosts. Final Honours Research Report, Newcastle Univ

Maddocks RF (1988) One hundred million years of predation on ostracods: The fossil record in Texas. In Hanai T, Ikeya N, Ishizaki K (eds). Evolutionary Biology of Ostracoda, Elsevier 637-657

Neale JW (1983) The Ostracoda and Uniformitarianism. 1. The later record: Recent, Pleistocene and Tertiary. Proc Yorks Geol Soc 44: 305-326

Omar AA (1992) Neogene biostratigraphy of Mersa Matruh and the Gulf of Suez. Unpubl Ph D Thesis, Minia Univ, Egypt

Renema W (2003) Larger foraminifera on reefs around Bali (Indonesia). Zool Verh Leiden 345 (31): 337-366

Renema W, Troelstra SR (2001) larger foraminifera distribution on a mesotrophic carbonate shelf in SW Sulawesi (Indonesia). Palaeogeogr Palaeoclimat Palaeoecol 175: 125-146

Reyment RA (1963) Studies on Nigerian Upper Cretaceous and Lower Tertiary Ostracoda. Part 2, Danian, Palaeocene and Eocene Ostracoda. Stockh Contr Geol, Stockholm 10: 286 p

Reyment RA (1966a) Preliminary observations on gastropod predation in the western Niger Delta. Palaeogeogr Plalaeoclimat Palaeoecol 2: 81-102

Reyment RA (1966b) Studies on Nigerian Upper Cretaceous and Lower Tertiary Ostracoda. Part 3, Stratigraphical, palaeoecological and Biometrical conclusions. Stockh Contr Geol, Stockholm 14: 151 p

Reyment RA (1967) Paleoethology and fossil drilling gastropods. Trans Kansas Acad Sci 70: 33-50

Reyment RA (1971) Introduction to quantitative paleoecology. Elsevier. Amsterdam 226 p

Reyment RA, Elewa AMT (2002) Predation by drills on Ostracoda. In Kelley P, Kowalewski M, Hansen T (eds). Predator-Prey Interactions in the Fossil record, Kluwer Academic/Plenum Publishers

Scott T (1902) Observations on the food of fishes. Rep Fishery Bd Scotl 20: 486-538

Wilson JJ (1975) Carbonate facies in geologic history. Springer, Berlin 471 p

4 Ostracod carnivory through time

I. Wilkinson[1], P. Wilby[1], P. Williams[2], D. Siveter[3] and J. Vannier[4]

[1]British Geological Survey, Keyworth, Nottingham, NG12 5GG, UK, ipw@bgs.ac.uk, [2]School of Earth and Environmental Sciences, University of Portsmouth, PO1 3QL, UK, [3]Department of Geology, University of Leicester, LE1 7RH, UK, [4]UMR 5125 du CNRS "Paléoenvironnements et Paléobiosphère", Université Claude Bernard Lyon 1, Campus Scientifique de La Doua, 2, rue Raphaël Dubois, 69622 Villeurbanne, France

4.1 Abstract

Carnivory in modern ostracods takes the form of predation, scavenging and parasitism. In the fossil record, carnivory is difficult to prove without the fossilisation of diagnostic functional morphological features or the preservation of the intimate association between the ostracods and the carrion or prey. Six examples of putative carnivory are known in geological deep time, the most persuasive being scavenging myodocopes of Ordovician, Carboniferous and Triassic ages (*Myodoprimigenia, Eocypridina* and *Triadocypris*), where swarms of ostracods are found associated with carcasses, and the early Silurian *Colymbosathon* in which characteristic soft part anatomy is preserved. Other putative scavengers, such as late Jurassic *Juralebris*, are unlikely to be carnivorous.

Keywords: Ostracoda, carnivory, predation, scavenging, parasitism.

4.2 Introduction

Ostracods are small, bivalved crustaceans that occupy all aquatic niches from the deep sea to temporary ponds and in moist leaf-litter around tree-lined lakes. They are represented by an estimated 33,000 living and fossil species and form the most diverse and prolific class of arthropods. Their mineralised carapace is readily fossilised, so that their record is long, the oldest known fossil carapaces of presumed Ostracoda being from the early Ordovician (about 490 million years old) (Hou et al. 1996). True ostracods (both Palaeocopa and Binodicopa) occur in the Tremadoc of Argentina (Salas, Vannier and Williams, in-press) and Estonia (Tinn and Meidla 2004).

Ostracods display a variety of feeding habits as filter-feeders, detrivores, herbivores and carnivores. Carnivory amongst modern ostracods is well established, having been observed in natural habitats and in aquarium experiments (e.g. Vannier et al. 1998). It takes three forms in ostracods: predation, scavenging and parasitism. Predation on larger animals is practiced by myodocopes such as *Gigantocypris* and *Macrocypridina* (Vannier et al. 1998; Moguilevsky and Gooday 1977). *Vargula* has been observed feeding on live polychaetes, but is also a scavenger (Vannier et al. 1998). Scavenging is the principle form of carnivory amongst ostracods, for example, in *Skogsbergia* (Cohen 1983) and in freshwater podocopes such as *Eucypris virens* and *Heterocypris incongruens*. Parasitism is poorly known, although the myodocope *Sheina orri* and, perhaps, *Vargula parasitica* have adopted this form of carnivory (Bennett et al. 1997).

Despite its widespread use as a feeding strategy amongst modern ostracods, carnivory is difficult to prove in the fossil record. It relies entirely on the exceptional preservation of distinctive, decay-prone, anatomical features and/or intimate association with food items. The importance and position of ostracods in ancient food webs is, therefore, difficult to assess. Six putative examples of fossilised scavengers are discussed herein. Although two of these are unlikely to be examples of carnivory, there is fossil evidence of scavenging ostracods in the Triassic, Carboniferous and Ordovician.

4.3 Carnivory in modern Ostracods

4.3.1 Predation

Predation by Recent myodocopes is normally restricted to small invertebrates such as worms, copepods and podocopid ostracods (Cohen 1982; Cohen and Kornicker 1987). One of the most voracious predators among ostracods is *Gigantocypris muelleri*, a large (up to 30mm long) and excellent swimmer that generally preys on fish fry, copepods and chaetognaths (Davenport 1990). Evidence for their diet being their gut contents, which were examined by Vannier et al. (1998) and Moguilevsky and Gooday (1977).

The act of predation has not been widely observed *in vivo*, and the most detailed account remains that of Vannier et al. (1998) who carried out numerous laboratory-based experiments. The introduction of live annelid polychaetes into an aquarium containing *Vargula hilgendorfii*, caused rapid swarming of the ostracods, which began biting at the mouth, gills, feeding tentacles, anus or ventral region of the body wall of the polychaetes (Fig. 1a), perhaps attracted to undigested food particles or secretions. Circular wounds on the epidermis and musculature resulted (Fig. 1b), and although these were initially not deep enough to be fatal, the release of blood and body fluids attracted yet more ostracods, which fed on the wounded areas, tearing more flesh. It required only moderate numbers (a few dozen) of cypridinids to kill and rapidly consume such prey (Vannier et al. 1998).

Swarms of ostracods on larger animals have also been recorded in freshwater lakes, although predation does not appear to have taken place. Seidel (1989) reported that swarms of the podocopid candonid *Cyclocypris serena* (which he erroneously called *C. ovum* – see Meisch 2000) were found clinging to toads (*Bombina variegata*) and newts (*Triturus vulgaris* and *T cristatus*). In one case, 263 individuals were counted on a single toad and, in another instance, 65 observed on a newt. In these instances, there was neither evidence that the ostracod swarms were attacking their "hosts" (they may have been feeding on the amphibian secretions) nor that they caused their death.

Predation by freshwater ostracods has been recorded, although not commonly. Amongst the Cyprididae, *Cypridopsis hartwigi* attacks ciliates, rotifers and insect larvae, and a swarm of between 10 and 15 individuals have been observed killing molluscs such as *Bullinus contortus* and *Planorbis glabratus* (Deschiens et al. 1953; Deschiens 1954). *Eucypris virens* feeds on the terminal setae of the phyllopod *Tanymastix* and it is also cannibalistic (Kiefer 1936). *Heterocypris incongruens* often attacks small prey such as *Daphnia magna*, cyclops, copepods, other ostracods, oligochaeta, cladocera and insect larvae (Ganning 1971; Meisch 2000).

Amongst the Candonidae, *Cypria ophthalica* has been observed attacking and killing the gastropod *Gyraulus crista* in aquarium experiments (Janz 1992). Normally, *C. ophthalica* lays its eggs on the shell of the gastropod, immediately prior to it spawning and shortly before the snail's death. The newly hatched ostracods then feed on the snail carcass (Janz 1992).

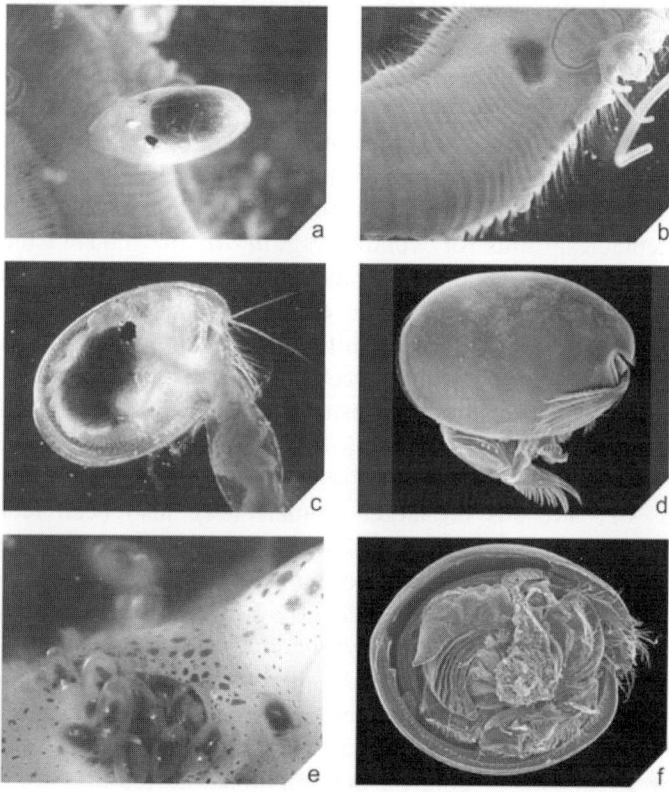

Fig. 1. A modern example of carnivory in ostracods: *Vargula hilgendorfii* (c. 3 mm long) predating on a live polychaete annelid. b. The wound on a live polychaete annelid after the attack of two specimens of *Vargular hilgendorfii*. c. *Vargula hilgendorfii* (c. 3 mm long) scavenging an insect larva, the a. food can be seen passing through the oesophagus and filling the stomach pouch. d. Lateral view of *Vargula hilgendorfii* (length carapace 2.5 mm) showing the furcae with claws and the second antennae for swimming. e. *Vargula hilgendorfii* (c. 3 mm long) scavenging on the carcass of a dead squid. The swarm attacked the eye first then penetrated into the body through the eye socket. f. *Leuroleberis surugaensis*, a typical cylindroleberidid, showing the soft parts (the gills and filtering setae are visisble). (carapace c. 5 mm long)

4.3.2 Scavenging

Scavenging is a common feeding strategy in Recent cypridinid myodocopes which are able to ingest relatively large amounts of food rapidly, between periods of fasting (Vannier et al. 1998). Morphological adaptation to scavenging includes the possession of a powerful furcal complex, which allows them to anchor firmly onto carrion and dismember it. Parker (1997) observed that the furca has several major functions in the feeding process of scavenging cypridinids such as *Azygocypridina lowryi*, including cutting and holding small food sections from an animal carcass and removing small fish scales. In baited trap experiments (Cohen 1983 1989; Collins et al. 1984; Stepien and Brusca 1985; Vannier and Abe 1993; Keable 1995; Parker 1997) swarms of cypridinids (sometimes in their thousands) were attracted to various food sources and in various states of decay. Swarms of several hundred individuals of *Vargula tsujii* gained access to internal tissues of dead or dying fish through natural openings (e.g. anus, gill chamber, genitalia) or lesions (Stepien and Brusca 1985; Vannier et al. 1998), although they alone did not cause the injury to the fish. They congregated along the base of their fins and around the operculae and anal openings, where they fed exclusively on mucus and skin, and it was only where isopods had caused serious injury that they fed on internal organs.

Vargula hilgendorfii behaves both as a predator, attacking for example, polychaete annelids, and as an opportunistic scavenger on carcasses of larger animals such as fish and squid. It is well adapted to this lifestyle morphologically; the fourth limb has strongly sclerotised setae on the endopodite and the furcal lamellae possess claws (Fig. 1d). These limbs are used to abrade and eventually tear open the integument of living prey and carcasses. Vannier et al. (1998) observed *V. hilgendorfii* clustering around injured areas of fish and penetrating the membrane covering the eye of a dead squid in order to consume the liquid contents (Fig. 1e).

Scavenging in the freshwater realm is exemplified by a number of members of the Cyprididae (Podocopa). Swarms of *Heterocypris incongruens* feed on dead gastropods (Meisch 2000) and carcasses of water birds (Reichholf 1983), and *Eucypris virens* has also been observed feeding on dead animals (Kiefer 1936).

4.3.3 Parasitism

Parasitism of fish by cypridinid myodocopes has been reported in several cases (Wilson 1913; Monod 1923; Harding 1966). The myodocope *Vargula parasitica* was said to congregate within the gills of hammerhead sharks (*Sphyrna zygaena*), sea bass (*Epinephelus adscensionis*) and jack fish (*Caranx crysos*) (Wilson 1913). It was argued that the ostracods had remained in the gills for some time, which was evidence of parasitism. The myodocope, *Skogsbergia squamosa,* was considered to be a parasite (Monod 1923), having been found on dead fish and securely attached to the head, mouth and back of live scorpion fish (*Scorpaena scrofa*) where, it was said, they probably fed off the blood of the host. Harding (1966) recorded the cypridinid *Sheina orri* firmly attached to the gills of fantail rays (*Taeniura lymna*) and an epaulette shark (*Hemiscyllium ocellatum*) and assumed they were parasitic. The parasitic lifestyle of these species was questioned by Cohen (1983), who suggested that attacks were carried out on dead or dying fish that had been injured as a result of trapping. Although this is probably the case for some records, Bennett et al. (1997) showed that parasitic ostracods exist; the gills of healthy specimens of *Hemiscyllium ocellatum*, offshore eastern Australia, were infested with *Sheina orri*. Evidence that *Sheina orri* led a parasitic lifestyle, attached to fish gills are: puncture marks and grooves in the gill epithelium; the ostracods always occurred between gill filaments and had been positioned there for some time, such that gill filaments were distorted around the carapace to form 'pockets'; all of the ostracods were consistently orientated with respect to the gill septum and arch, and generally in the lower part of the hemibranch; the preferred orientation of the ostracods within the 'pockets' precludes the notion that they were feeding on waterborne food particles moving through the gills; and the fact that the ostracods were attached to the host by means of hook-like claws on the mandible and fourth limb.

4.4 Carnivory in the fossil record

Despite the fact that carnivory in ostracods is widespread in modern oceans, examples in the fossil record are sparse and sometimes equivocal. Evidence of predation in deep time is unknown, although it is reasonable to suppose that it existed given the long history of nekto-benthonic myodocopes, at least back to the Silurian (Siveter 1984), and the known feeding habits of myodocopes such as *Gigantocypris, Macrocypridina* and *Vargula*. Parasitism is likewise unknown in the fossil record as, without the intimate ostracod-host relationship, parasitism cannot be directly proved. Scavenging is a feeding strategy adopted by many Recent marine invertebrates and ostracods are no exception (isopods for example which are sometimes found together with ostracods on the same carcass). Indeed, in certain settings ostracods form an important and prolific part of the scavenging guild (e.g. Keable 1995). In order to demonstrate more positively the occurrence of this lifestyle in ancient communities, the preservation of the scavenger in association with the scavenged carcass. It might be supposed that fossil evidence of scavenging by ostracods should be easily demonstrated, but it is in fact rare. There are six putative examples in the geological record that have been considered to demonstrate carnivory by scavenging, although two of these can no longer be sustained.

4.4.1 Early Cretaceous Pattersoncypris micopapillosa from Brazil

Bate (1971 1972) described an association between several hundred podocopid ostracods, *Pattersoncypris micopapillosa* (Figure 2a) and a teleost fish, *Cladocyclus gardneri*, from nodules within the Lower Cretaceous Santana Formation (Romualdo Member) on the flanks of the Chapada do Araripe, west of Recife, north-eastern Brazil. Bate considered *Pattersoncypris* to belong to Cyprididae/Cypridinae, but Smith (2000) pointed out its similarity with the modern species *Eucypris virens* and concluded that it should probably be placed into Cyprinotinae.

Bate drew attention to the powerful 'toothed' mandibles and the apparently close taphonomic relationship between the ostracods and the fish, concluding that "the animal was a scavenger, feeding on decaying plant and animal debris …" He argued that the *Pattersoncypris* specimens "were suddenly buried and rapidly asphyxiated, dying with their valves slightly agape." Considering the feeding habits of certain modern cyprids, this appears to be a reasonable argument, but there are several reasons why this conclusion cannot be supported. Ostracods occur in large numbers throughout the succession, in some cases forming coquina or thin ostracodal limestones, even when fish carcasses are absent. Therefore, the large numbers of carapaces associated with the fish cannot be unequivocally interpreted as swarming on carrion. The taphonomic relationship between the ostracods and the fish has not been fully resolved; none were found in an intimate relationship with the fish carcass, as is the case of other examples of scavenging discussed herein, and there is still a question whether they died at different times, or during the same widespread mortality event. In addition, Smith (1998 2000) demonstrated that the soft part anatomy of *P. micropapillosa* (Fig. 2b) showed close similarities with modern *E. virens* and concluded that not only can it be inferred that their limb functions were similar, but they also shared similar lifestyles. *Pattersoncypris*, he argued, would have been morphologically better suited to a nekto-benthonic, detritus feeding lifestyle rather than to a specialised scavenging one (Smith 2000, p. 95).

4.4.2 Jurassic Juraleberis jubata from Russia

Dzik (1978) documented the presence of several myodocope ostracods in the gut region of an Upper Jurassic pliosaur reptile from the early Volgian (Tithonian) of Pugatchov, Saratov District of Russia, which he considered had been inadvertently ingested by the reptile. Boucot (1990, p. 209) acknowledged this possibility but, in addition, suggested that scavenging was a conceivable alternative explanation for the association.

Following a detailed examination of the holotype by Vannier and Siveter (1995), a scavenging relationship now seems unlikely. Although its soft part anatomy is imperfectly known, *Juraleberis jubata* (Fig. 2c) shows affinities with *Leuroleberis* (Fig. 1f) and *Cycloleberis* and is assigned to Cylindroleberididae and questionably to Cyclasteropinae. In modern cylindroleberidids, the maxillae and fifth limb create water currents and food particles are collected by the maxillar setae and directed to the mouth by spatulate-like exopodites (Cannon 1933; Vannier and Siveter 1995). However, Cannon (1933) interpreted structures on the 4th, 5th (exopodite) and 6th limbs in Recent myodocopes, such as *Cyclasterope,* as filtering devices, although he made no *in vivo* obsevations. It now seems possible that these large ostracods were opportunistic filter feeders, able to grasp and tear open carcasses by using their furca, releasing food particles (especially if decay is advanced) so that the synchronized filtering device formed by the 3 pairs of appendages, can be used by the ostracod to gather the small food particles. This situation has been observed in association with truly scavenging *Vargula*; clouds of minute food particles are sometimes created when it feeds (Vannier, unpublished information). The presence of similar structures on the appendages of *Juraleberis* (Figure 2d) suggests that it belongs to this category of filter feeders.

4.4.3 Triassic Triadocypris spitzbergensis from Spitzbergen

One of the most compelling and widely accepted examples of scavenging in the fossil record has been documented by Weitschat (1983), who recovered exceptionally well-preserved myodocopes in close association with ammonoids from the Lower Triassic of Spitzbergen. The association is so well preserved that even the cilliate ectoparasites were present (Weitschat and Guhl 1994), and the ostracods retain anatomical features (Fig. 3a), including the first and second antennae, the epipodial fan of the fifth limb and the seventh limb. Similarities can be seen between the preserved soft parts of *Triadocypris spitzbergensis* and the modern nekto-benthonic *Vargula hilgendorfi*, although the feeding appendages are missing or poorly preserved (questionable setae of the 'biting apparatus' on the fourth limb are present) (Vannier et al. 1998). Unfortunately, a complete furca with furcal lamella (with claws) is absent, although a sclerotised plate (or sclerosome) is present (Weitschat 1983; Parker 1997). However, their close association with the ammonoids suggests a predatory or scavenging mode of life (Vannier et al. 1998, p. 405-406).

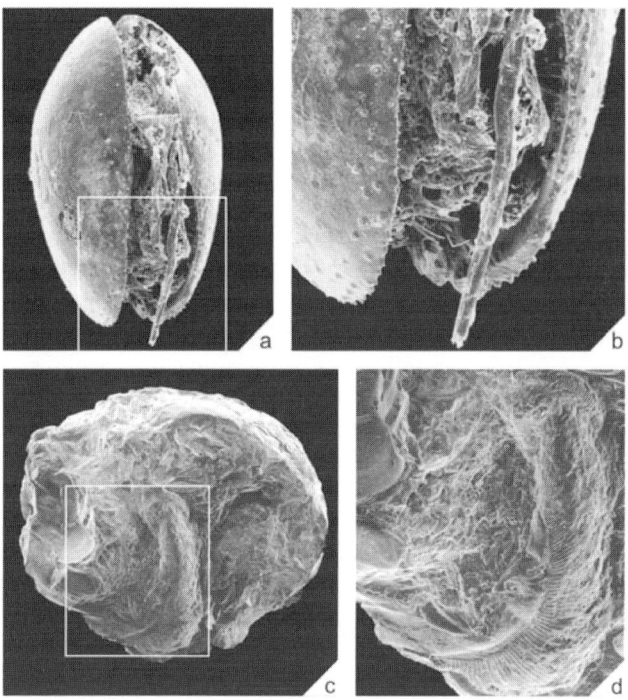

Fig. 2. Filter-feeding ostracods that were originally believed to be carnivorous: a. *Pattersoncypris micopapillosa.* A gaping carapace with preserved soft tissue from the Lower Cretaceous Santana Formation, Brazil, and originally considered to be scavenging teleost fish carcasses. Length: 1 mm (MPK13401). b. Details of the same specimen preserved soft tissue. c. *Juraleberis jubata,* an opportunistic filter-feeder.. Carapace (3.3 mm long) left lateral view with left valve removed to show the soft anatomy (holotype PIN 3775/1). Collected from the late Jurassic of Russia,and found in association with a pliosaur. d. Lateral detail of the epipodite of left 5th limb of the same specimen. (Taken from plate 22, 91, figs 1 and 2 of Vannier, J. and Siveter, D. J. 1995. *Stereo-Atlas of Ostracod Shells* 22, 86-95)

Swarming is one of the characteristics of scavenging ostracods and large numbers (50-100 individuals) of *Triadocypris spitzbergensis* were found restricted to the body chambers of the ammonoid *Keyserlingites,* where they apparently fed on soft tissues.

4.4.4 Carboniferous Eocypridina carsingtonensis from Central England

The Lower Hays Farm borrow pit near Carsington, Derbyshire, formerly exposed the Namurian Bowland Shale Formation. It yielded a large carbonate concretion containing marine faunas dominated by goniatites and bivalves, together with an association between a shark (*Orodus* sp.) and a large number of individuals of the myodocope ostracod *Eocypridina carsingtonensis* (Fig. 3b) (Wilkinson et al. 2004; Wilby et al. in press). At least 250 individuals of *E. carsingtonensis* crowd the upper surface of the shark and the displaced sections of its dermis, and also occur in smaller numbers amongst the teeth and beneath flaps of dermis behind the head (Fig. 3c-d). None occur beneath the shark or at any other level within the concretion.

The association was located near the northern edge of the WNW-ESE orientated Namurian Widmerpool Gulf where soupy mud-rich substrates accumulated in poorly oxygenated water, locally at depths of at least 100m (Holdsworth 1966; Trewin and Holdsworth 1973; Church and Gawthorpe 1994). As *E. carsingtonensis* has a well-developed, hook-like rostrum and a broad rostral sinus, it was probably capable of active swimming. However, unlike Recent pelagic non-mineralized halocypridids such as *Conchoecia* and, for example *Gigantocypris*, which typically have weakly calcified valves, those of *E. carsingtonensis* are moderately well mineralised suggesting a nekto-benthonic ecology, comparable to that of the extant, morphologically similar species *Vargula hilgendorfii* (Wilkinson et al. 2004). The association suggests that *E. carsingtonensis* was feeding on the shark, the relationship being predatory, scavenging or parasitic.

Although there are numerous examples of Recent myodocopes attacking small invertebrates (such as worms, copepods and fish fry), they alone could not cause serious injury to larger fish. Indeed, unrestrained fish would have simply evaded the ostracods or shaken them off. Consequently, a predatory relationship is unlikely. The only known examples of parasitism in myodocopes show intimate association with fish gills. This is inconsistent with the fossil association from Carsington, and the number and the size of the ostracods involved are very different, making a conclusion involving parasitism, unlikely.

Eocypridina carsingtonensis is interpreted as having been preserved whilst scavenging the carcass (Wilby et al. in press). Scavenging is a common feeding strategy and cypridinids are rapidly attracted, often in thousands, to a variety of food sources (including fish) in various states of decay. Certainly, it can be that the relatively large *Orodus* carcass would have exerted a powerful attraction on any opportunistic scavenging ostracods. Although it is unlikely that they would have been able to inflict serious damage to large fish such as *Orodus*, they may have gained access to internal tissues through natural openings (e.g. anus, gill chamber, genitalia) or lesions of a dead or dying fish to achieve their observed distribution. Recent carnivorous myodocopes have well-developed eyes (Vannier and Abe 1992) and some predatory mesopelagic species (e.g. *Gigantocypris* and *Macrocypridina*) show adaptations to low light levels (Land and Nilsson 1990). Opportunistic, scavenging myodocopes, however, do not appear to rely on vision to locate carrion (Stepien and Brusca 1985; Vannier et al. 1998), but chemoreceptors detect leaking body fluids and other organic constituents (Anderson 1977; Parker 1998; Stepien and Brusca 1985; Vannier et al. 1998). This appears to have been the case in *Eocypridina carsingtonensis* was clearly capable of locating the *Orodus* carcass and was prepared to travel relatively large distances into the basinal parts of the Widmerpool Gulf, in order to reach the shark (its absence from the background sediment indicates that it did not generally inhabit this part of the Gulf). Light levels would have been greatly reduced at depths in excess of 100m, but this was clearly not a problem; it may have behaved in a similar way to modern scavenging myodocopes, which are most active at night.

4.4.5 Early Silurian Colymbosathon ecplecticos from Herefordshire, England

The Lower Silurian Wenlock Series of Herefordshire, west Central England, includes volcanic ash with nodules (Briggs *et al* 1996; Orr et al. 2000). A number of animals, including soft part anatomy, are preserved in 3-dimensions within the nodules, including *Colymbosathon* (Fig. 3c), a cylindroleberidid myodocope that lacks the setose comb on the fifth appendage characteristic of modern forms (Siveter et al. 2003). It led a nekto-benthonic lifestyle at water depth of 150-200 m and is believed to have been a micro-predator and scavenger. As in modern scavengers, its furca is well developed, perhaps to hold and cut prey or carrion. Abe et al. (1996) showed that enzymes were present in the upper lip of Recent *Vargula* to aid digestion, and Siveter et al. (2003) speculated that processes on the labrum of *Colymosathon* may have been similar enzyme-secreting organs.

4.4.6 Late Ordovician Myodoprimigenia fistuca from South Africa

Gabbott et al. (2003) reported associations between ostracods and orthoconic cephalopods in the Upper Ordovician (upper Asgill) Cedarberg Formation of South Africa. An argillaceous unit at Keurbos, the Soom Shale Member, comprises finely laminated mudstones believed to have accumulated in a shallow water milieu in a sheltered embayment. Small numbers (typically <10) of the benthonic or nekto-benthonic myodocope ostracod, *Myodoprimigenia* (Fig. 3f), are preserved on the upper surface or within the body chambers of the orthocones (Fig. 3g-h) and, in one case, associated with the radula. Several explanations for the associations can be envisaged due to indifferent preservation (Gabbott et al. 2003). For example, a purely fortuitous taphonomic causation for the ostracod/cephalopod association might be argued, although this can be discounted as the delicate myodocopes show no signs of transportation and the sediment preserves evidence of neither current activity or bioturbation. The intimate association of ostracods with the orthocones and their rarity in adjacent sediment and elsewhere in the Soom Shale Member, suggests that the ostracods were scavenging cephalopod carrion. The association, therefore, represents the oldest known evidence of ostracod carnivory.

4.4.7 Cambrian bradoriids and the origin of carnivory

The origins of carnivory amongst Ostracoda are difficult to identify, as evidence in the Cambrian is wanting, although this trophic position may have been attained very early in the evolution of the group. Ostracods may have been represented in the Cambrian by some Bradoriida (Fig. 4a-b), although this probably polyphyletic taxon includes definite non-ostracod genera such as *Kunmingella* (Fig 4c) (Hou et al. 1996; Shu et al. 1999; Hou et al. 2004). Bradoriids such as *Kunmingella douvillei* are different from ostracods morphologically, but they may have occupied similar ecological niches (Shu et al. 1999; Vannier and Chen 2005). Carapace morphology amongst bradoriid arthropods varies considerably, perhaps reflecting, in some way, their preferred environmental niches and trophic position.

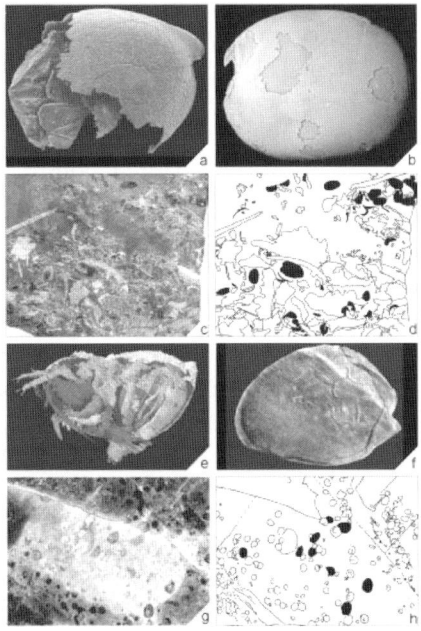

Fig. 3. Ancient scavenging myodocope ostracods. a. *Triadocypris spitzbergensis,* from the early Triassic of Flowerdalen, Spitzbergen, was found associated with ammonoids, and considered to be scavenging. Right valve, lateral view showing gill-like structures, epipodite of 5th limb and 7th limb (holotype, GPIHM 2558, 2.9 mm long). (taken from Wietschat, W. 1983. Stereo-Atlas of Ostracod Shells 10, 127 –138, late 10, 130, fig 1). b. *Eocypridina carsingtonensis* from the Carboniferous of Derbyshire, Central England, scavenged the shark *Orodus*. Left valve, lateral view (8 mm long) (GSM105459) (taken from plate 1 fig. 2 of Wilkinson et al. 2004, *Revista Española de Microplaeontología*). c. swarm of *E. carsingtonensis* in association with dermis from Orodus. d. diagrammatic representation showing the ostracods (black) and dermal tissue (white). e. *Colymbosathon ecplecticos* from the Early Silurian of Herefordshire, England. Left lateral view of the male holotype (OUM C.295670) (taken from Siveter et al. 2003, fig. A, Science, 300). f. *Myodoprimigenia fistuca,* from the Late Ordovician of South Africa, fed on the carcasses of orthoconic cephalopods. Right lateral view of a carapace (holotype) (on slab C409). g. Swarm of *Myodoprimigenia fistuca* on a cephalopod (slab C1002). Soom Shale, S. Africa.. h. Diagrammatic representation of swarming *Myodoprimigenia fistuca* (black) on cephalopod remains (white) (taken from Gabbott et al. 2003, figs 2c-d, 4c *Lethaia*)

Some bradoriids may have been motile epibenthic dwellers of the sediment-water interface in a similar fashion to modern ostracods (Vannier et al. 2005), but the suggestion that bradoriids may have been micro-scavengers or micro-predators of unknown meiofaunal organisms (Vannier et al. 1998), is conjectural. The protruding antennae of *Kunmingella* appear powerful and might have been used for feeding. Their high abundance supports the notion that bradoriids were important recyclers on the Early Cambrian seabed (Shu et al. 1999), and coprolite evidence within the Chengjiang (Vannier and Chen 2005) and Burgess Shale Lagerstätten (Williams and Siveter unpublished data) indicates they were themselves, an important food source for larger animals.

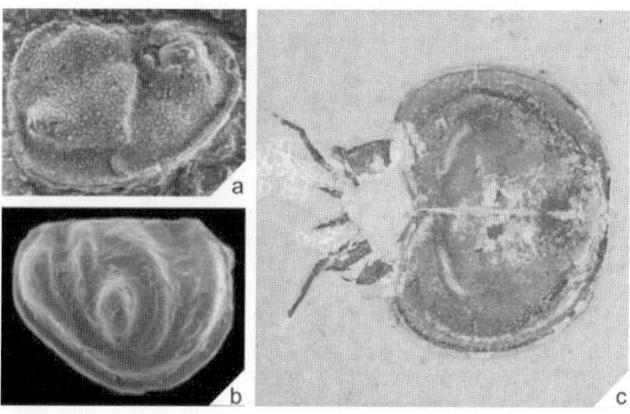

Fig. 4. Figure 4. Bradoriids may have been micro-scavengers or micro-predators and some may have been motile epibenthic species that occupied the same niches that ostracods were later to adopt. Some bradoriids may have been ostracods e.g. Fig. 4a. *Altajanella costulata* Melnikova (1.19mm long) from the Tandoshka Formation, Gorny Altay, Upper Cambrian, and Fig. 4b. *Vojbokalina magnifica* Melnikova (1.37mm long) from the Sablinka Formation, Leningrad Region, Russia, Middle Cambrian. However, *Kunmingella douvillei* (Fig. 4c, length of whole specimen including appendages is 7mm) is a bradoriid from the Chengjiang Lagerstaette, Lower Cambrian, Yunnan, China, that is no longer placed within the Ostracoda on the basis of the preserved soft part anatomy (Photograph by Derek Siveter; taken from Hou et al., 2004, The Cambrian Fossils of Chengiang, China: The Flowering of Early Animal Life, Blackwell Science Ltd, Oxford)

There is no evidence of carnivory in bradoriids or in Ordovician palaeocopes. Myodocopes are, therefore, the oldest known carnivorous taxa amongst the Ostracoda. Carnivory may have been one mechanism that enabled ostracods successfully to colonise the water column and search for food, and, once pelagic, to enter environments from which the palaeocopes were excluded.

4.5 Conclusions

The origins of carnivory amongst Ostracoda are difficult to identify, but this trophic position appears to have been attained very early in the evolution of the group. Predation may have begun amongst the benthonic and nekto-benthonic species of bradoriids and this lifestyle was adopted by early myodocopids.

Although there are numerous species of carnivorous ostracods in Recent marine and freshwater milieux, there are few convincing examples in the fossil record. Without soft part anatomy, functional morphology of carnivorous taxa cannot be identified and we are forced to rely on secondary evidence. There are, however, several cases of carnivory by scavenging, all by myodocope ostracods. These are the *Myodoprimigenia*/orthocone association in the Soom Shale Member, South Africa, the *Eocpridina*/shark association in the Bowland Formation, Central England, and the *Triadocypris*/ammonoid association in the Lower Triassic of Spitzbergen. Although not recovered in association with carrion or prey, the soft part anatomy of *Colymbosathon*, from the Lower Silurian of Herefordshire, England, suggests it was probably a predator or scavenger.

4.6 Acknowledgements

IPW and PRW publish with permission of the Executive Director of the British Geological Survey (N.E.R.C.).

References

Andersson A (1977) The organ of Bellonci in ostracodes: an ultrastructural study of the rod-shaped, or frontal, organ. Acta zoologica (Stockholm) 58 197-204

Bate RH (1971) Phosphatized ostracods from the Cretaceous of Brazil. Nature 230, 397-398

Bate RH (1972) Phosphatized ostracods with appendages from the Lower Cretaceous of Brazil. Palaeontology 15, 379-393

Bennett MB, Heupel MR, Bennett SM, Parker AR (1997) *Sheina orri* (Myodocopa: Cypridinidae), an ostracod parasitic on the gills of the epaulette shark, *Hemiscyllium ocellatum* (Elasmobranchii: Hemiscyllidae). International Journal for Parasitology 27(3), 275-281

Boucot AJ (1990) Evolutionary Paleobiology of behaviour and coevolution 725pp. [Elsevier, Amsterdam]

Briggs DEG, Siveter DJ, Siveter DJ (1996). Soft-bodied fossils from a Silurian volcaniclastic deposit. Nature 382, 248-250

Cannon HG (1933) On the feeding mechanism of certain ostracods. Transactions of the Royal Society of Edinburgh 57, 739-764

Church KD, Gawthorpe RL (1994) High resolution sequence stratigraphy of the late Namurian in the Widmerpool Gulf (East Midlands, UK). Marine and Petroleum Geology 11, 528-544

Cohen AC (1982) Ostracoda. In Parker, S.P. (ed.) Synopsis and classification of living organisms 181-202. [McGraw-Hill, New York]

Cohen AC (1983) Rearing and postembryonic development of the myodocope ostracode Skogsbergia lerneri from coral reefs of Belize and the Bahamas. Journal of Crustacean Biology 3, 235-256

Cohen AC (1989) Comparison of myodocope ostracods in the two zones of the Belize barrier reef near Carrie Bow Cay with changes in distribution 1978-1981. Bulletin of Marine Science 45, 316-337

Cohen AC, Kornicker LS (1987) Catalog of the Rutidermatidae (Crustacea: Ostracoda). Smithsonian Contributions to Zoology 449, 1-11

Collins KJ, Ralston S, Filak T, Bivens M (1984) The susceptibility of Oxyjulis californica to attack by ostracods on three substrates. Bulletin of the Southern Californian Academy of Science 83, 53-56

Davenport J (1990) Observations on swimming, posture and buoyancy in the giant oceanic ostracods, Gigantocypris mülleri and Macrocypridina castanea. Journal of the Marine Biological Association of the UK 70, 43-55

Deschiens R (1954) Mécanisme de l'action léthale de Cypridopsis hartwigi sur les mollusques vecteurs des bilharzioses. Bulletin de la Société de Pathologie Exotiques 47, 399-401

Deschiens R, Lamy L, Lamy H. (1953) Sur un ostracode prédateur de Bullins et de Planorbes. Bulletin de la Société de Pathologie Exotiques 46, 956-958

Dzik J (1978) A myodocope ostracode with preserved appendages from the Upper Jurassic of the Volga River region (USSR). Neues Jahrbuch fur Geologie und Palaontologie Monatshefte 7, 393-399

Gabbott SE, Siveter DJ, Aldridge RJ, Theron JN (2003) The earliest myodocopes: ostracodes from the late Ordovician Soom Shale Lagerstätte of South Africa. Lethaia 36, 151-160

Ganning B (1971) On the ecology of Heterocypris salinus, H. incongruens and Cypridopsis aculeata (Crustacea, Ostracoda) from Baltic brackish-water rockpools. Marine Biology 8, 271-279

Harding JP (1966) Myodocopan ostracods from the gills and nostrils of fishes. In Barnes H (ed.) Some contemporary studies in marine science 369-374 [Allen and Unwin, London]

Holdsworth BK (1966) A preliminary study of the palaeontology and palaeoenvironment of some Namurian limestone 'bullions'. Mercian Geologist 1, 315-337.
Hou XG, Siveter DJ, Williams M, Walossek D, Bergstöm J (1996) An early Cambrian bradoriid arthropod from China with preserved appendages: its bearing on the origin of the Ostracoda. Philosophical transactions of the Royal Society of London, Series B 351,131-1145
Hou XG, Aldridge RJ, Bergström J, Siveter DJ, Siveter DJ, Feng XH (2004) The Cambrian Fossils of Chengiang, China: The Flowering of Early Animal Life 233 pp, [Blackwell Science Ltd, Oxford]
Janz H (1992) Eine fakultative Beziehung zwischen *Cypria ophthalmica* (Jurine) (Ostracoda) und *Gyraulus crista* (L.) (Gastropoda) und ihre mögliche biologische Bedeutung. Stuttgarter Beiträge zur Naturkunde, Serie A (Biologie) 476, 1-11
Keable SJ (1995) Structure of the marine invertebrate scavenging guild of a tropical reef ecosystem: field studies at Lizard Island, Queensland, Australia. Journal of Natural History 29, 27-45
Kiefer F (1936) Über die Krebstiere, insbesondere die Ruderfusskrebse des Eichener Sees. Beiträge zur naturkundlichen Forschung in SW-Deutchland 1, 157-162
Land M, Nilsson D-E (1990) Observations on the compond eyes of the deep-sea ostracod Macrocypridina castanea. Journal of Experimental Bioogy 148, 221-233.
Meisch C (2000) Freshwater Ostracoda of western and central Europe. Süsswasserfauna von Mitteleuropa 8/3, 522 pp [Spektrum Akademischer Verlag, Heidelberg]
Moguilevsky A, Gooday AJ (1977) Some observations on the vertical distribution and stomach contents of *Gigantocypris muelleri* Skogsberg 1920 (Ostracoda, Myodocopina). In: Loeffler H, Danielopol D (eds) Aspects of ecology and zoogeography of Recent and fossil ostracods, Proceedings of the 6th International Symposium on Ostracoda, Saafelden, Austria 263-270 [W. Junk bv, The Hague]
Monod T (1923) Notes carcinologiques. (Parasites et commensaux). Bulletin de L'Institut Océanographique 427, 1-23
Orr PJ, Briggs DEG, Siveter DJ, Siveter DJ (2000) Three-dimensional preservation of a non-biomineralized arthropod in concretions in Silurian volcaniclastic rocks from Hereford, England. Journal of the Geological Society, London 157, 173-186.
Parker AR (1997) Functional morphology of the myodocopine (Ostracoda) furca and sclerotized body plate. Journal of Crustacean Biology, 17, 632-653
Parker AR (1998) Exoskeleton, distribution, and movement of the flexible setules on the Myodocopine (Ostracoda: Myodocopa) first antenna. Journal of Crustacean Biology 18, 95-110
Reichholf J (1983) Ökologie und Verhalten des Muschelkrebses Heterocypris incongruens Claus, 1892. Spixiana 6 205-210

Seidel B (1989) Phoresis of Cyclocypris ovum (Jurine) (Ostracoda, Podocopida, Cyprididae) on Bombina vatiegata (L.) (Anura, Amphibia) and Trituris vulgaris (L.) (Urodela, Amphibia). Crustaceana 57, 171-176

Shu D-G, Vannier J, Luo H-L, Chen L, Zhang X-L, Hu S-X, (1999). The anatomy and lifestyle of Kunmingella (Arthropoda, Bradoriida) from the Chengjiang fossil Lagerstätte (early Cambrian; southern China). Lethaia 42, 279-298

Siveter DJ (1984) Habitats and modes of life of Siluian ostracodes. Special Papaers in Palaeontology 32, 71-85

Siveter DJ, Vannier JMC, Palmer D (1991) Silurian Myodocopes: Pioneer pelagic ostracods and the chronology of an ecological shift. Journal of Micropalaeontology 10, 151-173

Siveter DJ, Sutton MD, Briggs DEG, Siveter DJ (2003) An ostracode crustacean with soft parts from the Lower Silurian. Science 300, 1749-1751

Smith RJ (1998) Biology and ontogeny of Cretaceous and Recent Cyprididae Ostracoda (Crustacea). Unpublished PhD thesis, University of Leicester

Smith RJ (2000) Morphology and ontogeny of Cretaceous ostracods with preserved appendages from Brazil. Palaeontology 43, 63-98

Stepien CA, Brusca RC (1985) Nocturnal attacks on nearshore fishes in southern California by crustacean zooplankton. Marine Ecology – Progress Series, 25, 91-105.

Tinn O, Meidla T (2004) Phylogenetic relationships of Early Middle Ordovician ostracods of Baltoscandia. Palaeontology 47 199-221

Trewin NH, Holdsworth BK (1973) Sedimentation in the Lower Namurian rocks of the north Staffordshire Basin. Proceedings of the Yorkshire Geological Society 19, 371-408

Vannier J, Abe K (1992) Recent and early Palaeozoic myodocope ostracodes: functional morphology, phylogeny, distribution and lifestyles. Palaeontology 35, 485-517

Vannier J, Abe K (1993) Functional morphology and behaviour of Vargula hilgendorfii (Ostracoda: Myodocopea) from Japan, and discussion of its crustacean ectoparasites: preliminary results from video recordings. Journal of Crustacean Biology 13, 51-76

Vannier J, Chen J-Y (2005) Early Cambrian food chain: new evidence from fossil aggregates in the Maotianshan Shale biota, SW China. Palios 20, 3-26

Vannier J, Siveter DJ (1995) On Juraleberis jubata Vannier and Siveter gen. et sp. nov. Stereo-Atlas of Ostracod Shells 22 (21), 86-95

Vannier J, Abe K, Ikuta K (1998) Feeding in myodocope ostracods: functional morphology and laboratory observations from videos. Marine Biology 132, 391-408

Vannier J, Williams M, Alvaro JJ, Vicaïno D, Monceret S, Monceret E (2005) New Early Cambrian bivalved arthropods from southern France. Geological Magazine 142, 751-763

Weitschat W (1983) Ostracoden (O. Myodocopea) mit Weichkörper-Erhaltung aus der Unter-Trias von Spitzbergen. Paläontologische Zeitschrift 57, 309-323.

Weitschat W, Guhl W (1994) Erster Nachweis fossiler Ciliaten. Paläontologische Zeitschrift 68, 17-31

Wilby PR, Wilkinson IP, Riley NJ (in press) Late Carboniferous scavenging ostracods: feeding strategies and taphonomy. Transactions of the Royal Society of Edinburgh, Earth Sciences

Wilkinson IP, Williams M, Siveter DJ, Wilby PR (2004) A Carboniferous necrophagous myodocope ostracod from Derbyshire, England. Revista Española de Micropaleontología 36 195-206

Wilson CB (1913) Crustacean parasites of West Indian fishes and land crabs, with descriptions of new genera and species. Proceedings of the United States National Museum 44, 189-277

5 Trophic relationships in crustacean decapods of a river with a floodplain

P. Collins[1,2,3], V. Williner[1,2] and F. Giri,[1,3]

[1]Instituto Nacional de Limnología (CONICET-UNL), José Maciá 1933, 3016 Santo Tomé; [2]FByCB-FHUC, Universidad Nacional del Litoral; [3]FCyT, Universidad Autónoma Entre Ríos, Argentina,
pcollins@arnet.com.ar, wvero@arnet.com.ar, fegiri@infovia.com.ar

5.1 Abstract

This review deals of trophic relationships of crustacean decapods in rivers with a floodplain. To do so the feeding aspects of some represents of four families (Sergestidae, Palaemonidae, Trichodactylidae and Aeglidae) of neotropicals decapods were analysed. Freshwater prawns and crabs are considered to be not only predators but also preys being important elements in the food chain of large rivers. This group is found everywhere in the Parana systems, ranging from rivers to wetlands, from pelagic to semi terrestrial areas, from free vegetation, such as beaches, to vegetated areas. These crustaceans take part in the matter-energy exchange between aquatic and terrestrial systems. This occurs because they constitute a source of food for terrestrial and aquatic animals. Approximately, 85% of the feeding ecology of prawns (Sergestidae and Palaemonidae) and 45 % of crabs (Trichodactylidae and Aeglidae) of the Paraná River and its main tributaries has been studied. From the data obtained from stomach content analyses, it is evident that the diet of these decapods is mainly omnivorous, using different trophic levels (plant remains, algae, zooplankton, insect larvae, and oligochaetans, among others). Resource use may vary according to diverse cycles (ontogenetic, ecdysis cycle, daily, annual etc.) and also to the inter and intraspecific competitive pressure.

Keywords: Natural feeding, alluvial valley, freshwater, crabs, prawns, stomach contents analysis.

5.2 Introduction

Trophic relationships among Crustacea Decapoda are strongly influenced by the association between communities and environments. Our example deals with decapods interaction at different trophic levels in rivers with a floodplain. Freshwater prawns and crabs are considered to be not only predators but also preys being important elements in the food chain of the large rivers. This group is found everywhere in rivers, from its principal channel (e.g. planktonic shrimps) to the terrestrial area where cave-dwelling crabs may travel many kilometers on the savanna under certain circumstances of hydric stress (Fernández and Collins 2002). The abiotic and biotic factors characteristic of these large rivers, e.g. hydric and thermal cycles, join to form a complex interrelationship. It must be taken into account that freshwater decapod ancestors and current sea taxa live in a more stable habitat. In South America, the Paraná River flows through an important floodplain of high environmental diversity. Therefore, and due to its particular characteristics, some areas have been declared RAMSAR sites (Giraudo and Cordiviola 2002). Its decapod fauna (Table 1) is less diverse than the Amazon or other rivers in warmer regions, but one feature of this system is that densities are very high (Collins 2000a, Collins et al. 2006). The Paraná River system, together with the biogeographical regions of the Amazon – Pantanal – Matto Grosso shares about 70% of freshwater families. However, many of the components of the former are endemic, e.g. *Palaemonetes argentinus* and *Macrobrachium borellii* (Collins et al. 2004).

Table 1. Genus of Crustacean Decapod at the Paraná River and their floodplain

Infraorder	Family	Genus	Vulgar name
	Sergestidae	Acetes	Planctonic shrimp
Caridea	Palaemonidae	Macrobrachium	Prawn
		Palaemonetes	Prawn
		Pseudopalaemon	Prawn
Brachyura	Trichodactylidae	Trichodactylus	Crab
		Dilocarcinus	Crab
		Sylviocarcinus	Crab
		Zilchiopsis	Crab
		Poppiana	Crab
		Valdivia	Crab
Anomura	Aeglidae	Aegla	Pancora or Crab

This decapod group is probably found in a variety of habitats ranging from rivers to wetlands, from pelagic to semi terrestrial areas, from free vegetation, such as beaches, to vegetated areas. Furthermore, the same species can change habitats according to different internal (ontogenetic and ecdysis cycles) and external (hydric, daily, annual cycles) conditions, thus influencing its intra and interspecific relationships (Collins et al. 2006).

There can, however, be a wide range of changes in the physicochemical parameters of the environments where they live. For example, conductivity ranges from 60 µS cm^{-1} to 7,000 µS cm^{-1}; temperatures can fall below 10 ºC or rise above 40 ºC; or a certain environment can even dry out (Fernández and Collins 2002; Collins 2005a; Collins inedited).

In the diverse environments existing along the Paraná River and their tributaries, Crustacea Decapoda can survive and adjust to the variations of abiotic conditions, including those associated to the hydric cycle, namely high and low water periods and even drying in certain cases. For these adjustments to be effective, feeding plays a significant role, having to supply the nutrients that development requires at each ontogenetic stage in all environments and seasons.

It is important to consider that this system started at Pliocene 3 to 4 million years ago (Paoli et al. 2000) and in this period these crustaceans had to adjust to unstable conditions, even when coming from a more stable system like the sea, or migrating from other freshwater systems, undergoing all the typical changes of freshwaterization.

5.2.1 Physical Environment: A river with a floodplain

Large rivers with extensive floodplains can be found in South America. The Paraná River, from the Plata system, is the second in importance. It has a basin of approximately 2.8 million km^2. In general, it has a soft slope and its main channel is anastomosed, forming several secondary streams, islands with internal ponds and a complex system of shallow lakes, the dynamics of which follow the hydric cycle (Fig. 1). The cycle has a regular flow, with high and low water periods occurring in winter and summer respectively. Water discharge ranges from 500 million m^3.s^{-1} to 65,000 million m^3.s^{-1}, their mean values being 16,000 m^3.s^{-1}. This variation causes different degrees of communication between the lotic and lentic environments throughout the year (pre-isolated, isolated, post-isolated and flooded periods), the identity of the shallow lakes being lost during floods. These differences are characteristics of each lentic environment and depend on lake volume, distance from river, amplitude-longitude of flooding pulse and time of water residence (Fig. 1) (Bonetto and Waiss 1995).

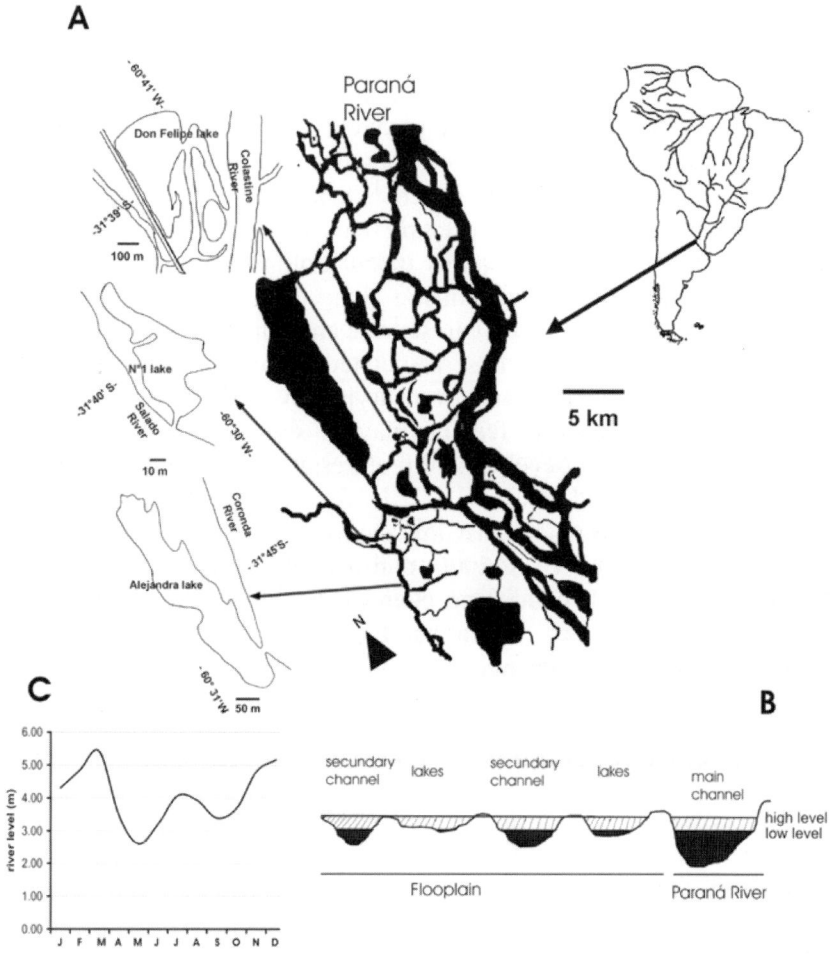

Fig. 1. Schematic representation of rivers with floodplain in South America (e.g. Paraná River) including details of three shallow lakes (A); transverse section of the Paraná River showing the main channel and floodplain with lakes and secondary channels during flooding (high level) and falling (low level) phases (B); normal water-cycle level of the Paraná River (C)

5.2.2 Potential predators

When considering trophic interrelationships, those groups which can pose a direct or indirect risk of minor or extreme injury to these crustaceans are included.

First of all, it must be pointed out that cannibalism occurs within this group, mainly involving those individuals which have just molted, or those of a smaller size (Williner and Collins 2000).

As also happens among certain other species, some prawns attack other less aggressive species: in the case of two sympatric species, *M. borellii* causes more frequent damage to *P. argentinus*. The analysis of stomach content indicated that it is possible to establish trophic relations and place the species in a trophic web. This network showed that decapods are an important component of the aquatic community due to their density and their role in the energy transfer (Fig. 2). Several authors refer to Decapoda as the components of the fish trophic spectrum (Bonetto et al. 1963; Cabrera et al. 1973; Oliva et al. 1981). Furthermore, crustaceans take part in the matter-energy exchange between aquatic and terrestrial systems. This happens when they are preyed upon by bird species (Beltzer 1983a, b; Beltzer and Paporello 1984; Beltzer 1984; Navas 1991; Bó and Darrieu 1993; Lajmanovich and Beltzer 1993; Navas 1993, 1995). In addition, it can be seen in literature that crabs of the Trichodactylidae family are part of the trophic spectrum of some mammals and amphibians (Massoia 1976; Bianchini and Delupi 1993; López et al. 2006). Moreover, crabs from the Aeglidae family have been mentioned as preys of caimans (Bond Buckup and Buckup 1994). Likewise, some aeglids have been found in the stomachs of fish (Ferriz et al. 2000), and the Neotropical river otter (Gori et al. 2003). Some crabs have also been commonly preyed upon by snakes (Gallardo 1977; Williams and Scrochi 1994) and even by monkeys (Port-Carvalho et al. 2004).

5.2.3 Potential preys

Even minor components are included when considering the potential trophic offer. This is the case of fungi, for example, since their intake supplies nutrients to the group, whether directly or indirectly. This could be attributed to the fact that these elements can be deliberately caught or occasionally accompany the intake of other preys. Furthermore, some bacteria can contribute to the decomposition of some elements when the food has been transported to the interiors of the digestive tract.

Also, in order to evaluate potential preys, the various communities (planktonic, benthic, and associated vegetation) have to be evaluated.

Natural fauna supplies and satisfies decapods with macronutrients (proteins, lipids, and carbohydrates) and micronutrients such as vitamins, cholesterol, phospholipids, and minerals (Collins 2004). Common preys provide them with a high ratio of proteins (49-59 %) and planktonic elements such as cladocerans, copepods, and phytoplankton play an important role at certain moments of their ontogenetic and seasonal cycles.

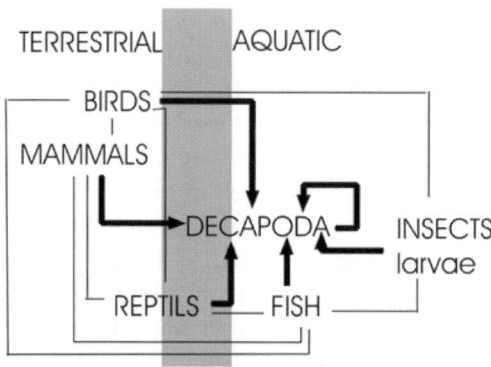

Fig. 2. Potential predators of Crustacean Decapods from aquatic and terrestrial environments of the Paraná River and their possible direct (thick line) and indirect (fine line) relationships

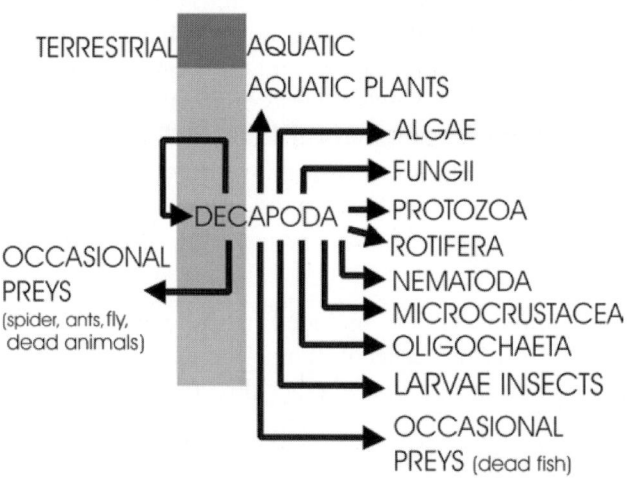

Fig. 3. Potential preys of Crustacean Decapods from aquatic and terrestrial environments of the Paraná River and their possible direct (thick line) and indirect (fine line) relationships

As these organisms supply them with necessary vitamins and microelements, it is important to include them in the trophic analysis of this group, though in certain cases their volumes may not seem significant, or their intake is considered accidental (Collins 1999a).

Among the biotic components that may appear in the system, we should consider mesotrophic bacteria and their variation according to the concentration of suspended solids (Emiliani 1985). Phytoplankton is abundant and diverse, evidencing differences between the main channel, the secondary stream, tributaries and ponds. This group is homogenized horizontally as well as vertically during the high water periods. Dominant groups are diatoms, green and green-blue algae. The most significant changes occur in the Chlorophyceae and Cyanophyceae groups during the course of the year (Garcia de Emiliani 1990; Izaguirre et al. 2001; Unrein 2002). Zooplankton is mainly represented by rotifers, cladocerans and copepods manifesting seasonal changes in density and composition and reaching values of 150,000 ind.m^{-3} (Paggi 1980; Jose de Paggi 1984).

Benthic fauna of the main channel is scarce and presents a homogeneous structure. The qualitative and quantitative composition of this community is differentiated with higher values towards the floodplain. The same happens in lentic environments, near or far from the main course (Drago et al. 2003). The main taxa represented correspond to Nematoda, Oligochaeta, Copepoda and Diptera (Chironomidae larvae) reaching density values of 90,000 ind.m^{-2} (Marchese 1984; Bertoldi de Pomar et al. 1986). In turn, benthic microfauna at the water-sediment interface is constituted by ciliates, flagellates, amoebas, platyhelminths, nematodes, gastrotrichs, rotifers, oligochaetes, cladocerans and ostracods, reaching values of 196 ind.cm^{-2} (Ojea 1994). Aquatic vegetation develops and is associated with islands, shallow lakes and the riparian area in the floodplain, showing the highest growth in the months when the temperature exceeds 15 °C (Lallana 1980). This vegetation is related to an abundant trophic offer and refuges, due to its variety in fauna taxa and microhabits, including nematodes, oligochaetes, cladocerans, copepods, ostracods, amphipods, insects and molluscs (Poi de Neiff and Carignan 1997). Sarcodines, rotifers, hirudinidans, coelenterates and other invertebrates and vertebrates such as fish, amphibians, reptils, birds and mammals are also frequent.

Among macroinvertebrates, densities reach values above 115,000 ind.m^{-2} at certain moments of the year (Paporello de Amsler 1987). Fluctuations in hydric levels in addition to limnic properties (transparency, nutrients, granulometry, and current velocity) and temperature determine the differences in the quantitative and qualitative composition of the mesofauna in the vegetated areas on riparian and flooding areas of the river (Bruquetas de Zozaya 1986).

5.3 Feeding ecology

Approximately, 85% of the feeding ecology of prawns (Sergestidae and Palaemonidae) and 45 % of crabs (Trichodactylidae and Aeglidae) of the Paraná River and its main tributaries has been studied. From the data obtained from those analyses, it is evident that the diet of these decapods is mainly omnivorous, using different trophic levels. However, trophic-habit characterization is very wide, e.g. some prawns are considered scavenger-predators, such as *Macrobrachium nattereri* in the Amazon basin. Others are considered carnivorous, such as *M. malcolmsonii* or bottom feeders, such as *M. gangeticum* in the rivers of India and *M. idella idella* in Tanganika Africa (Jayachandra 2001). According to this characterization, in some species a higher proportion of fauna elements can be observed among the stomach contents, whereas others select vegetation (algae and macrophytes) (Table 2). In turn, resource use may vary according to diverse cycles and to inter and intraspecific competitive pressure (Collins 2005b). For example, *M. borellii* is an omnivorous prawn, exhibiting a more significant presence of animal items as preys. They mainly feed on dipterans and oligochaete larvae (Collins and Paggi 1998; Collins 2005b). A similar characteristic can be found in other species of *Macrobrachium* (*M. amazonicum, M. carcinus, M. idella, M. rosembergii*) and genus *Palaemon, Processa* and *Penaeus* (Collart 1988; Lewis et al. 1966; Jayachandra and Joseph 1989; Ling 1969; Guerao 1994, 1995; Wasemberg and Hill 1987). Similarly, *M. jelskii*, a prawn which is distributed in the system of large rivers in South America (Collins 2000b; Melo 2003), shares an homologous trophic spectrum (Collins 2001).

Acetes paraguayensis, another omnivorous shrimp, feeds mainly on phytoplankton and zooplankton (rotifers and microcrustaceans) (Collins and Williner 2003) such as *Palaemon adspersus* and *P. serratus*, as well as *Pseudopalaemon chryseus* and *P. amazoniensis* of the Amazon basin (Figueras 1986; Kensley and Walker 1982). *A. paraguayensis* also feeds on larvae of chironomids and oligochaetes. Therefore, their natural diet is made up of elements from the benthic and lotic littoral communities. This reveals that this species trophically links lentic and lotic habitats, as well as benthic and pelagic communities, playing an important role in the energy transfer between shallowlakes and rivers (Collins and Williner 2003).

Among crabs, the omnivorous *Dilocarcinus pagei* incorporates a great quantity of plant material, for which it is considered a grazer or browser (Williner and Collins 2002). Likewise, *T. borellianus* shows similar characteristics to *Aegla uruguayana* and *A. platensis*. However, these two crabs present a higher percentage and diversity of animal elements

(Williner and Collins 1999; Bueno and Bond-Buckup 2004; Williner and Collins 2005).

Feeding habits of many Decapoda change according to the stage of their life cycle, due to their larval development. There is no ontogenetic variation in the diet of those in which larval development is abbreviated, i.e., those which are born at an advanced larval stage or almost as adults. This occurs, for example, in *M. borellii* and all crabs of this system. In the early stages, after eggs hatch, they have the characteristics of adults and use the same habitat corresponding to the littoral-benthic community. However, an example of a prawn that undergoes ontogenetic changes is *P. argentinus*, which during the stages after egg hatching develops planktonic habits (Boschi 1981) (Fig. 4). This is the same situation for *M. rosembergii* of Asia, *M. gangeticum* of Ganges River or *M. idella idella* of African freshwater environments. On the other hand, the ontogenetic shift can happen from omnivorous to carnivorous as is the case for *M. malcolmsonii* of the rivers of India (Ibrahim 1962; Rao 1967; Jayachandra and Joseph 1989; Roy and Singh 1997)

Further light on the trophic ecology of this group came from a study of the elements found in the digestive tract. Oligochaetes and insect larvae were the resources mostly used by all prawns and crabs, except *D. pagei* and *A. uruguayana*. Insect larvae belong to the ephemeropteran, tricopteran and dipteran orders, the latter being found mainly in the Chironomidae family (*Chironomus* sp., *Parachironomus* sp.), whereas in oligochaetans the most frequent genera were *Dero* sp. and *Pristina* sp. These groups are common among aquatic vegetation and benthos which indicates that these environments are used for a trophic purpose. Regarding the concept of optimal foraging, those groups show in the energy equations a positive gain / cost ratio, what it is recognized for several groups of aquatic animals (Popchenko 1971; Bouguenec and Giani 1989; Collins and Paggi 1998). Another positive item in the energy relationship is represented by the non parasitical nematodes found in the stomachs of prawns such as preys, mainly *M. jelskii*, *M. amazonicum* and *P. argentinus* (Table 2).

Microcrustaceans are elements which are frequently found in the stomachs of prawns and with a slightly less frequency in crabs (Table 2), including ostracods (e.g. *Cypridopsis* sp.), cladocerans (e.g. *Macrotrix* sp., *Chydorus* sp., *Bosmina* sp.), calanoid copepods (e.g. *Notodiaptomus* sp., *Diaptomus* sp.) and ciclopoid copepods (e.g. *Macrocyclops* sp., *Eucyclops* sp.). These include species that are part of the planktonic as well as benthic community and associated to vegetation (Collins and Paggi 1998). This makes it possible to visualize their ability to use a great variety of environments, also exploiting the water column, as is reflected by the pelagic shrimp *A. paraguayensis*.

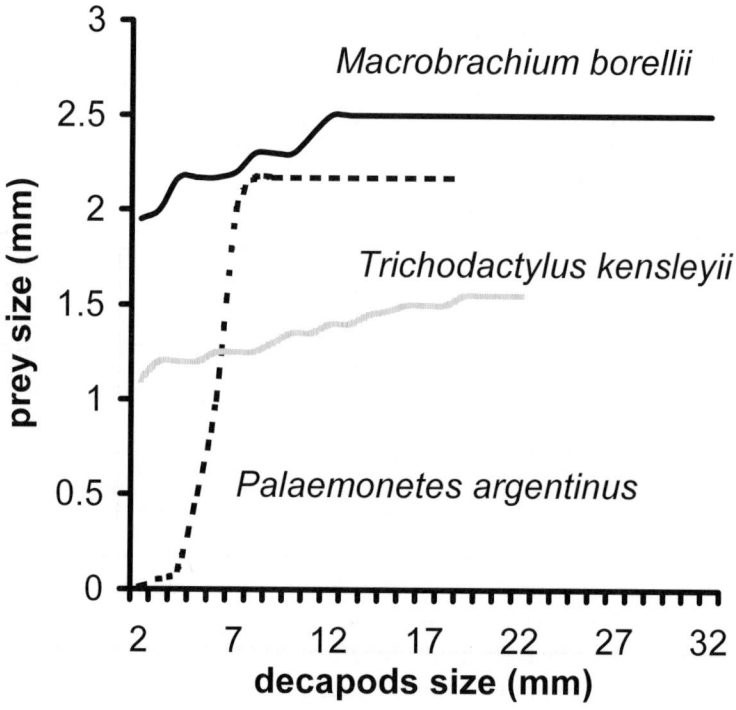

Fig. 4. Relationships between decapod size (cephalothorax length in prawns, width carapace in crabs) and mean of the prey size found in their stomach contents. (According to Collins and Paggi 1998; Collins 1999; Collins and Williner 2005)

Decapods, as a food resource, are common in aeglids and in the most aggressive prawn *M. borellii* although it can also be seen in *P. argentinus* (Table 2). The protozoa (e.g. *Diflugia* sp., *Chlamidaster* sp.) and rotifers (e.g. *Brachionus* sp., *Keratella* sp.) found in the stomachs of crabs and prawns are all mainly associated with the presence of some substrate, plant and/or belong to the epibenthos, or in some cases they also use typical species of plankton (e.g. *Lecane* sp., *Bosminopsis* sp.).

Table 2. List and frequencies of major food items found in prawns and crabs that inhabit the floodplain of the Paraná River and their tributaries (* low frequency; ** medium frequency; *** high frequency). UA: unicellular algae; FA: filamentous algae; C: Cyanophiceae; B: Bacillariophiceae; PR: plant remains; F: Fungii; PROT: Protozoa; ROTIF: Rotifera; N: Nematoda; OS: Ostracoda; CLAD: Cladocera; COPCAL: Copepoda Calanoidea; COPCICL: Copepoda Ciclopoidea; D: Decapoda; CH L: Chironomidae larvae; I L: other Insecta larvae; A: Acari; T: Tardigrada; OLIG: Oligochaeta; S: sand. 1. Collins and Williner 2003, 2. Collins and Paggi 1998, 3. Collins 2005b, 4. Collins and Williner 2001, 5. Collins and Williner 2002, 6.Collins 1999b, 7. Williner and Collins 2002, 8. Williner and Collins 1999, 9. Collins and Williner 2005, 10. Williner 2003

c	Sergestidae	Palaemonidae				Brachyura			Anomura
	A. paraguayensis [1]	M. borellii [2,3]	M. jelskii [4]	M. amazonicum [5]	P. argentinus [3,6]	D. pagei [7]	T. borellianus [8]	T. kensleyi [9]	A. uruguayana [10]
FA	**	**	***	***	***	***	*	**	**
UA	**	*	***	***	***	**	*	*	*
C		*	*	**	*				
B	**		**	*	***	***	*	**	**
PR	**	***	*	**	***	***	***	**	***
F	*					*	*		***
PROT	***					**	*	**	*
ROTIF	***		**	***	***	**	*	**	*
N		*		*					*
OS						*			
CLAD	**	**	**	**	**	*	**	*	*
COPCAL	**	**	**	**	**		**	*	*
COPCICL		**			**	*	**		*
D		**			*				*
CH L									
I L	*	***	***	***	***	*	***	***	**
A									*
T									***
OLIG	**	***	***	***	***	*	***	***	**
S	***	***	***	***	***	***	***	***	***

Among algae, those which have a filamentous morphology (e.g. *Basicladia* sp., *Oedogonium* sp., *Zignema* sp.) are very commonly found in the stomachs of both crabs and prawns. On the other hand, algae with a unicellular morphology (e.g. *Coelastrum* sp., *Ankistrodemus* sp., *Euastrum* sp.) show variations in quantity as well as in frequency, blue-green algae (e.g. *Gloeotrichia* sp., *Nostoc* sp.) being the rarest group. However, diatoms (e.g. *Gomphonema* sp., *Navicula* sp., *Bacillaria* sp.) correspond to a group which is very common in some crabs and prawns (Devercelli and Williner 2006). Among the plants found in all species, vascular macrophytes show the highest values, both in frequency and quantity (Table 2).

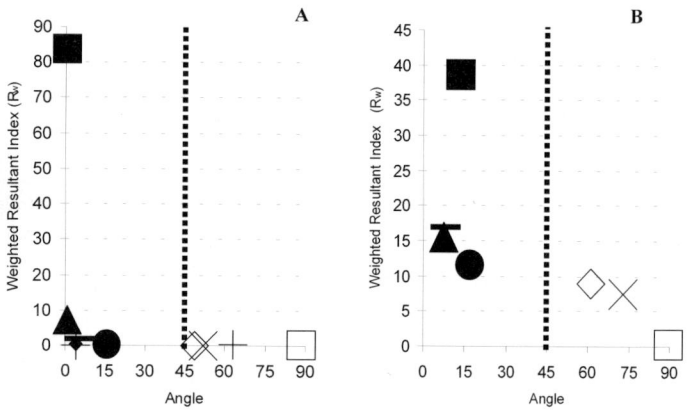

Fig. 5. Weighted result index (RW) of food items from the stomach content of *Aegla uruguayana* (A) and *Trichodactylus borellianus* (B) according to the angle of preponderance. (A) More influence of the volume of the item: ■ aquatic plant, ▲ oligochaetes and chironomid larvae, ◆ naupli, ostracods and terrestrial insect, _ cladocerans, ● copepods. More influence of the frequency: ◇ rotifers, × tardigrades, + amoebas, □ filamentous algae, unicellular and colonial algae, diatoms, fungi and ciliates. (B) More influence of the volume of the item: ■ aquatic plant, ▲ oligochaetes, _ chironomid larvae, ● insect larvae. More influence of the frequency: ◇ cladocerans, × copepods, □ filamentous algae, unicellular algae, diatoms, fungi, amoebas, rotifers

The Index of Relative Importance (IRI) and the Weighed Resultant Index (RW) (frequently used in birds and fish) were applied to quantify the consumption of the different items in order to assess and compare the importance of each group as a trophic resource (see Appendix 1). Considering the frequency, amount and size of each prey, these indexes make it possible to evaluate each food item present in the stomach contents. The remains of plant, oligochaete and chironomid larvae are the most important resources, while rotifers and protozoa are of a lower value (Figs. 5, 6).

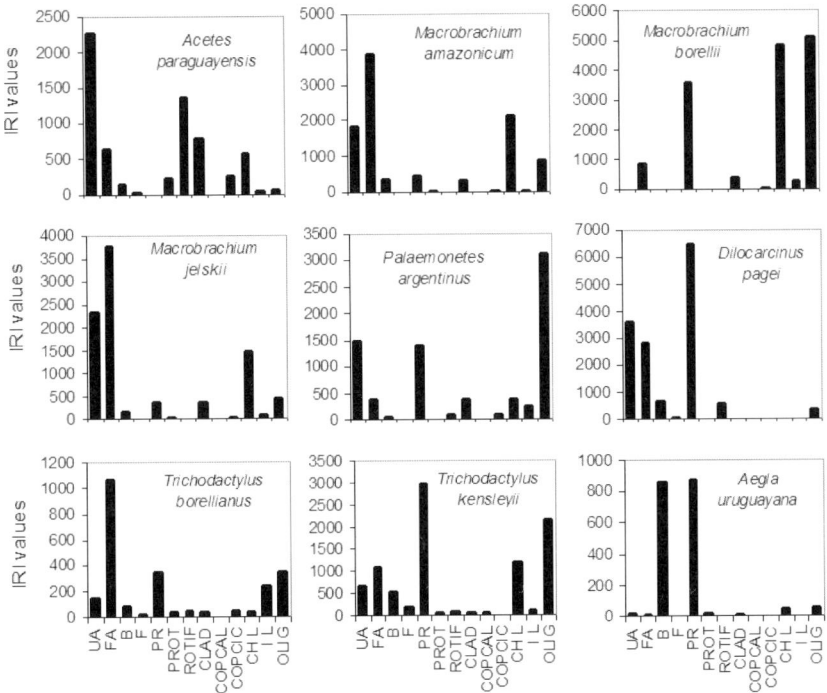

Fig. 6. Index of Relative Importance (IRI) of major food items found in prawns and crabs that inhabit the floodplain of the Paraná River and their tributaries (mean values of annual and daily cycles and of several sites). UA: unicellular algae; FA: filamentous algae; B: Bacillariophiceae; F: Fungii; PR: plant remains; PROT: Protozoa; ROTIF: Rotifera; CLAD: Cladocera; COPCAL: Copepoda Calanoidea; COPCICL: Copepoda Ciclopoidea; CH L: Chironomidae larvae; I L: other Insecta larvae; OLIG: Oligochaeta. (modified from Collins and Paggi 1998; Collins 1999b, 2005b; Williner and Collins 1999, 2002; Collins and Williner 2002, 2003)

5.3.1 Selectivity

There is a certain preference towards slow moving and large preys rather than small and very evasive preys, due to optimal energetic balances. In turn, the former are most frequently chosen for their high protein content and essential aminoacids (Collins 1999a).

The selection of food items fluctuates according to the cycles and movements of preys, e.g. vertical and horizontal movement of zooplancton, availability affected by circadian, seasonal and annual cycles, being the hydric cycle the most important structuring factor (Fig. 7) (Collins and Paggi 1998; Collins 1999b; Collins et al. 2006).

Crabs, in this region, are considered to be efficient phytoplancton grazers. Moreover, other decapods can be considered effective predators of benthic macroinvertebrates which move slowly. However, they also catch their prey in open waters.

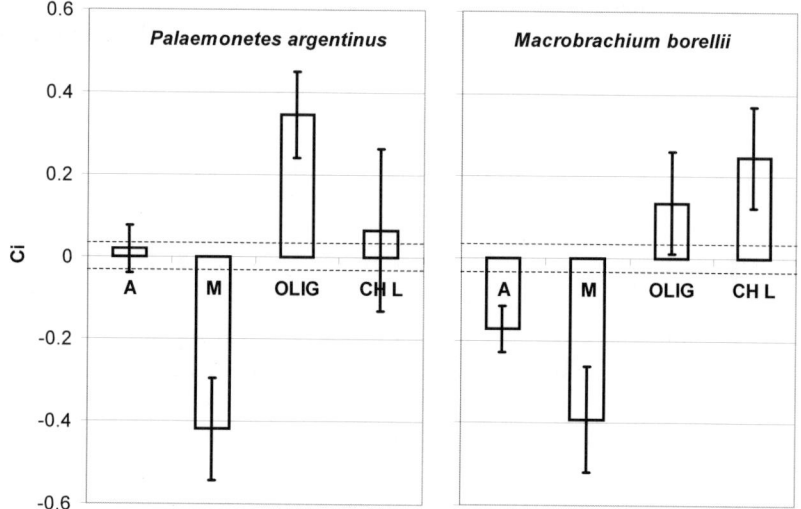

Fig. 7. Pearre Food Selection Index (C_i) of four groups found in two prawns that inhabit the Paraná River and their floodplain. Mean values and standard deviation (line) of annual cycles. Algae (A): unicellular, filamentous algae; microcrustacean (M): Cladocera, Copepoda Calanoidea; Copepoda Ciclopoidea; Oligochaeta (Olig) and Chironomidae larvae (CH L) (modified from Collins and Paggi 1998; Collins 1999b)

5.3.2 Circadian rhythms

The rythmic manifestation of any activity is influenced by exogenous and endogenous factors which act to improve the energetic equation of the activity at the lowest possible risk. Environmental conditions, succession of days and nights, the presence of prey and predators, together with the characteristics of the preys, associated with the higher or lower number of hard parts constitutes external agents that adjust the frequencies in the trophic activity (Aréchiga and Rodríguez-Sosa 1997). Therefore, the trophic activity of freshwater decapods occurs over a 24-hour-period, but not with the same intensity. Qualitative and quantitative variations of the ingested material indicate periods of higher or lower activity. Such moments cannot only be observed due to the fullness of digestive tracts (Fig. 8) but also through oscillations in enzymatic secretion (Cuzon et al. 1982; Giri et al. 2002). Variations in activity occur differentially according to age or stage of development, being in juveniles more irregular than in adults, and this characterizes the species and the population. These cycles can vary according to the elasticity of each species and their ability to respond to external pressure, adjusting to a new periodicity in which the risks of attacks are lower (Fig. 9) (Giri et al. 2002; Collins 2005b).

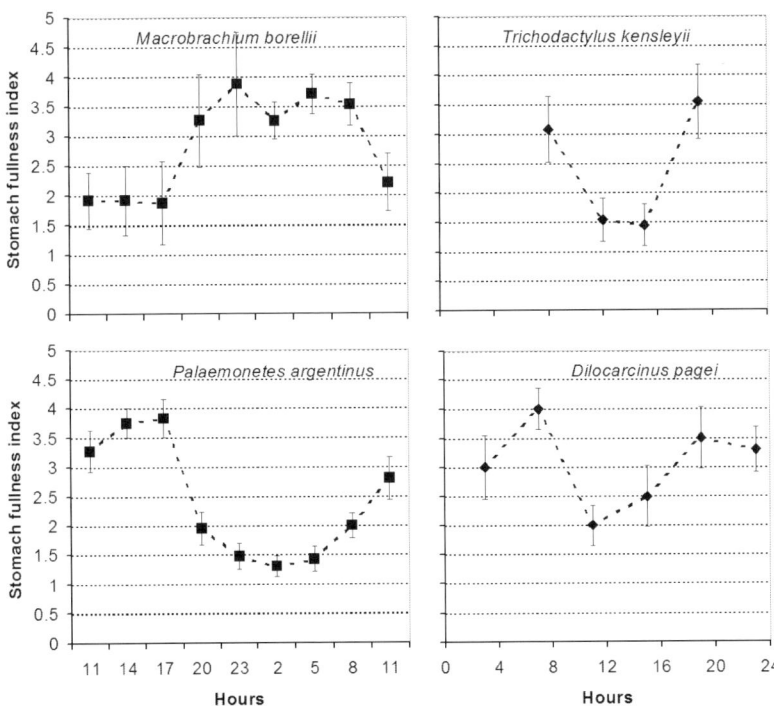

Fig. 8. Stomach fullness index of prawns and crabs during several daily cycles (mean values and standard deviation). Ranging from 0 (empty) to 5 (stomach fully distended with food (modified from Williner and Collins 2002; Collins, 2005b; Collins and Williner, 2005)

The size of the stomach together with the presence or absence of internal structures which imprint a certain pace on the food process, in addition to other factors, are some of the elements which may act as structuring factors. In turn the cycle presented by cells F, R and B in the hepatopancreas complete the complex interaction of elements. The production of digestive enzymes, possible sex differences, ecdysis and hormonal factors among others, act internally modeling the expression of trophic activity (Collins 1997a; Cuzon et al. 1982; Boschi 1981; Shyama 1987; Collart 1988).

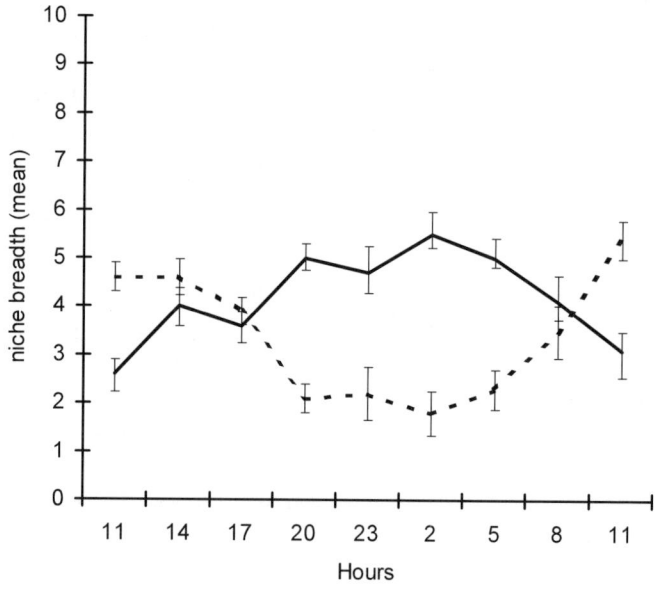

Fig. 9. Food niche breadth (B) of *P. argentinus* (- - -) and *M. borellii* (———) in a circadian cycle (for more detail see Collins 2005b)

One of the factors to be determined in order to define a cycle is the time taken to digest and evacuate the excess of prey. From the evaluation of the consumption of insect larvae in *P. argentinus* it can be seen that the process occupies 39% of one whole day and it varies according to size (Giri et al. 2002). This period could also be modified by the type of prey consumed due to the presence of hard structures, as mentioned above, which may lead to a slowing down of the digestive processes. In the case of this prawn, for example, a full stomach requires 10 hours to become completely empty. This indicates that the same individual could fill its digestive tract at least twice a day (Giri et al. 2002).

Among marine shrimps, it was observed that the evacuation rate occurs an hour after ingestion in *Penaeus esculentus* whereas in *P. monodon* only half of the stomach contents are emptied in this period. This variability in the digestion rate is also related to the characteristics of the prey ingested, in addition to the higher or lower quantity of hard or indigestible parts (Hill 1976; Hill and Wasemberg 1987; Marte 1980).

The manifestation of a cycle is given by behavioral pressures (e.g. competence, agonistic behavior, predator press), and are also shaped by physiological and morphological characteristics of the species which act on the populations so that a certain activity exhibits a rhythm. In the former, we refer to actions which tend to avoid visual predators and the coupling of the rhythm of potential preys. Another element considered to be behavioral is the pressure exerted by competition, as is shown by two sympatric prawns (*P. argentinus* and *M. borellii*). This last species shows aggressive behavior towards *P. argentinus* (Collins 2005b). This character displacement can be observed when the trophic offer is not very abundant and the population size, using this resource, is large. In a floodplain, this can be evident when the lakes are isolated from the main channel and the water surface begins to become smaller. Besides, the effect of the ecdysis cycle is associated to starvation after molting, which masks the normal rhythm. When all the structures in charge of the manipulation of preys become harder, they start feeding again (Collins 1995).

In general, decapods increase their trophic activity from sunset to the hours immediately preceding and following dawn. This rhythm results from the interaction between a certain stage of the cycle and a stimulus which in the case of *M. borellii* may possibly be the photoperiod (Collins 1997a). The most frequent consumption of certain items at a certain moment during the day also indicates the habitat it frequents. For example, in the crab *D. pagei*, animal items, such as oligochaetes, amoebas and rotifers, are more common at night and associated to the intake of sand and sediment particles. Other animals can be taken in together with plants, such as ostracods. The same is so for cladocerans and copepods, being frequent members and associated to the littoral and benthic community. The remains of plants are more frequent during the day, indicating that these crabs move along the water column to feed and find shelter (Williner and Collins 2002). On the other hand, similar tendencies have also been observed in prawns, except among *P. argentinus*.

For decapods, in rivers with a floodplain, the effects of tides are not relevant in shaping daily cycles, while the effects of the moon have yet to be studied.

5.3.3 Annual rhythms

Trophic activity is not of the same intensity throughout the year. This is due to the presence and abundance of certain potential preys which are affected by thermal and hydric cycles.

Along the Paraná River and in its floodplain, the highest level of food consumption occurs between spring (September-October) and autumn (April-May) when both the most important (hydric and thermal) cycles show high values. In this period, growth and development are more active together with the occurrence of reproductive events (Collins and Paggi 1998). In winter, when macrophytes decrease their coverage due to the cold weather and the low levels of the river, microcrustacea and plankton are used as an important alternative resource. On the other hand, in spring there is also an increase in the consumption of algae and plants supplying and supplementing diets with vitamins and other essential substances. This allows for an optimal and more frequent ecdysis together with reproduction.

Sergestid shrimps are worthy of note. They are commonly found in estuarine and marine environments, but *Acetes paraguayensis* habits in some neotropical rivers. Its feeding habits have been analyzed during migration and high water periods, corresponding to summer and early autumn. From the analysis of stomach contents it was concluded that *A. paraguayensis* preys on littoral and lotic communities.

5.4 How does Crustacea Decapoda obtain its food?

The food present in the stomachs of Crustacea Decapoda is macerated by a complex system of internal and external structures (Collins 2000a, Giri and Collins 2003, 2004; Williner and Collins 2005). Preys are caught in sediment, aquatic vegetation, and water column and/or scraping from any substrate by the chelipeds -second pereiopods. Afterwards, food passes to the mouthpieces where it is broken, macerated and mixed with mucus. Then the material enters the stomach, where it is further crushed by chitinous structures and sand (e.g. *Dilocarcinus*, *Aegla*), or only by sand taken in together with food (e.g. palaemonids).

The structure of the digestive system and the feeding habits of decapods can be divided into two large groups. Firstly, one in which food is finely mashed in the cardiac chamber of the stomach by chitinous structures (gastric mill) able to break and mash (large crabs and aeglas). Secondly, another group, the members of which show a stomach free from ossicles and teeth that only have a line of setae of a filtering function which lead the remains or pieces of food to the following segments of the digestive tract (hepatopancreas and intestine) (prawns). In the latter, food returns to the external mouthpieces so that these mash them again, and this happens as many times as is necessary so that food reaches a minimum particle size so that it can pass the filters towards the pyloric stomach. The stomach is divided in two chambers: in the first one (cardiac stomach) the pieces of food are mashed again by the gastric mill and/or the erosive action of the sand clusters taken in with the food. Then, particles go through certain setae which act as filters to the pyloric stomach, where digestion continues. Particles of a certain size, in turn, pass to the mid- intestine gland or hepatopancreas, where in their tubules the digestion ends and the R-cells absorb nutrients.

The great amount of suspended detritus in the water column suggests that it is intaken together with the prey or during manipulation. Sand grains may assume the role of a gastric mill in those species which lack one (e.g., palaemonids).

Regarding the functionality of the feeding process, it was observed that the trophic niche width is similar among groups. The obtaining of food depends on the relationship between the incorporation of energy and the expense of metabolic consumption, growth, reproduction, maintenance of body processes, food search and prey manipulation. The predation process consists of a series of linked events likely to be divided into three stages: search, capture and intake. For example, in the group of better swimming ability such as prawns, this activity increases before the recognition of the presence of prey settling on some substrate after the prey has been caught. During the first phase, decapods perceive the movement of food by means of mechanoreceptors, which is followed later by chemical receptors. Visual recognition may not be considered important or efficient in these environments, due to the fact that the natural habitat is unfavorable in this sense due to the high content of suspended material (Giri and Collins 2004). For the recognition of dead or motionless food only chemical receptors may be effective.

Choosing the prey for prawns is related to body size and pelagic or benthonic habits shown by species during their development. In general, there is a more efficient use of food by the smaller than by bigger eaters observed in the variation of the food conversion rate as well as in the growth rate (Collins and Petriella 1996, Collins 1997b, 1999a, 2004). This may also depend on the moment of the life cycle of the individual and the type of prey captured, being size, presence of hard or protective structures and the prey's ability to move important factors (Giri and Collins 2004).

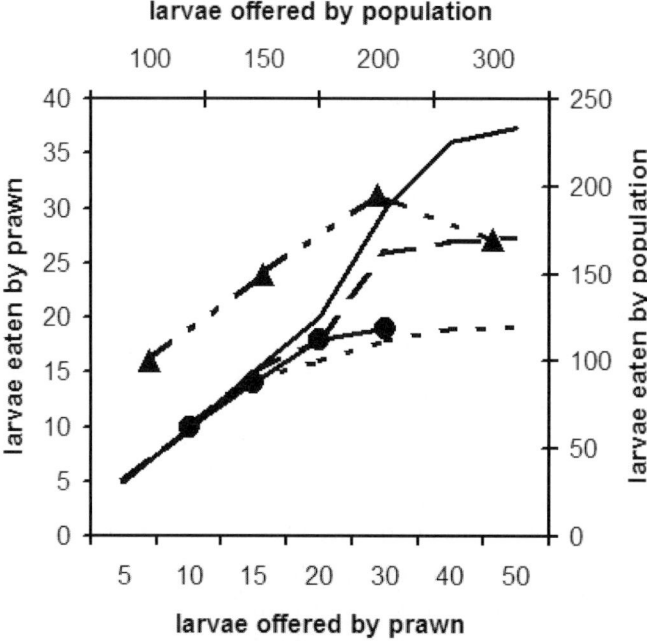

Fig. 10. Relationship between larvae offered and larvae eaten by two prawns. Prawns grouped in a population (triangle). *M. borellii* small (point line), medium (dashed line), large (solid line). Adult of *P. argentinus* (circle). (Modified from Collins 1998; Giri and Collins 2003)

After a period of starvation, the use of resources is active in the first hour. Predation decreases with time, due to the satiation process. Daily predation fits a functional response type 2 (Begon et al. 1995), increasing the consumption rate together with the increase in prey density. However, the predation rate decreases reaching a plateau in which the consumption rate remains constant regardless of prey density.

A rhythm can be observed in feeding, which may be due to the succession of days and nights with the corresponding oscillation of external factors of ecological interest, as well as the internal factors already mentioned (Fig. 10).

The fullness of the digestive tracts through a greater capture of preys is followed by an increase of enzymatic secretion. Renewal rate determines a significant time in the cycle, which may account for the variations in time (hours and days) (Giri and Collins 2003).

References

Aréchiga H and Rodríguez-Sosa L (1997) Coupling of environmental and endogenous factors in the control of the rhythmic behaviour in decapod crustaceans. Journal of Marine Biology and Association of the United Kingdom 77: 17-29

Begon M, Harper JL and Townsend CR (1995) Ecologia: individuos, poblaciones y comunidades. Omega, Barcelona 886 pp

Beltzer AH (1983a) Alimentación de la garcita azulada (*Butorides striatus*) en el valle aluvial del río Paraná medio (Ciconiiformes: Ardeidae). Revista de Hydrobiologia Tropical 16 (2): 203-206

Beltzer AH (1983b) Nota sobre fidelidad y participación trófica del "Bigua común" (*Phalacrocorax olivaceus*) en ambientes del río Paraná medio (Pelecaniformes: Phalacrocoracidae). Revista de la Asociacion de Ciencias Naturales del Litoral 4 (2): 111-114

Beltzer AH (1984) Ecología alimentaria de *Aramides ypecaha* (Aves: Rallidae) en el valle aluvial del río Paraná Medio (Argentina). Revista de la Asociación de Ciencias Naturales del Litoral 16 (1): 73-83

Beltzer AH and Paporello G (1984) Alimentación de aves en el valle aluvial del río Paraná. IV *Agelaius cyanopus cyanopus* Vieillot, 1819 (Passeriformes: Icteridae). Iheringia Serie Zoologica 62: 55-60

Bertoldi de Pomar C, Copes C, Ezcurra de Drago I and Marchese M (1986) Carácterísticas limnológicas del río Paraná y sus principales tributarios en el tramo Goya – Diamante. Los sedimentos de fondo y su fauna. Revista de la Asociación de Ciencias Naturales del Litoral 17 (1): 79-97

Bhatnagar GK and Karamchandani SJ (1970) Food and feeding habits of *Labeo fimbriatus* (Bloch) in River Narmada near Hoshangabad (m.P.) Journal of Inland Fisheries Society of India 2: 30-50

Bianchini JJ and Delupi LH (1993) Mammalia. In Ageitos de Castellanos Z (dir.) Fauna de Agua Dulce de la República Argentina. Vol. 44 (2) (actualización). PROFADU, Buenos Aires 1-79

Bó NA and Darrieu CA (1993) Aves ciconiformes. In Ageitos de Castellanos, Z. (dir.) Fauna de Agua Dulce de la República Argentina. Vol. 43 (1B). PROFADU, Buenos Aires 1-59

Bond Buckup G and Buckup L (1994) A familia Aeglidae (Crustacea, Decapoda, Anomura). Arquivos de Zoologia 32 (4): 159-346

Bonetto AA and Wais IR (1995) Southern South American streams and rivers. En: Ecosystems of the World 22 River and stream ecosystems In C. E. Cushing, K. W. Cummins and G. W. Minshall (eds). Elsevier, Ámsterdam 257-293

Bonetto AA, Pignalberi C and Cordiviola E (1963) Ecología alimentaria del amarillo y moncholo, *Pimelodus clarias* (Bloch) y *Pimelodus albicans* (Valenciennes) (Pisces, Pimelodidae). Physis 24 (67): 87-94

Boschi EE (1981) Decapoda Natantia. Serie Fauna de Agua Dulce de la República Argentina 61 pp

Bouguenec V and Giani N (1989) Les Oligochetes aquatiques en tant que proies des invertrebres et des vertebres: une revue. Acta Ecologica Applicata 10 (3): 177-196

Bruquetas de Zozaya I (1986) Invertebrados que pueblan áreas anegables de la cuenca del riachuelo (Prov. de Corrientes, Argentina); I: Variaciones temporales en una charca somera vegetada. Revista de la Asociación de Ciencias Naturales del Litoral 17 (2): 217-229

Bueno AAP and Bond-Buckup G (2004) Natural diet of *Aegla platensis* Schmitt and *Aegla ligulata* Bond-Buckup & Buckup (Crustacea, Decapoda, Aeglidae) from Brazil. Acta Limnologica Brasileira 16 (2): 115-127

Cabrera DE, Baiz ML, Candia CR and Christiansen HE (1973) Algunos aspectos biológicos de las especies de ictiofauna de la zona de Punta Lara (río de la Plata). 2ª parte. Alimentación natural del bagre porteño (Parapimelodus valenciennesi). Armada Argentina, Servicio de Hidrografía Naval H.1029, 7-47

Collart OO (1988) Aspectos ecológicos do camarão *Macrobrachium amazonicum* (Heller, 1862) no baixo Tocatins (Pa-Brasil). Memoria de la Sociedad de Ciencias Naturales La Salle 48 (Supl): 341-353

Collins PA (1995) Variaciones diarias de la actividad trófica en una población de *Palaemonetes argentinus* (Crustacea Decapoda). Revista de la Asociacion de Ciencias Naturales del Litoral 26 (1): 57-66

Collins PA (1997a) Ritmo diario de alimentación en el camarón *Macrobrachium borellii* (Decapoda, Palaemonidae). Iheringia Série Zoologica 82:19-24

Collins PA (1997b) Cultivo del camarón *Macrobrachium borellii* (Crustacea: Decapoda: Palaemonidae), con dietas artificiales. Natura Neotropicalis 28 (1): 39-45

Collins PA (1999a) Role of natural productivity and artificial feed in enclosures with the freshwater prawn, *Macrobrachium borellii* (Nobili, 1896). Journal of Aquaculture in the Tropics 14 (1): 47-56

Collins PA (1999b) Feeding of *Palaemonetes argentinus* (Nobili) (Decapoda: Palaemonidae) in flood valley of river Paraná Argentina. Journal of Crustacean Biology 19 (3): 485-492

Collins PA (2000a) Mecanismos de coexistencia en poblaciones de Palaemónidos diulciacuícolas (Crustacea, Decapoda, Caridea). Tesis Doctoral Universidad Nacional de La Plata 319 pp

Collins PA (2000b) A new distribution for *Macrobrachium jelskii* (Miers, 1877) in Argentina. Crustaceana 73 (9): 1167-1169

Collins PA (2004) Cultivo alternativo en el valle aluvial del río Paraná: camarones dulciacuícolas en jaulas flotantes. En: CIVA2003: Comunicaciones y foros de discusión. Universidad de Zaragoza España. ISBN: 84-609-0184-x 427-433

Collins PA (2005a) Distribución altitudinal del cangrejo *Trichodactylus kensleyi* (Rodríguez, 1992) en Misiones, Argentina. Natura Neotropicalis 36-37: 23-26

Collins PA (2005b) A coexistence mechanism for two freshwater prawns in the Paraná river floodplain. Journal of Crustacean Biology 25 (2): 219–225

Collins PA and Paggi JC (1998) Feeding ecology of *Macrobrachium borellii* (Nobili) (Decapoda: Palaemonidae) in the flood valley of the River Paraná, Argentina. Hydrobiologia 362: 21-30

Collins PA and Petriella AM (1996) Crecimiento y supervivencia del camarón *Macrobrachium borellii* (Decapoda Palaemonidae) con dietas artificiales. Neotropica 42 (107-108): 3-7

Collins P and Williner V (2001) Espectro trófico del camarón dulciacuícola *Macrobrachium jelskii* en el Parque Nacional Río Pilcomayo. V Congreso Latinoamericano de Ecología, Jujuy

Collins PA and Williner V (2002) Espectro trófico natural del camarón *Macrobrachium amazonicum* en el Parque Nacional Río Pilcomayo Argentina. XXIV Congreso Brasileiro de Zoologia, Itajai, Brasil, 83 pp

Collins PA and Williner V (2003) Feeding of *Acetes paraguayensis* (Nobili) (Decapoda: Sergestidae) from the Parana River, Argentina. Hydrobiologia 493: 1-6

Collins PA and Williner V (2005) Ecología trófica y ritmo nictimeral del cangrejo *Trichodactylus kensleyi* en Misiones, Argentina. 3° Congreso Argentino de Limnología – CAL III. Chascomús, Argentina, 20 pp

Collins PA, Williner V and Giri F (2004) Crustáceos Decápodos del Litoral Fluvial Argentino, In Aceñolaza FG (ed) Temas de la Biodiversidad del Litoral Fluvial Argentino. INSUGEO, Miscelánea 12: 253-264

Collins PA, Giri F and Williner V (2006) Population dynamics of *Trichodactylus borellianus* (Crustacea Decapoda Brachyura) and interactions with the aquatic vegetation of the Paraná River (South America, Argentina). Annals of Limnology 42 (1): 25-31

Cuzon G, Hew G and Cognie D (1982) Time lag effect of feeding of growth of juvenile shrimp, Penaeus japonicus. Aquaculture 29: 33-44

Devercelli M and Williner V (2006) Diatom grazing by *Aegla uruguayana* (Decapoda: Anomura: Aeglidae): digestibility and cell viability after gut passage. Annals of Limnology, in press

Drago EC, Escurra IE, Oliveros OB and Paira AR (2003) Aquatic habitats, fish and invertebrate assemblages of the Middle Paraná River. Amazoniana 17 (3/4): 291-341

Emiliani F (1985) Estudios limnológicos en una sección transversal del tramo medio del río Paraná, XIII: Bacteriología. Revista de la Asociación de Ciencias Naturales del Litoral 16 (2): 217-225

Fernandez D y Collins PA (2002) Supervivencia de cangrejos en ambientes dulciacuícolas inestables. Natura Neotropicalis 33 (1y2): 81-84

Ferriz RA, Villar CA, Colautti D and Bonetto C (2000) Alimentación de *Pterodoras granulosus* (Valenciennes) (Pises, Doradidae) en la baja cuenca del Plata. Revista del Museo Argentino de Ciencias Naturales 2 (2): 151-156

Figueras AJ (1986) Alimentación de P*alaemon adspersus* (Rathke, 1837) y *Palaemon serratus* (Pennant, 1777) (Decapoda Natantia) en la ría de Vigo (N.O. España). Cahiers de Biologie Marine 27 (1): 77-99

Gallardo JM (1977) Reptiles de los alrededores de Buenos Aires. Eudeba, Buenos Aires 213 pp

Garcia de Emiliani MO (1990) Phytoplankton ecology of the Middle Paraná River. Acta Limnologica Brasileira 3: 391-417
Giraudo AR and Cordiviola E (2002) Jaaukanigás, un nuevo sitio Ramsar en la Argentina. Naturaleza y Conservación 5 (10): 34-35
Giri F and Collins PA (2003) Evaluación de *Palaemonetes argentinus* (Nobili 1901) (Decapoda: Natantia) como controlador biológico de larvas de mosquito *Culex pipiens* s.l. (Linnaeus 1758) (Diptera: Culicidae) en condiciones de laboratorio. Iheringia Serie Zoológica 93 (3): 237-242
Giri F and Collins PA (2004) Eficiencia de captura del camarón dulceacuícola *Palaemonetes argentinus* (NOBILI, 1901) sobre larvas de mosquito *Culex pipiens* s.l. (Linnaeus 1758). Hidrobiologica 14 (2): 84-90
Giri F, Williner V and Collins PA (2002) Tiempo de evacuación del camarón dulceacuícola *Palaemonetes argentinus* alimentado con larvas de mosquito *Culex pipiens* s.l. FABICIB 6: 37-41
Gori M, Carpaneto GM and Ottino P (2003) Spatial distribution and diet o the Neotropical otter *Lontra longicaudis* in the Ibera Lake (northern Argentina) Acta Theriologica 48 (4): 495-504
Guerao G (1994) Feeding habits of the prawns *Processa edulis* and *Palaemon adspersus* (Crustacea, Decapoda, Caridea) in the Alfacs Bay, Ebro delta (NW Mediterranean). Miscellanea Zoologica 17: 115-122
Guerao G (1995) Locomotor activity patterns and feeding habits in the prawn *Palaemon xiphias* (Crustacea: Decapoda: Palaemonidae) in Alfacs Bay, Ebro delta (Northwest Mediterranean). Marine Biology 122: 115-119
Hill BJ (1976) Natural food, foregut clearence-rate and activity of the crab *Scylla serrata*. Marine Biology 34:109-116
Hill BJ and Wasemberg TJ (1987) Feeding behavior of adult tiger prawns, *Penaeus esculentus*, under laboratory conditions. Australian Journal of Marine and Freshwater Research 38:183-190
Ibrahim KH (1962) Observations on the fishery and biology of the freshwater prawn, *Macrobrachium malcolmsonii* H. Mine Edwards of River Godavari. Indian Journal of Fisheries 9A(2): 433-467
Izaguirre I, O´Farrell I and Tell G (2001) Variation in phytoplankton composition and limnological features in a water-water ecotone of the Lower Paraná Basin (Argentina) Freshwater Biology 46: 63-74
Jayachandra KV (2001) Palaemonid Prawns: Biodiversity, Taxonomy, Biology and Management. Science Publishers, Enfield, 624 pp.
Jayachandra KV and Joseph NY (1989) Food and feeding habits of the slender river prawn, *Macrobrachium idella* (Hilgendorf, 1898) (Decapoda, Palaemonidae). Mahasagar 22: 121-129
Jose de Paggi S (1984) Estudios limnológicos en una sección transversal del tramo medio del río Paraná: distribución estacional del plancton. Revista de la Asociación de Ciencias Naturales del Litoral 15 (2): 135-155
Kensley B and Walker I (1982) Palaemonid shrimps from the Amazon basin, Brazil (Crustacea: Decapoda: Natantia). Smithsonian Contributions to Zoology 362: i-iii, 1-28

Lajmanovich RC and Beltzer AH (1993) Aporte al conocimiento de la biología alimentaria de la pollona negra *Gallinula chloropus* en el Paraná Medio, Argentina. El Hornero 13 (4): 289-291

Lallana VH (1980) Productividad de *Eichhornia crassipes* (Mart.) Solms. En una laguna isleña de la cuenca del río Paraná medio. II. Biomasa y dinámica de población. Ecología 5: 1-16

Legendre L and Legendre P (1979) Ecologie Numerique II. Masson, Paris 247 pp

Lewis JB, Ward J and McIver A (1966) The breeding cycle, growth and food of the fresh water shrimp *Macrobrachium carcinus* (Linnaeus). Crustaceana 10: 48-52

Ling SW (1969) The general biology and development of *Macrobrachium roembergii*. FAO Fish Rep. 57: 589-606

López JA, Arias MM, Peltzer PM and Lajmanovich RC (2006) Dieta y variación morfométrica de *Leptodactylus ocellatus* (Linnaeus, 1758) (Anura: Leptodactylidae) en tres localidades del centro-este de Argentina. Boletin Español de Herpetología, in press

Marchese M (1984) Estudios limnológicos en una sección transversal del tramo medio del río Paraná. XI: Zoobentos. Revista de la Asociación de Ciencias Naturales del Litoral 15 (2): 157-174

Marte CL (1980) The food and feeding habits of *Penaeus monodon* collected from Makato river, Akland Philiphines. Crustacena 38: 225-236

Massoia E (1976) Mammalia. In Ringuelet RA (dir) Fauna de Agua Dulce de la República Argentina. Vol. 44. FECIC, Buenos Aires 1-128

Melo GAS (2003) Manual de identificação dos Crustacea Decapoda de água doce do Brasil. São Paolo Ed. Loyola, Centro Universitario São Camilo, Museu de Zoología, Universidade de São Paolo 429 pp

Mohan MV, Sankaran TM (1989) Two new indices for stomach content analysis of fishes. Journal of Fish Biology 33: 289-292

Navas J (1991) Aves gruiformes. In Ageitos de Castellanos Z (dir) Fauna de Agua Dulce de la República Argentina Vol. 43 (3). PROFADU, Buenos Aires 1-80

Navas J (1993) Aves Podicipediformes y Pelecaniformes In: Ageitos de Castellanos, Z. (dir.) Fauna de Agua Dulce de la República Argentina. Vol. 43 (1A). PROFADU, Buenos Aires 1-79

Navas J (1995) Aves ciconiformes. In Ageitos de Castellanos Z (dir) Fauna de Agua Dulce de la República Argentina.Vol. 43 (1C). PROFADU, Buenos Aires 1-53

Ojea N (1994) Arcellidae, Difflugidae, Centropyxidae y Nebelidae (Rhizopoda, Testacelobosia) del valle aluvial del río Paraná medio. Revista de la Asociación de Ciencias Naturales del Litoral 24-25: 64-67

Oliva A, Ubeda CA, Vignes EI and Iriondo A (1981) Contribución al conocimiento de la ecología alimentaria del bagre amarillo *(Pimelodus maculatus* Lacépède 1803) del río de la Plata (Pisces, Pimelodidae). Comunicaciones del Museo Argentino de Ciencias Naturales Ecología 1 (4): 31-50

Paggi JC (1980) Campaña limnológica "Keratella I" en el tramo Paraná medio (Argentina): zooplancton de ambientes leníticos. Ecología 4: 77-88

Paoli C, Iriondo M and Garcia N (2000) Características de las cuencas de aporte. In Paoli C & Schreider M (eds) El río Parana en su tramo medio. Centro de Publicaciones UNL 29-68.

Paporello de Amsler G (1987) Fauna asociada a las raíces de *Eicchornia crassipes* en cauces secundarios y tributarios del río Paraná en el tramo Goya – Diamante. Revista de la Asociación de Ciencias Naturales del Litoral 18 (1): 37-50

Peterson BJ and Fry B (1987) Stable isotopes in ecosystem studies. Annual Review from Ecology and Systematics 18: 293-320

Pillay TVR (1952) A critique of the Methods of study of food of fishes. Journal of the Zoological Society of India 4: 185-200

Pinkas L, Olipham MS and Iversor ILK (1971) Food habits of albacore, bluefin tuna and bonito in California waters. Fishery Bulletin 152: 1-105

Poi de Neiff A and Carignan R (1997) Macroinvertebrates on *Eichhornia crassipes* roots in two lakes of the Paraná river floodplain. Hydrobiologia 345: 185-196

Popchenko VI (1971) Consumption of Oligochaeta by fishes and invertebrates. Journal of Ichthyology 11 (1): 75-80

Port-Carvalho M, Ferrari SF and Magalhães C (2004) Predation of crabs by Tufted Capuchins (*Cebus apella*) in Eastern Amazonia. Folia Primatologica 75: 154-156

Rao RM (1967) Studies on the biology of *Macrobrachium rosembergii* (de Man) of the Hooghly estuary with notes on its fishery Proceedings of the Indian National Science Academy 33B (5-6): 252-279

Roy D and Singh SR (1997) The food and feeding habits of a freswater prawn *Macrobrachium choprai*. Asian Fisheries Science 10: 51-63

Shyama SK (1987) Studies on moulting and reproduction in the prawn *Macrobrachium idella*. Mahasagar 20: 15-21

Swinnerton GH and Worthington EB (1940) Note on the food of fish in Haweswater (Westermorland). J Anim Ecol 9: 183-187.

Unrein F (2002) Changes in phytoplankton community along a transversal section of the lower Paraná floodplain, Argentina. Hydrobiologia 468: 123-134

Wassemberg TJ and Hill BJ (1987) Natural diet of the tiger prawns *Penaeus esculentus* and *P. semisulcatus*. Aust J Mar Freshw Res 38: 169-182

WIlliams JD and Scrochi G (1994) Ofidios de agua dulce de la Republica Argentina. Fauna de agua dulce de la Republica Argentina. Vol. 42 Reptilia, fascículo 3, Oficia Lepidosauria 55 pp

Williner V (2003) Ecología trófica del cangrejo *Aegla uruguayana* Schmitt 1942 (Decapoda: Anomura: Aeglidae). VIII Jornadas de Ciencias Naturales del Litoral – I Jornadas de Ciencias Naturales del NOA 177 pp

Williner V and Collins P (1999) Estudio preliminar sobre la ecología trófica del cangrejo *Trichodactylus borellianus* Nobili, 1896 (Crustacea, Decapoda, Trichodactylidae). 64° Reunión de Comunicaciones Científicas, ACNL 8 pp

Williner V and Collins P (2000) ¿Existe jerarquización en las poblaciones de Palemónidos del valle aluvial del Río Paraná? Natura Neotropicalis 31(1y2): 53-60

Williner V and Collins PA (2002) Daily rhythm of feeding activity of the freshwater crab *Dilocarcinus pagei pagei* in the Río Pilcomayo National Park, Formosa, Argentina. Modern Approaches to the Study of Crustacea 171-178

Williner V and Collins PA (2005) Ecología trófica de cangrejos Anomuros: una aproximación metodológica. 3rd Congreso Argentino de Limnología – CAL III. Chascomús 56 pp

Appendix 1

Trophic analysis in freshwater prawns and crabs. Some methodological considerations.

In trophic studies, there are different possibilities in approaches to understanding feeding relationships. This review is based on the data related to the analysis of stomach content. Another tool is the use of a stable isotope to map the routes that the nutrients follow in freshwater ecosystems (Peterson and Fry 1987). This methodology provides important information and represents the future step in better understanding the trophic relationships of neotropicals decapods.

The study of stomach contents in decapod crustaceans from freshwater present certain difficulties due to several factors, among which we can mention the small size of individuals and therefore the small size of stomachs, as is the case in prawns. Likewise, the maceration degree of the food items, due to the mechanical breakdown in the mouth and cardiac stomach, makes it difficult to recognize the ingested food.

This appendix adds to the knowledge of some aspects that should be considered important when starting a trophic study. They emerge from the analysis of the natural diet of some prawns and crabs in different freshwater systems.

A first point to consider is that it is necessary to set a sampling schedule regarding intake moments. Even though these decapods eat their food throughout the day, there is usually a time span in which such activity is at its highest peak.

If the aim is not to analyze the daily feeding cycle and we want to make sure that most stomachs have a high degree of fullness, setting such a span is necessary. The degree of foreguts fullness was assessed visually by a subjective scale ranking. Each foregut was assigned to 0 (empty) from 3 (fullness). Visual assessment was made possible by the fact that the foregut is a thin-walled translucent organ. Thus, the sample size to be studied in order to be representative can be optimized. The method to fix and preserve the organisms avoiding that the ingested material is regurgitated is also particularly important. On this topic, we have observed that specimens anaesthetized by cooling in a refrigerator with ice and preserved in 70% alcohol or 10% formalin in the field is the best technique as almost no food is regurgitated.

Another factor for the optimization of the number of organisms to analyze is the use of a Diversity Index (Shannon and Weaver or Simpson) (Legendre and Legendre 1979) which plots the diversity curve accumulated by the stomach while its contents are observed. Thus, the moment the curve becomes asymptotic; it is evident that this number of stomachs can be considered sufficient to achieve representativeness. However, it is important to study a higher number of stomachs than the quantity yielded by the application of the index mentioned above.

On the other hand, before starting observations, it is necessary to consider if the contents of the stomach is to be quantified on a full stomach as happens in prawns and crabs of a small size. Otherwise, it would be necessary to establish a certain number of aliquots due to the fact that the volume of the ingested material is important as happens in big stomachs. In both cases, counts and quantification are done through microscopic observation at different magnification (10X, 20X, 40X) according to the size of identifiable fragments. For example, in *Aegla* species, the size of identifiable fragments ranges from 50 to 1,500 µm (Williner and Collins, 2005).

To express the results of the observation of stomach contents, several indices can be applied. The interesting indices are those that take into account the volumetric data, for they provide more information than those using only frequency data. Disadvantages, however, are linked to the observation time and the calculus implied. Some indices frequently applied are: e.g. Index of Relative Importance (IRI) (Pinkas et al. 1971), Weighted Resultant Index (RW) (Mohan and Sankaran 1988), points method (Swinnerton and Worthington 1940), Occurrence method (Pillay 1952), and index of preference Bhatnagar and Karamchandani (1970). Detailing, Weighted Resultant Index (RW) uses the occurrence percentage values and volume percentage and to know which contributes bigger proportion in the conformation of the same one an angle of these parameters it is calculated. The Index of Relative IRI uses the volumetric content, the numeric content and the frequency of occurrence of prey item.

6 The role of predation in shaping biological communities, with particular emphasis to insects

Panos V. Petrakis[1] and Anastasios Legakis[2]

[1]National Agricultural Research Foundation, Institute of Mediterranean Forest Ecosystem Research, Lab. of Entomology, Terma Alkmanos, 1152 Ilissia, Athens, Greece, pvpetrakis@fria.gr, [2]University of Athens, Dept. of Biology, Zoological Museum, 15784 Athens, Greece, alegakis@biol.uoa.gr

"It is an almost universal rule that each animal either has enemies which seek to feed upon it, or that it seeks itself to feed upon other animals. In the first case, it has to escape its enemies or it cannot long continue live. This it does either by its swiftness of flight, by its watchfulness, or by hiding itself from view"

[*A. R. Wallace 1879*]

6.1 Abstract

Predation in animal communities is the consumption of tissues belonging to another organism. The content of the term predation varies among trophic levels, mode of action and species, and thus several additional terms have been used such as parasitism, herbivory, cannibalism, and larceny. This treatise deals with predation *sensu stricto* and cannibalism. It was found that in the majority of biotic communities predation by arthropods constitutes the most frequent significant effect as far as competition is concerned, especially in the absence of vertebrate predators. The important predator-prey interaction is addressed *via* the modelling of the community, a common approach to prediction making on population densities of predator and prey species.

Many models have been developed apart from the classical ones of Holling and Lotka-Voltera. LES, Rosenzweig-MacArthur and Luck are newly developed models that predict two paradoxes, namely the *"paradox of enrichment"* and the *"biological control paradox"*. Models that take into accounts the spatial component such as the *reaction-diffusion* models allow the investigation of many aspects of the behaviour of the predator-prey system within biotic communities. In terms of the relative importance of predation and competition *sensu stricto* in shaping insect communities the debate is far from settled. Typically, predation is the energy passed from prey to predator while competition is the modification of the trajectory of energy transfer. In this respect, competition cannot be traced in the fossil record unlike predation, which leaves some detectable traces. In effect, past competition can only be hypothesized. Although the size of predators can be closely related to the size of the attacked prey there are prey species corresponding to two or more predator sizes. These predators belong to two or more assemblages preying on the same prey species.

Cannibalism was interpreted as a type of predation. In this view, cannibalism is predation on conspecifics and in this respect it can emerge when competition is low and predators are scarce. In these cases cannibalism is the only mechanism that controls the population densities of organisms.

Nevertheless, many authors believe that cannibalism lacks any adaptive significance. The usage of the term '*natural enemy*' allowed the inclusion of competitors and predators in a single expression, in the sense that competitors are enemies. In this sense, a prey newcomer cannot invade any biotic community unless it has a difference in its anti-predatory traits from already existing prey species. An advanced form of this interaction is the hypothesis of '*differential diversification*' of sister clades that determines the outcome of biotic interaction in evolutionary and ecological time scales.

Biodiversity is promoted and maintained by "*keystone predators*". These predators act on numerous or common organisms preventing the dominance of one or a few species. It seems that the higher the rank of the trophic level the greater the difficulty to receive energy from the ecosystem. Thus, increased biodiversity levels are exploited in many ways to attain a firm grasp of energy input in the biological community. Possibly, this is the main cause of the extreme diversity of insect orders. On the other hand, the biodiversity is spatially distributed according to various ecosystem components.

Chemical communication, though a primitive way of communication, is heavily employed both in the detection of prey and in the avoidance of predation. The emission of chemical messages and the ability to sense them is so important as to reflect the evolution of the emitter. There is no work on the evolutionary message conferred to the receiver but the alterations needed to emit the chemical message are more sophisticated than the alterations necessary to sense the new blend. The use of chemical messages is tightly connected to the behavioural ecology of insects. Usually, chemistry and morphology are both engaged in the predatory interactions of insects. A number of hypotheses have been proposed to explain these interactions.

Keywords: Predation, cannibalism, insects, biotic interaction, chemical detection, predator avoidance, optimal forage theory.

6.2 Predation and its types in insects

Predation is conceived as the consumption of individuals or tissues of them belonging to species of the same or, more usually, lower trophic levels. If the prey species belongs to autotrophs the predation is specifically called "herbivory". In the case of predation that involves individuals from the same species it is called cannibalism. The terms *'predator-prey'* usually refer to the classical case where the prey species belongs to a lower trophic level. Polis et al. (1989) have suggested a set of terms for the description of the interaction of two species.

Sometimes the term *'predation'* is used to describe consumption of special organs of other organisms such as seeds. In such cases it is usually referred as *'seed predation'* (Janzen 1971). In the case where the tissues of the prey are renewable without any effect to the reproductive output for the prey organism it can be variably named as *'collection'* or *'larceny'* (Inouye 1980). Predators that live at the expense of the tissues of other organisms are called *'parasites'* and if their emergence from the host assumes its death they are called *'parasitoids'*. Traditionally predators are considered to attack many prey individuals while parasites are exploiting in large numbers only one prey individual. Intermediate to these two extremes are parasitoids, which typically attack one individual. Nevertheless, exceptions to this can be found in several parasitoid-prey systems, such as the *Pieris brassicae – Pteromalus puparum* system. This system exhibits a rather complicated behaviour since pupae are attacked by several individuals while neonate larvae are attacked by only one individual, that is, they receive only one egg (van Alphen and Jervis 1996). Parasitoids are divided further to *'idiobionts'* and *'koinobionts'* depending on whether they permit the development of the host from the stage attacked or not. In this way *Petromalus puparum* (Hymenoptera, Pteromalidae) which attacks larval stages of *Pieris brassicae* (Lepidoptera, Pieridae) is an idiobiont while *Cotesia glomerata* (*=Apanteles glomeratus*) (Hymenoptera, Braconidae) which usually attacks newly hatched larvae of the same butterfly allowing them to feed and develop together with the parasitoid inside them are koinobionts (van Alphen and Jervis 1996). Since in parasitoids predation always assumes a certain type of reproductive behaviour, this type of predation will not be studied in this work. Interested readers should consult the chapters in the excellent review edited by Jervis and Kidd (1996).

When treating predation in insect communities it is necessary to define some important terms commonly used in the scientific literature. A '*patch*' is usually a subset of the space where a population or a segment of it, usually referred as '*local population*', lives (Hanski and Gilpin 1991). Patches usually comprise islands, both true islands and discontinuities of a large continuous landscape. In this form, the term '*patch*' is used both by paleobiologists and neontologists (Jackson et al. 1996). In the same style, the term '*metapopulation*' is the set of local populations of many potentially or actually interacting populations that move among the different patches. '*Metacommunity*' is used to describe more than one species inhabiting one or more patches. In this sense the metacommunity comprises a subset of the species pool of an area (Hanski and Gilpin 1991; Jackson et al. 1996).

In a recent review of field experiments concerning predation, competition and prey communities, Sih et al. (1985) found that in experiments with large significant effects, predation by arthropods in general constitutes a large proportion (98.0%) of specific interactions when the respective percentage for vertebrate predators is smaller (91.3%). Moreover, experiments carried out in terrestrial systems gave much less significant effects (55.8%) than experiments in lentic systems (70.5%). In general, the role of predation was more emphasized in marine and lentic systems than in terrestrial habitats including flowing waters (lotic). The reason usually invoked for this discrepancy is the structural simplicity of lentic and marine systems. Menge and Sutherland (1976) predicted that predation should be more important at lower trophic levels in the sense that experimental manipulations should be more intense (= large effects) in primary than in secondary predators.

6.3 Prey-predator interaction seen through models

One of the most important methods to study the population dynamics of predators and their prey is through modelling. Unless the experimental design involves closed system units and a plenty of time and resources (e.g. for competition among aquatic organisms see Morin et al. 1988; for theoretical treatment of animal colonization see Cantrell et al. 1996), modelling permits the experimentation at time and parameter scales not available in real situations. The argument that models are idealized mathematical constructions with simplifying assumptions for our intelligence to "... cope with the blooming, buzzing confusion of the natural world..." (Odenbaugh 2005) can be criticised from the viewpoint that a model is constructed to study (not mimic) the behaviour of natural systems. In many cases, models are the only means to study natural systems while MacArthur's quote '*a model is a lie that helps us see the truth*' (Dr Alan J.A. Stewart pers. comm.) rephrases the way that learning proceeds in humans. Odenbaugh (2005) has investigated the ability of ecological models to be inaccurate and simultaneously successful in ecological themes quite complex such as populations, communities and ecosystems. Dostalkova et al. (2002) proceeded further and considered three points that have to be taken into account in all models. Moreover, they constructed their own model incorporating all three considerations. The considerations are: (i) Juveniles must be treated as entirely different species from adults since they do not move between patches (except the first instar crawlers of coccids and many Carabidae and Staphylinidae (Coleoptera) predatory species) and they do not reproduce. (ii) Food availability is more restrictive to juveniles than adults since the former are not very mobile. (iii) Cannibalism of eggs and larvae is very common among insect and arachnid predators and therefore the quality of a patch is more decisive for the reproductive output of adults (Wise 2006).

An extended account of modelling schools of thought, concerning insects either as predators or prey, is given in the multi-authored book edited by Jervis and Kidd (1996). Another account concerning general concepts of ecological models is given in the book of Maynard Smith (1984).

Lately, there is a growing body of biological evidence that when predators have to search for their food, the efficiency of searching is based on the ratio of prey items to predator individuals (Arditi et al. 1991a and b; Kuang and Beretta 1998). Functional response models are the models predicting the response, in the form of a function, of the predator to the prey density (Xiao and Ruan 2001). In particular, the response of the density of predator populations per unit time and change per unit prey density is usually referred as functional response, and the most useful functions describing it are those of Holling II or Michaelis-Menten (Jost et al. 1999) with equations redefined at the origin as:

$$\frac{d}{dt}x(t) := x(t)\cdot(a - b\cdot x(t)) - \frac{c\cdot x(t)\cdot y(t)}{m\cdot y(t) + x(t)} \quad (6.1)$$

$$\frac{d}{dt}y(t) := y(t)\cdot\left[-d + f\cdot\frac{x(t)}{(m\cdot y(t) + x(t))}\right] \quad (6.2)$$

$$\frac{d}{dt}x(0) := 0 \qquad \frac{d}{dt}y(0) := 0 \quad (6.3)$$

where x(t) and y(t) are the densities of prey and predator populations at time *t*, and *a*, respectively, and *b, c, d, m, f* are positive constants; in particular *d* is the death rate of the predator, *a/b* is the ratio corresponding to the carrying capacity of the environment for the prey, *a* is the prey intrinsic growth rate, *c* is capturing rate for the predator, *m* is the half saturation constant and *f* is the conversion rate. The model in system (3-1) has two equilibrium points, one at (0,0) and one on the x-axis (K,0) (Kuang and Beretta 1998). The equilibrium properties of the system (1) were comprehensively studied by Xiao and Ruan (2001) for various types of the parameters *a, b, c, d, m, f*. Interestingly, the ratio *a/b*, which is the carrying capacity of the environment for the prey, is also either a saddle or a global attractor as marked on the diagrams of Fig. 1.

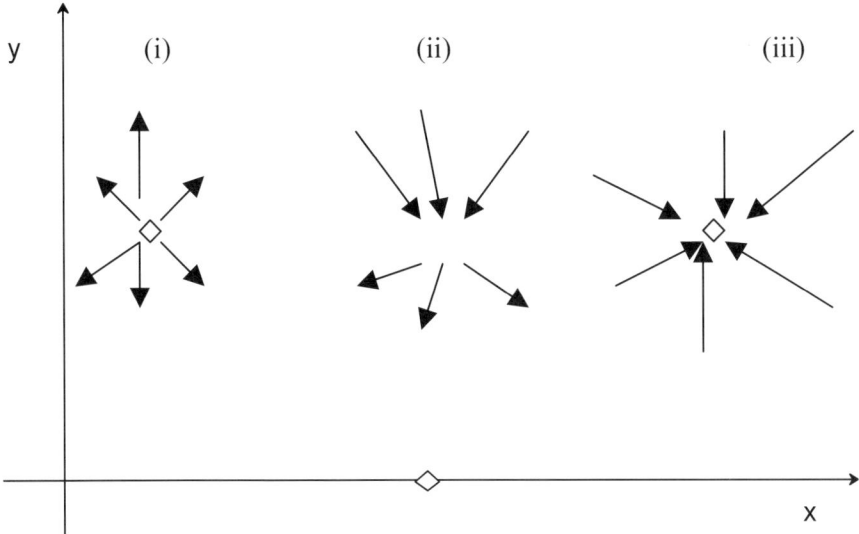

Fig. 1. Diagrams showing the topological structure of the system (3-1) at the point (a/b, 0) represented by a rhomboid. When (i) $f > d$ the equilibrium is said to be saddle (ii) if $f = d$ the point is called saddle point and (iii) if $f < d$ the point is referred as stable node

Attempts have been made to functionally relate many classical biological models, such as the dynamic Lotka-Volterra, to LES (= Life Energy System) models. In fact, the equivalence of both models was examined by Huang and Zu (2001). These authors found that the models always lead to similar predictions but the interpretation of the Lotka-Volterra models is more complicated and not straightforward. In contrast, LES models always have parameters amenable to biological interpretation.

The real progress in ecological modelling has been made when two points were recognized as '*paradoxes*'. Kuang and Beretta (1998) believe that this is […*A milestone progress*…]. The first is coming from the Rosenzweig-MacArthur model (Maynard Smith 1974) and is called the '*paradox of enrichment*'. According to this, the enrichment with predators of a predator-prey system results in higher equilibrium densities for the predator but not in smaller prey densities and this destabilizes the community equilibrium. This is a general predator-prey model with equations quite similar to the equations of (6-1 to 6-3):

$$\frac{d}{dt}x(t) := a \cdot x(t) \cdot \left(1 - \frac{x(t)}{K}\right) - \frac{c \cdot x(t) \cdot y(t)}{(m + x(t))} \qquad (6.4)$$

$$\frac{d}{dt}y(t) := y(t) \cdot \left[-d + f \cdot \frac{x(t)}{(m + x(t))}\right] \qquad (6.5)$$

where x and y are time functions of prey and predator density respectively, and a, K, c, m, f, d are all positive constants that stand for the intrinsic growth rate of prey, the carrying capacity of the ecosystem for the prey, the capturing rate of the predator, the half capturing saturation constant, the conversion rate and finally the death rate of the predator, respectively. The other paradox is called the "*biological control paradox*" and was introduced by Luck (1990). According to this model, the existence of two equilibrium points, one low and one corresponding to stable prey density, is impossible. Many ecologists consider this as a problem (Arditi et al. 1991a and b; Arditi and Saiah 1992; Gutierrez 1992) contributing to the discrepancy of the model predictions with field observations.

The most important models that have a spatial component and consider the movement of an organism or its tactic behaviour, are the reaction-diffusion models (Murray, 2001; Fagan et al., 2002) They were initially used for species invasion, after the finding of Kareiva (1983, in Murray 2001) who convincingly stated that the dispersion of an animal can be adequately described with a reaction-diffusion model and have a constant diffusion coefficient, which he obtained experimentally from various insect species. The relevance to predation of this type of modelling is that it incorporates the chemotactic behaviour of the studied insects, which is the common way for insect predators to find prey (Muray 2001: page 405). Mathematically, the chemotactic motion is one step more elaborated than the simple diffusive motion since the motion is done upwards or, more rarely as in the case of repellent compounds, downwards a concentration gradient. However, we know that, in nature, in some very simple cases of chemical orientation –e.g. in Lepidoptera - the chemotactic motion is carried out in a zigzag trajectory, in order for the moving organism to sample all the discrete puffs of the emitted plume (Elkinton and Carde 1994). One simple version of the model is given in the system of two non linear equations (6-6, 6-7).

$$\frac{\partial}{\partial t} n := D \cdot \frac{\partial^2}{\partial x^2} \left[n - \chi 0 \cdot \frac{\partial}{\partial x} \left(n \cdot \frac{\partial}{\partial x} \alpha \right) \right] \quad (6.6)$$

$$\frac{\partial}{\partial t} \alpha := h \cdot n - k \cdot \alpha + D_a \cdot \frac{\partial^2}{\partial x^2} \alpha \quad (6.7)$$

where n is the number of moving insects, a is the food to which the oriented insect moves, h and k are positive constants while $h \cdot n$ is the spontaneous production of the attractant, i.e. a chemical blend. The expression $-k \cdot a$ represents the activity of the attractant, if the attractant is not produced by the moving insects, such as aggregation pheromones used by bark beetles to recruit the bulk of the population on the bole of a susceptible tree.

Murray (2001) states that, in general, diffusion is a stabilising force whereas attractive chemotaxis -i.e. upward movement- is a destabilising force. The chemo-attractivity of the medium has been commonly expressed as an index $I = D_2/D_1$ where D_2 and D_1 are the diffusion coefficients of the medium and the insect, among many other indices. The researchers of reaction-diffusion-chemotaxis models are usually cell biologists, protozoologists or of a similar discipline, mainly interested in the movement properties of the modelled organisms.

6.4 Predation in relation to competition, parasitism, cannibalism and size

6.4.1 Competition

In terms of energy flow, predation is the transport of the energy existing in one organism, commonly called prey, to another, called predator. However, in the same terms, competition is the impact exerted on the trajectory of energy transportation, not the energy itself, for the benefit of one of the competitors. All competing members exert their own impact on the trajectory in order to exploit the limited available energy. In this sense predation and competition are two different impact modes on energy transfer incorporating all aspects of competition (Keddy 1989). Since these are among the most important biotic interactions, there is an ongoing discussion on which is more important in shaping biological communities (the discussion in Stewart 1996 and Denno et al. 1995 is particularly revealing). In many circumstances where an assembly rule for a particular community was investigated, competition was not a convincing explanation (Strong et al. 1984). Many authors tried to explain the lack of current examples of competition dominance by invoking selective forces based on competition that acted in the past; this was named by Connel (1980) '*the ghost of competition past*'. Lawton (1984) who studied insect communities on bracken, found that not all British species occurred on bracken at Skipwith Common, York, UK. Lawton speculated that if the realized feeding niches of all missing insect species were too similar to those of the existing insects, then, invasion was impossible; the proportion of British species excluded from the niche matrix is expected to increase as the cells of the matrix are populated. However, no such compliance was evidenced at the woodland and the open site at Skipwith. In the same lines, Strong (1984) emphasized that natural enemies exert a great influence in maintaining populations of insects on *Heliconia*. Strong believes that this factor keeps populations of insects far below the densities that are necessary for '...neo-Malthusian forces to operate among species...'. Actually, the theory of Connell (1980) is impossible to falsify. It is impossible to exorcise the 'ghost of competition past' with the current technology and theories (Strong 1984). Competition leaves no traces in the fossil record and in effect no falsification can be provided for the '*competition past*'. Only the traces left by predators on fossil prey are typified forms readily seen, such as drill holes in the shells or eggs, or exit marks of parasitoids (Jackson et al. 1996).

Some authors, especially those engaged in the research of the philosophical background of biology, consider predation as a form of competition (Bonner 1988). Influenced by Darwin, Bonner says that if an organism is efficient for the intake of enough food, the same is expected for its productivity. In this respect, predation is much more efficient than competition since the predator is selected for more efficient exploitation of prey; prey is selected for the ability to escape from the reach of predators or for colonising new environments where there are no predators. This escape to new environments releases many biological issues such as speciation, biodiversity increase and outbreeding (Bonner 1988). A very special mode of competition –and predation- is the '*arms race*' (Dawkins and Krebs 1979). The meaning of this term is that between two competitors, or in our case, in a predator-prey pair, the emergence of an adaptation in one of them causes the emergence of a counter adaptation to the other. If this is followed by evolutionary events, such as speciation, the phylogenetic picture essentially shows two trees with matching terminal branches (Brooks and McLennan, 1990).

Interestingly, many authors consider competitors as natural enemies and under this name they are treated as predators, though not in the classical sense. The distinction between the types of natural enemies that are or are not predators is quite clear. In a recent review, Vamosi (2005) treats competitors as enemies on the principle that they exert an adverse effect on the shared resources. In these lines he quotes the important work of Holt (1977; Holt and Lawton 1994). Vamosi (2005) also cites the paradigm of the damselfly *Enallagma* studied by McPeek in a series of papers. These damselflies live in several north American lakes. Two types of lakes can be typified. Fish lakes are full of fish predating damselflies. The other type is typically referred as dragonfly lakes and in these, damselflies are predated by dragonflies. In fish lakes the diversity of damselflies is much higher ($N=34$ species) than in dragonfly lakes ($N=4$ species). The apparent difference between the damselflies of the two lakes is that in the dragonfly lake the immature stages have an increased caudal lamella, which makes them efficient swimmers that escape from dragonfly predation. However, in this way they become more apparent to fish. To escape from fish predation the respective damselflies use crypsis as strategy. McPeek stated that apart from the higher damselfly diversity in fish lakes, dragonflies exert a detectable selection pressure on *Enallagma*.

6.4.2 Parasitism

Parasitism and predation in animals is widespread, and among the animal phyla there are nine entirely parasitic and twenty-two (predominantly) predatory ones (Lafferty and Kuris 2002). Many clades within each phylum are entirely parasitic or predatory. This implies that whether a group of organisms will evolve towards parasitism or predation depends on the actual or future allowed size. For example, the parasitoid status evolved once in parasitic Hymenoptera and was followed by the widest radiations in the entire evolutionary history (Godfray 1994; Lafferty and Kuris 2002). Importantly, hymenopteran parasitoids are all small in size and in a few cases they evolved trophic strategies intermediate of the two principal strategies. Indeed, many parasitic wasps and flies need to act as predators and eat animal flesh before they complete their egg development. According to Godfray (1994), they have evolved to the parasitoid status from predators of insects, fungi and scavengers of insects.

6.4.3 Size and predation

Central to ecology is that many ecological inferences can be drawn from morphological data, which can be easily seen in museum collections (Hespenheide 1975). The size of an organism is critical in many ecological issues and about this many 'rules' have been constructed that demand an underlying mechanism. Bergmann's rule is perhaps the best example. According to this rule the higher the latitude the larger the size of individuals of mammals, birds, reptiles and amphibians. The underlying reason for this rule is much harder to understand and the temperature gradients do not seem to suffice (Bonner 1988).

However, invertebrates and especially insects do not seem to follow this rule. In general, insects evolved many ways to overcome size constraints. Examples are the many feeding modes such as chewers, suckers, borers and gall formers, broadly in the same route that has been followed by primaeval bacteria, which adopted the style of "if you can't lick 'em, join 'em" becoming thus parasites, mutualists or symbiotes of other multicellular organisms (Bonner 1988). Stewart (1996) and Strong et al. (1984) quote a number of features emerging from the feeding mode of insects. For example, the incidence of competition is much higher in sap-feeding insects than in chewers and leaf miners and the asymmetric interactions are lower. Admittedly, there is no plausible explanation for these patterns.

The size of the members in the predator-prey system determines the capture success of the predator. Claessen et al. (2002) have found that if the ratio of the body lengths of prey to predator fall within a specific range, referred as '*predation window*', where δ is the lower limit and ϵ the upper limit, then the capture is possible. Although the research team discovered this in piscivorous fish, insect piscivores such as the hemipteran *Belostoma* were not taken into account. Aquatic and semiaquatic Hemiptera are usually outside –though not entirely- the predation window while they confer some mortality to fishes in lentic environments. In some instances of predation *Belostoma* may go outside the predation window, while *Notonecta, Corixa* and *Sigara* can capture larger prey including fish, amphibian and aquatic reptile hatchlings and pulmonate mollusks.

In a study of estuarine real food webs with four crustacean predators it was shown that size relations determine the strength of the various interactions in the food webs, in predator-prey systems. The strength of these relations determines the stability characteristic of the respective food webs (Emmerson and Raffaelli 2004). Although the definition of various stability patterns is not within the objectives of this chapter something must be said here. The account is largely based on the book of Yodzis (1989). *Local stability* is the preferred term –by Yodzis- though in the ecological literature prevails the commonly used term is *neighbourhood stability*. An *equilibrium point* in a density graph where the x-axis is the density axis and the abscissa is the population growth rate f, can be locally stable or not and it is defined as those point at which $df/dN = 0$. It is *locally stable* if small density departures from equilibrium densities (N_e) re-drive the system towards this equilibrium. This area of "small density departures" is usually called *domain of attraction* of the equilibrium. In insects that show a range of social behaviours it is very common to have a picture of the decline in per capita birth rate usually called *Allee effect*. The Allee effect usually introduces local stability in cases where equilibrium points were unstable (Yodgis, 1989:12-13).

The Allee effect can be of great importance in many biological phenomena such as '*biological invasions*'. In particular Petrovskii et al.(2002) have investigated the patchy invasion in a predatore-prey system. The authors showed that determinism via predator-prey modeling with Allee effect made possible such patchy invasions. Until now the researchers invoked environmental stochasticity tp show that such a possibility exists. The addition of the Allee effect to the integrodifference equations describing the invasion of an organism, in particular *Drosophila pseudoobscura* (Diptera: Drosophilidae) decreases the [...overall rate of spread...] of an organism and introduces a critical range that must be surpassed for a successful invasions (Kot et al., 1996).

According to Hespenheide (1975) there is a close relation of prey size, prey type and the niche width of the predator. First, it was shown that the sizes of individual insects –beetles in this case- follow a lognormal frequency distribution, that is the if the sizes are transformed to logarithms, they become normalised. Beetles were selected as an appropriate bird stomach item because they are very common insects, they are digested slowly and for this they can be identified some time after eating. The arguments and the points made by Hespenheide can be better seen in Fig. 2.

In this figure it can be seen that identical prey ranges can be produced by different avian predators such as vireos (slope 0.0275) corresponding to line *a* and swallows (slope 0.0060) corresponding to line *b*.

Simultaneously, the effect of two different lines ensures that many prey size ranges are consumed by the predators of a local fauna. The different lines in the prey-predator plane were interpreted by Hespenheide (1975) as the two components of niche width, namely the "within-phenotype = WPC" and "between-phenotype =BPC" component. WPC is considered the range of prey taken by one individual predator (=one phenotype). BWC is the total range of prey sizes taken by considering all the sizes of predators.

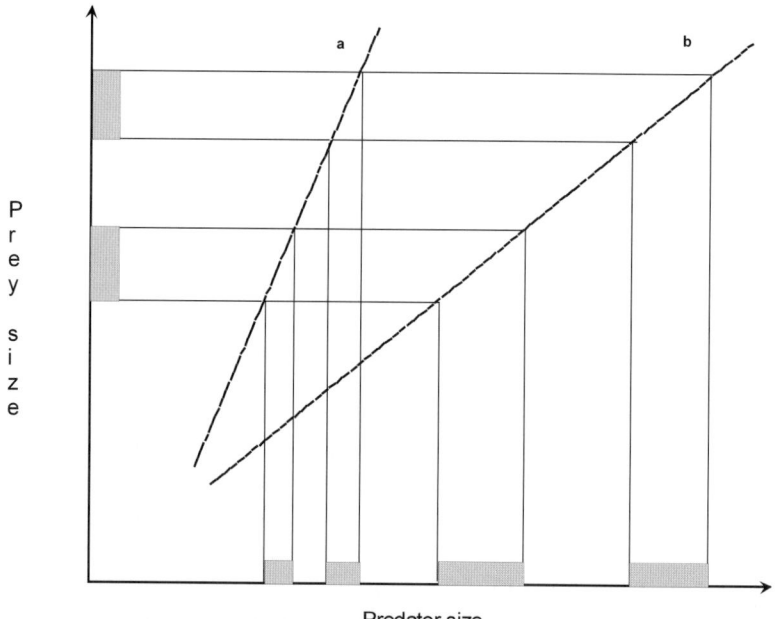

Fig. 2. The lines *a* and *b* correspond to two different relationships between the sizes of predator and the respective prey. Eventually both lines correspond to two different prey types taken by different predator assemblages. It can be seen that a particular range of prey items corresponds to two different ranges of predator sizes. Bigger predators are given by line *b* (after Hespenheide, 1973).

Given this puzzle it is unclear what is the evolutionary trend of ecosystems. Given the general macroevolutionary trend known as Cope's Rule (Jablonski, 1996) –i.e. organisms evolve towards larger sizes- we cannot say what is the trend for prey species or for predators. Even in cases where the entire assemblage is evolved toward bigger sizes (Bonner, 1988), as in theropod dinosaurs there is no evidence that the size relations to prey species have changed so the range of prey taken become more extended. Size dependent predation was also invoked as a mechanism behind habitat segregation of darkling beetles (Coleoptera, Tenebrionidae) in Negev Desert, Israel (Groner and Ayal, 2001). It was found that birds prefer large tenebrionids and for this large sizes of these beetles is found in densely vegetated habitats. Laboratory experiments designed to answer whether this is an affect of reduced efficiency of avian predation in dense vegetation cover or it is a bird preference *per se*, showed that the birds observed in the field and used in the laboratory experiments (white stork, Ciconia ciconia and stone curlew Burhinus oedicnemus) showed a clear preference for large sized tenebrionids. In addition, the anti-predatory defences of darkling beetles can be ordered according to their size. Large beetles are refuge-dependent, medium sized are sheltered in enemy-free spaces of the habitats while small beetles are predator independent. All these size related escape strategies are also reflected in the size-distributions of the tenebrionid beetles.

6.4.4 Cannibalism

For many authors, cannibalism is the mechanism acting before any shortage of resources can be detrimental to the population. However, for cannibalism to act, the size of the predator can be sufficiently larger than the intraspecific prey (Fox 1975). In the anthocorid *Xylocoris flavipes*, cannibalism is exerted by adults and large nymphs on smaller nymphs but not on eggs. The opposite situation was never observed and the eggs were never eaten. Possibly this behaviour is adaptive in the sense that the more durable life stage is left to exploit resources that are not available at the moment of cannibalism but are expected to emerge (Arbogast 1979; Polis 1981). In the reduviid true bug *Pisilus tipuliformis* many chemical stimuli were found to elicit predatory behaviour while cannibalism was observed in mating and parental behavioural sequences most of them interpreted as competition relaxing processes (Parker 1965). In particular, the effect of cannibalism at mating has received adaptive interpretations such as the reduction of future mating and the increase in female fecundity (Polis 1981; Johns and Maxwell 1997). Nevertheless, Arnqvist and Henriksson (1997) found no plausible adaptive explanation of the cannibalistic behaviour of fishing spiders and the authors consider it as a by-product of the selection on correlated traits such as predation, territoriality and aggression.

A reply to the lack of adaptation in cannibalism came from Kreiter and Wise (2001) who worked on the same group of spiders with the species *Dolomedes triton*. They found that the fecundity of the females is food limited and the increased mobility of females is towards the finding of natural prey. The cannibalism exhibited by the females of this species is an adaptation to counterbalance the shortage of food in order to keep egg production at the normal levels. Similarly Eggert and Sakaluk (1994) showed that in sagebrush crickets *Cyphoderris strepitans* (Orthoptera, Halgidae) the females eat the fleshy hind wings and the haemolymph oozing from the wounds they inflict. Whilst males with no hind wings are not dying, they obtained significantly fewer matings. The hind wings help males in transferring spermatophores to the mounted females.

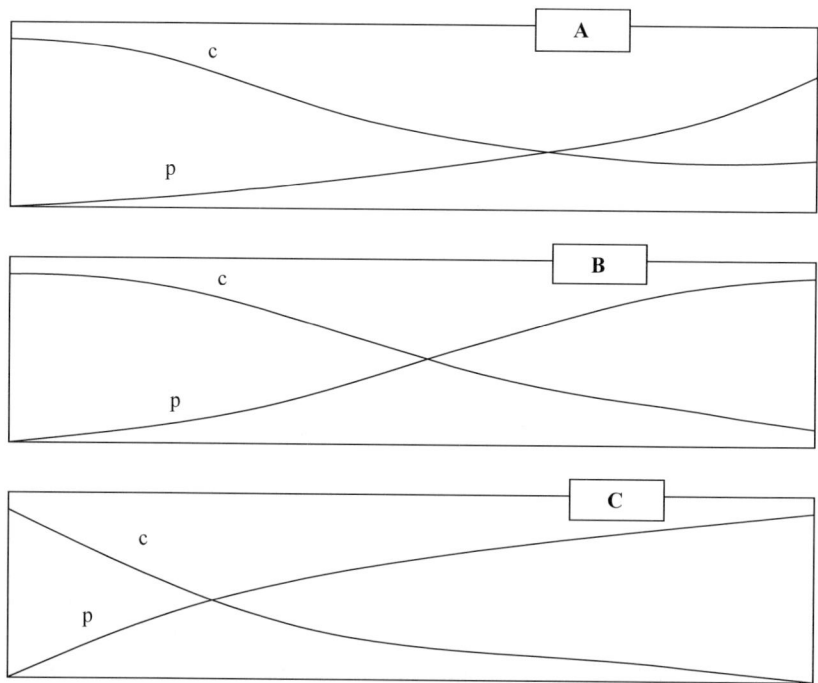

Fig. 3. Relative proportions of cannibalism [*c*] and predation [*p*] on competitors of a species. A: The aggressive species is more common than its competitors, B: the species and its competitors are equally abundant and C: the species is rare with respect to its competitors (after Fox 1975)

In wolf spiders, the species *Schizocosa oreata* shows size-based cannibalism on males - i.e. large females eat smaller males - and especially on those males with poor condition of the tufts of bristles on the first pair of front legs. Evidently, males with symmetric tufts are considered as appropriate partners for mating (Persons and Uetz 2005). This behaviour is adaptive since it promotes mate choice for larger mate size and many secondary sexual characters. If food limitation is predictable then cannibalism may be particularly advantageous according to Fox (1975 but see Wise (2006) for spider population regulation). For this, the predatory mite *Typhlodromus occidentalis*, which lives in Mediterranean climates with harsh and dry summers, is more cannibalistic than other populations of the same species living in the Pacific northwest. Fox (1975) presented in a figure the basic predictions he made about the advantages of cannibalism. In Fig. 3 these predictions can be better visualised and the particular shape of the abundance curves may vary. A salient feature of the predictions of arguments depicted in Fig. 3 is that the frequencies at which the rates of cannibalism and predation are equal vary with the relative abundance of the aggressive species. For instance, in panel A it can be seen that when the competitors are in low density then cannibalism may be high, while it decreases with increasing densities of competing species.

6.5 The control of prey population by predators

According to the majority of entomologists, insect communities are principally structured by predation forces exerted by natural enemies on prey populations. The predator-prey interaction is among the most powerful force that is capable of driving natural selection (Endler 1991; Brandt and Mahsberg 2002). A predator can control the population of a prey species if it can perform without problems the following behavioural sequence: (1) detection of the prey; (2) identification; (3) approach to a single or a number of prey individuals; (4) subjugation; (5) consumption and (6) normal post-ingestive physiology. All these steps have to be reflected to the important life history parameters of the prey species such as birth and death (=mortality) rate. From the point of view of a population ecologist three questions must be answered (Dempster 1983): (i) what is the '*key factor*' that causes the density fluctuations of a population (for explanations concerning the application of this method the work of Southwood and Henderson (2000) must be consulted), (ii) what is the governing factor of these fluctuations and (iii) what governs the abundance of these organisms — i.e. the mean level — and causes some species to be common and other species to be rare. Among entomologists, lepidopterists consider that (i) regulation and (ii) limitation are the two main hypotheses that pertain to the natural control of prey populations (Dempster 1983). Lepidoptera are homogenous with regard to habits and living modes. The form of the logistic equation suggested by Verhulst (Dempster 1983) is the model that can support the discussion of population control. The differential equation for the population growth rate is

$$\frac{d}{dt}N(t) := N(t) \cdot r \cdot \left(1 - \frac{N(t)}{K}\right) \qquad (6.8)$$

where *N(t)* is the population density of the prey at time *t* which is given in days or weeks, *K* is the '*carrying capacity*' and *r* is the intrinsic rate of increase of the population,. The model was examined by many workers (see for example in Jervis and Kidd 1996) and has a useful property. As

N(t) approaches *K*, the population differential growth declines to zero. This property of the Verhulst equation introduces its ability to show density dependent effects. Weather variations are a blurring factor in many population processes and are very difficult to be detected. For instance, it was found that *Ladoga camilla* is controlled by birds. Birds predate on the advanced larval instars and pupae of the white admiral butterfly and the rate of predation depends on summer temperatures. It remains unclear if weather *per se* controls the population or if it is correlated with the actual factors that exert control (Dempster 1983). The same dilemma is found when the fecundity of insects is measured through the size of females or directly from counting the complement of eggs in gravid females. Dempster states that weather variations is may influence the oviposition behaviour of prey females either directly, by restricting the time with conditions suitable for oviposition, or indirectly, by restricting food resources and through this the egg complement of gravid females.

Indeed, among the 24 works on Lepidoptera populations cited by Dempster (1983), eight studies could not demonstrate the existence of density-dependent regulation mechanisms such as starvation, virus-induced dispersal, delayed fecundity, cannibalism or disease, even though they lasted for more than ten generations. In three out of the 24 studies listed in Dempster (1983) population regulation came from natural enemies such as egg parasitoids, viruses and polyphagous parasitoids.

In other insect groups the situation is quite different. Especially in insects that occupy aquatic habitats, such as Odonata (=dragonflies and damselflies), predation is commoner as a regulating factor (Johnson, 1991). In addition, they may exert predation and regulate the populations of other insects such as mosquitoes, a feature exploited by mosquito control programs (see this volume). Since Odonate predators are capable of conferring mortality to mosquito populations, they are capable of controlling mosquitoes. When the enrichment of a habitat with odonates is wanted, the intra-guild and the intraspecific predation must be taken into account. The predation of *Enallagma* by dragonflies has also been noted above. Johnson (1991) states that in semivoltine species, that is species, which complete a life cycle in more than one year, the coexistence of many instars is quite common. Thus, intra-guild cannibalism is promoted, especially the type based on differences in size. For the same reasons intraspecific cannibalism is also common where different instars of the same species coexists. In many studies it is observed that intraspecific predation is the result of an attempt of the occupant to defend a perching site, a refuge from predators or a food supply. Such a behaviour is called *winning* since it is the occupant that usually wins.

It has been observed that in fishless water bodies, odonates can attain large densities and as a result the available resources are severely depleted. A mechanism that reduces the competition in such environments is intra-guild predation (Johnson 1991). Caution must be taken when the scarcity of odonates in fishless ponds is justified on the basis of intra-guild or intraspecific predation. At Schinias, Attica, Greece, the population and biodiversity of odonates was abruptly reduced in 2004 (personal observation). We attribute this to the more intense spraying program against mosquitoes, together with spraying in the coastal pinewood against the pine scale Marchalina hellenica (Hemiptera, Margarodidae) closely to the rowing Olympic centre on the occasion of the 2004 Olympic games.

In nature, the existence of many predators regulating one or more prey species seems to be the rule. In insects this fact is further supported by speciation, which is much more intense in insects occupying trophic levels corresponding to primary and secondary predators. Southwood (1985) in concert with this emphasizes that phytophagous insects comprise less than a quarter of all insect species while the rest is comprised by insectivorous species and other feeding types. The amount of nitrogen needed by insects is high (>18%; Southwood 1985). It is mandatory for an insect predator to have several prey species as food sources to secure that there always be available food. Of course many insects have followed other life modes manipulating the duration and timing in life history events. For insects that have many predators the impact of each predator has to be evaluated and a simple measure of it is the change in prey density *per capita* of the predator under study (Sih et al. 1998). In this sense, the impact of a single predator on a certain prey species cannot be easily isolated from the impact of others and probably this does not deserve the required substantial efforts. The existing data for food webs indicate that only in eight out of ninety-two examples of food webs was there a correspondence of one predator to one prey. In the rest, each prey species was predated by two to three predatory taxa (Schoener 1989; Sih et al. 1998). Sih et al. (1998) suggested that the classical approach for the detection of *multiple predator effects* (=MPE) is inefficient. Researchers adopted the construction of a 2x2 factorial experimental design for two predators with four treatments: (i) no predators, (ii) predator A only, (iii) predator B only, (iv) predators A and B. These authors refute the additive model on the basis that it allows prey to be killed twice. Instead they adopted the *multiplicative risk* model suggested by Soluk and Collins (1988). Sih et al. (1998) have found many studies that showed genuine MPE's and among them there was a substantial amount showing predatory risk enhancement. As an example they quote water striders that skate to the banks of the pond in order to avoid predation by sunfish. However, on the bank they are predated by spiders and this proves that the spatial refuge from predation is not effective. Actually Sih et al. quote this as a counter example saying that no enhancement was produced, since the water striders exhibit other generalised responses to multiple predators. They reduced the particularly risky mating activity among others. If the reduction of matings is reflected to the reduction of inseminations of females and the production of fertile eggs then it is an emerging MPE. In evolutionary time, which certainly is not checked, the reduction of matings would have the adverse effect of an insufficient reshuffling of genetic material and the ceasing of sexual selection with unpredictable effects to the survival of the species.

Sih et al. (1998) consider these conflicting prey defences as the main cause of risk enhancement. The most important feature of multiple predators is the reduction of the final mortality effect by interactions of predators. They quote several examples from the field of biological control examining the effect that the addition of a predator may have on the prey species, which is usually a pest. If the added predator affects a life stage — i.e. an instar — not yet affected by existing predators (or parasitoids) then the reduction of the prey is rapid and drastic.

The effect of multiple predators is not easily estimated in natural populations and only recently it has started to gain new insights. A future direction of this research is the study of the behavioural ecology of prey species, in particular, the behavioural sequences that are responses to single and multiple predators. The effect of these predators and the antipredatory responses of the prey in reducing the predation risk are of particular value.

6.6 The relation of predation to biodiversity

The initial increase of biodiversity with predation can be understood in the special case where the predation is exerted on autotrophs, usually called herbivory. The special type of herbivory called grazing causes interesting effects in the plant cover. It acts more intensely on the temporally dominating plants. In this way the domination of the site by this species or a few additional ones is prevented (Miles 1980). The widely held opinion that low grazing increases plant diversity is also responsible for the changes in the process of succession. On a spatial micro-scale many successional and seral stages coexist creating a habitat where there are no dominant plant(s) (Inouye 1980). The same in general qualitative terms exists when the prey species belongs to more advanced trophic levels. Although most evidence for the increase of biodiversity with predation comes from intertidal communities (Giller 1984) there are many cases where the predator is an insect that changes the competition status, thus preventing the dominance of a few species and exclusion of the rest. In this way, high biodiversity is created or maintained by a special type of predators usually called *'keystone predators'* (Pianka 1978).

It is important also to note that predation can also reduce biodiversity. For instance, competition among various species in a community may be alleviated due to predation. As a result, the affected species make more extensive use of the resources e.g. through more intense reproduction. This creates the dominance of these species, which in effect reduces biodiversity. For instance, Allan (1982) found that in some freshwater habitats, such as streams, the removal of the most important predator (the trout *Salmo trutta*) left intact the rest of the subordinate species. This was explained on the basis that the competitive bonds among the species in the lower level(s) were not particularly strong and for this the predation by trout was not an organizing force of the stream community.

The arguments about the impact of predation on biodiversity are diverse. For instance, many authors believe that in the xylobiont (=xylophagous) fauna it is important to separate the effects of, at least two, scales. The small scale, which requires a substantial amount of dead wood, and a large scale, which requires the occurrence of diverse habitats, among them forests, in order to add to the structural complexity of the landscape (= the ecosystems of a site). This ensures that the biodiversity of at least three families of insects will be maintained. The families, Cerambycidae, Buprestidae and Lucanidae contain xylophagous, especially saproxylic insects with particularly strong competitive relations since the available wood resources are scarce (Moretti and Barbalat 2004). Some Scarabaeidae beetles also exhibit this type of behaviour. Some species of this group have gone extinct from many places in Greece in parallel with the disappearance of mature and unmanaged forests. An example is the trichiin *Osmoderma eremita*, which presumably due to its strong repellent odour, has no important predators. Today, this species is one of the rarest beetles in Greece (Ranius et al. 2005). This insect was recently rediscovered in a set of stands of mature trees, remnants of old forests, in Greece.

The impact of predation on the diversity of ecological communities was also examined by modellers. Dennis and Patil (1979) pointed out that predation is a major species interaction together with competition and symbiosis. This covariance between the two noises may be taken to represent the interaction of the two species. A visualization of the species interaction, according to Dennis and Patil (1979), is achieved when the abundance vector of the community is close to the hyper-line $n_1=n_2=...=n_{St}$ where n_i are the population densities of the constituting species. A departure of the points from the hyper-line means a high likelihood of having low diversity. In this way the authors found a direct way to relate the interaction of two species with biodiversity. Apart from the mathematical simplicity of the above derivation, the relationship is not easily estimated in real ecosystems as the two following case studies show.

In a deciduous forest in Prespa national park, north-western Greece, (Petrakis 1989) it was found that there is a general spatial correlation of diversities among components at three trophic levels (plants, passerine birds and insects). In this work, two sets of variables were defined for the description of biotopes. They were responsible for the horizontal (horizontality) and vertical heterogeneity (verticality) of the biotope. Birds are easily recognized as four different groups while insects were distributed in accordance to the differences of biotopes. Bird and insect species of Mediterranean origin were in general widely distributed in comparison to central -European ones. In accordance to biotope structure, in a vertical and horizontal sense, bird and insect diversities are distributed in accordance to habitat heterogeneity. Birds were more faithful to that pattern since they do not rely solely on plants but they predate also on insects.

In an east Mediterranean ecosystem dominated by phrygana and maquis it was found that the importance values of plants were correlated with phytophagous hemipteran insects (Petrakis 1991). Hemipteran predator densities were also highly correlated with many aspects of plant diversity probably because they are correlated with their prey. The number of hemipteran insects was high, probably due to the existence of predators that keep populations of some insects at moderate levels preventing the dominance of a single species. Phenological investigations showed that at the onset of some phases one or two species became temporarily numerous. However, the natural enemies of these species are acting so as to reconstitute the community's equitability. The various phenological phases of insects in ecological time and their circadian differences sustain a diverse number of predatory species, mainly insects and arachnids. While there were only a few attempts to model the population densities of various insects, it is evident that diversity at one level – e.g. the plants - is reflected at the next trophic level –i.e. phytophagous insects. In particular, the work of Spencer (2000) on invertebrate communities has shown that predators are scarce below the level at which the conservatism of energy among trophic levels (≤ 0.10) predicts (Whittaker 1975). More precisely, Spencer modelled several communities and showed that the proportion of species is higher than that of individuals if the size of predators is equal or larger than the size of their prey. On the same grounds after experiments wit microcosms invaded by macroinvertebrates, mainly insects, Ahlering and Carrel (2001) showed that indeed the proportion of individuals is lower than that of species but this happened without regard to the size of predators (Fig. 4). They supported their findings by stating that predators are always specialising to certain prey types and this causes the lowering of the proportion of individuals of predators.

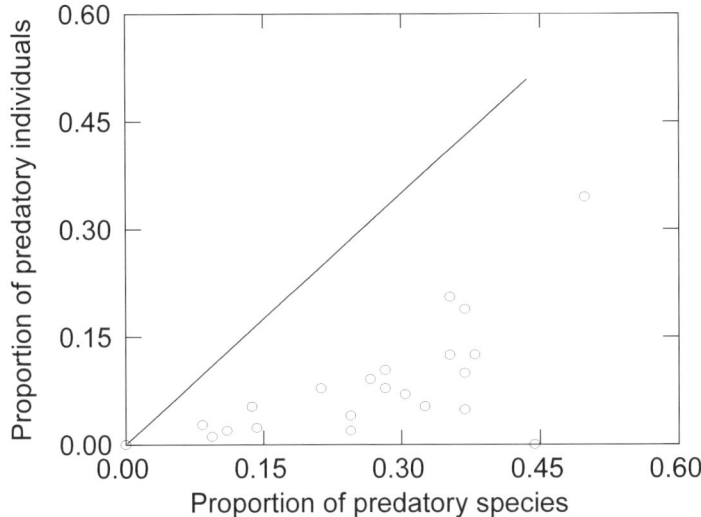

Fig. 4. Diagram showing the proportion of predatory species vs predatory individuals in a set of 22 experimental detrital microcosms. The diagonal line (y=x) shows equal proportions of species and individuals (Ahlering and Carrel, 2001)

Actually predation is not very different from other methods of obtaining energy from the environment. In the primaeval soup the photosynthesis – the capture of sunlight and carbon dioxide as raw material for sugars- is not essentially different from the engulfing directly particulate and dissolved food. For economy reasons, all organisms that retained one machinery lost the other. The insects actually adopted the second machinery and became predators, either phytophagous at the second trophic level or zoophagous at the third or higher trophic level. Once the predatory status of organisms was established there were a lot of new selective forces that acted towards the increase of diversity of local assemblages. Bonner (1988) stated the same argument for eukaryotes prior to multicellularity.

6.7 The chemical ecology of the prey-predator systems

Chemicals in prey-predator systems serve both as prey defence compounds or as prey detection cues. Reviewers of these systems retain this view from an important point. They consider food webs as information flows in addition to the energy flow, which is the traditional view (Dicke and Grostal 2001). Arthropods and particularly insects are especially suited to convey such chemical information for several reasons: (1) arthropods are important components of food webs, (2) among living organisms they comprise the largest group in terms of individual numbers and species diversity, (3) they are the best models for biological studies (small size, short life span, ease of maintaining them in captivity), (4) they heavily rely on chemical information for the detection of prey or avoiding predators, (5) they are particularly well equipped to detect and send chemical messages, and finally (6) since they are important for agriculture and public health there is always strong interest from authorities for their study and incorporation into practical programs. One widely accepted terminology that has been proposed, according to the above consideration by Dicke and Sabelis (1988), is given in Table 7-1.

It is interesting that, as a rule, insects are affected by primitive means of information exchange such as chemical communication. However, they have incorporated all other advanced ways of sensing the environment, such as vision and acoustics. As many authors have pointed out, all predators and parasitoids use, for detection, any chemical product from the prey available, such as feces, pheromones or exuviae (Vet and Dicke 1992; Godfray 1994; van Alphen and Jervis 1996; Carrasco 2005; Weiss 2006). The use of the chemical signal from the prey as infochemical (semiochemical) is also widespread in such a degree it caused the evolution of adaptations in both prey and predators (e.g. for fruit flies [Diptera, Tephritidae] see Greany et al. 1977; Carasco et al. 2005; Miller et al. 2005). These authors proved that the fruit fly *Anastrefa ludens* uses a complex set of compounds to detect the suitability of mango fruits for oviposition. The differentiation of chemicals caused by oviposition events of other flies allowed the decrease of competition for oviposition substrate. The predators *Enoclerus lecontei* and *Thanassimus undulatus* (Coleoptera, Cleridae) of the bark beetles *Ips pini* and *Orthotomicus latidens* (Coleoptera, Scolytidae) use the same components of the pheromone blend emitted by the bark beetles in the same dose-dependent way. That is, the response of the bark beetles is dose-dependent and so is the response of the predators. Moreover, the predators differentiated with respect to the dose-response attraction in the presence of β-phellandrene, which leaves unaltered the behaviour of the first but affects the latter. Bark beetles are able to manipulate the relative concentrations of the enantiomeres of the compounds in the pheromone blend and in this way they always exert selection pressures causing changes in the genetic system of their predators over short time (Raffa 2001). The main advantage for this manipulation is based on the ability of bark beetles to maintain a response window wider than the production window (see for an example Schlyter et al. 2001). On the same grounds, bark beetles manipulate the chemical signals used as kairomones by predators in an attempt to escape from predation. The result of this manipulation is ambiguous since the escape results in a decrease of competition of already existing predators and parasitoids (Pettersson 2001). Raffa and Dahlsten (1995) investigating the responses of predators to subtly altered pheromone blends emitted from *Ips pini* (Coleoptera, Scolytidae) found that, locally, the predators and bark beetles evolved adaptations and counter-adaptations to increase or decrease the detection rate respectively. The predators are strongly attracted to prey volatiles from non-adapted bark beetles. Volatiles that emanate from adapted bark beetles are not properly sensed by predators. In this way the predation exerted on bark beetles is reduced. However, the extent to which this is the result of adaptation and not a consequence of a random alteration of the volatile profile is not clear. Possibly this cannot be safely proved on the basis of available data.

Prey species usually try to escape predation or at last to relax the risk of predation by a variety of strategies, the most common being (i) time shifts in various life history stages, (ii) spatial escape by using habitats not normally visited by predators, (iii) cryptic behaviour of the prey species so that it is not apparent to potential predators, and (iv) masking, that is the decoration of body with foreign materials. However, because many predators use chemical compounds emitted from the host as cues for prey detection — i.e. kairomones — prey species evolved a class of chemically diverse compounds that are used for defence. In many cases the evolution of blends used for defence and kairomone-based detection is gradual, in an 'arms race' way. This is the case of a special class of kairomones emitted by plants and sensed by phytophagous insects, which are produced by the plant as a means to repel predators and are then used by ants to sense the precious tissue or storage medium of the plant (Caroll and Janzen 1973; Brooks and McLennan 1991). Some ants have proceeded further and use chemical compounds to estimate the health status of a plant. In ants that attend Homoptera this is vital. Ants detect the health of the plant and in case of plant sap exhaustion they move the honeydew producing Homoptera to more healthy parts of the plant or on other plants (Way

1963; Holldöbler and Wilson 1990). Chemical compounds of the plant that cannot be sensed by, the often sessile, homopterans, indicate the vigour of a plant. Instead, they are sensed by their attending ants that regulate the intake of plant sap. If it is not economically feasible to change the feeding position of the attended scale, the ants may decide to consume the homopteran scales. Caroll and Janzen (1973) quote examples from trees that produce nectar bodies directly harvestable from the ants. We have observed ants of the species *Crematogaster scutellaris* that attended the pine scale *Marchalina hellenica* to built occasionally squared covers to keep a stock of the feeding scale, which they guarded, to protect it from other, presumably arthropod, predators and harsh weather. The tritrophic system of plants — Homoptera — ants is an excellent example of this type of chemical relations.

Some groups of insects that have abandoned visual predation are heavily using infochemicals to detect prey. It is known that many herbivores are able to detect the altered volatile profile of a plant and also detect previous exploitation of the tissues by a herbivore (e.g. Dickens 1999; Broeckling and Salom 2003). Many insects, especially cryptic ones like miners and borers, are predated by parasitoids that rely on chemical signals emitted either by the prey or by the combination of prey and prey-damage on the tissue. In a study on the predators of scolytids and the way they are relaxed by prey, Aukema and Raffa (2004) investigated the role of the two major predators of *Ips pini*, namely *Thanassimus dubius* (Coleoptera, Cleridae) and *Platysoma cylindrica* (Coleoptera, Histeridae). It was found that the location of prey within the pine host by the beetle predators is greatly facilitated by the pheromone emitted by *I. pini*. Importantly, *T. dubius* senses the pheromone of the prey better than the scolytid itself (Raffa and Klepzig 1989). In particular, the predator is attracted to the prey odours four times more intensively than the image of the prey. Since in the bark beetle aggregation dilutes the predation risk, the insect uses aggregation pheromones to attract as many male and female insects as possible in the host tree. The authors state that predator-swamping was increased by the additive effects of the two predators but it is unlikely that swamping is behind the origin of aggregation behaviour in bark beetles as a predator avoidance strategy. The evidence gathered by the workers on bark beetles suggests that *I. pini* shifts the concentrations of chemical compounds of the pheromone blend in order to escape detection by the predators and simultaneously keeps the intraspecific communication by using the compound lanierone as a synergist of the two components.

The chemicals used by insects either as defences or as detection agents are also useful for phylogeny reconstruction. The cuticular profile of compounds of some insects (Peschke and Eisner 1987; Pearson et al.

1988) is a widely known example of such chemicals. Another important example is the defensive compounds of Carabidae (Forsyth 1972; Eisner et al. 1977; Moore 1979; Dettner 1987; Will et al. 2000) secreted by the pygidial gland of these beetles. Will et al. (2000) suggested that the inclusion of formic acid in the reservoir of the pygidial gland is indicative of a newer tribe in Carabidae. Formic acid producing tribes have radiated in all regions of the globe, a fact suggesting that this compound is a very effective defensive compound. The lack of primitive –or basal taxa- from the tropics is explained as an extinction on the basis of predation exerted by birds and ants because of the lack of formic acid. These authors support their arguments by stating that more potent chemicals exist in situations where high predation from birds and ants is expected. Many other defensive compounds are not of evolutionary importance –i.e. markers- and are invented occasionally to solve predatory problems arising in particular circumstances. An example of this situation is the bombardier beetle (Coleoptera, Carabidae, *Brachinus* sp.) (Eisner et al. 1977; Trautner and Geigenmuller 1987). The insect produces by the pygidial gland a mixture of quinones and hydrogen peroxide which are kept separated. When the beetle is disturbed, presumably the predator empties the components in the same chamber where an exothermous enzymic reaction takes place. The temperature rise of the mixture is rapid –i.e. up to $100^{o}C$- and the products, which are the pressure provider oxygen and quinones, are ejaculated to the exterior.

The chemical ecology of predation in insects is tightly bound to the behavioural ecology of these insects and the antipredator defences employed in these relations. For instance, tritrophic interactions and direct feeding of plants by natural enemies are considered important evolutionary forces affecting the mortality of herbivores and the fitness of their natural enemies (Ode 2006). Usually, the behavioural sequence includes steps, which correspond to one type of behaviour. A widely known example is the long-range sex pheromone blend, which releases other behaviours when the insect is approaching the sender (Carde and Baker 1984).

The number of cases where insects engage morphology and chemistry are many. Aposematism is a typical example of this class of relations. Aposematism is interpreted as a warning signal, which can be morphological, chemical (odour) or behavioural, and is sent by certain prey items to the predator in order to modify its behaviour (Mappes 2005). Since the processing of information is low within the brain of animals with respect to the information encountered in the environment, even in the more advanced forms, the animal is forced to select the more easily recognizable prey items (Dukas 2002). Aposematism is thus forced to emerge and maintained at low levels; as a result negative frequency-dependent selection is produced (Mappes 2005).

Many theories have been recently suggested to explain such phenomena and predator-prey interactions. Among them, the following were discussed in recent treatments of predation in insects.

6.8 Predator Confusion Hypothesis

According to this hypothesis (Dukas 2002), a predator must divide its attention among many of the visible members of a swarm. This results to a decrease of the probability of capturing a certain individual. In the case of predators that focus on an individual the outcome of the division of the attention is impossible. However, it is unlikely that a predator focuses attention on a certain individual of the swarm since all nearby individuals are identical.

6.9 Search Image Behaviour

This type of behaviour corresponds to a sort of selective searching in which a predator focuses on one out of many prey types if these types are equally cryptic and equally available. As a result of this behaviour a prey frequency-dependent predation usually emerges. The fact that a predator focuses on the more common prey, maintains the prey polymorphism and in effect an increase of biological diversity.

6.10 Sensory Exploitation Hypothesis

The hypothesis appeared also under the name *'selective attention hypothesis'*. It is known that animals can process only a limited portion of the information in the reach of their sensory system. According to this, females focus on selected patterns of the male courtship signals. The hypothesis is usually invoked when one tries to explain some structured choruses of calling males. The predation exerted on these choruses by predators perceiving them by acoustic sensing may be followed by chemically based predation especially when this happens in reduced light where only chemical perception is possible.

6.11 Predator Interference Hypothesis

The hypothesis (Tscharntke 1997) stems from the observation that the suppression exerted by birds on insects is higher on the natural enemies of phytophagous insects (=pests) than on the insects themselves. As a result, pest populations usually increase. Less effective predators are usually unaffected by top predators, which cannot be equally competitive. The fact that intermediate predators are of prime importance in the absence of higher order predators is also termed '*mesopredator release*' (Holt and Lawton 1994).

6.12 Pest Release Hypothesis

The feeding activity of some birds on bark beetles causes severe damage to the searched tree by removing the bark or drilling the trunk thus creating the appropriate conditions for future bark beetle ovipositions and fungus infestations. Defecation of already existing larvae by birds and further transport of pathogens to uninfected habitats causes an increase of bark beetles. To increase this effect many insects in oviposition lay also some frass together with eggs while others construct special envelops for their eggs called '*scatoshells*' (Weiss 2006).

6.13 Optimal foraging theory

This theory predicts that the way that an animal forages is always under selection for decisions that maximise the reproductive success (Krebs and Davies 1987; Stephens and Krebs 1986). In this context, the *Predator Confusion Hypothesis*, *Aposematism, Sensory Exploitation Hypothesis,* and the various ways of escaping from predators are all consequences of the optimal foraging theory. Although optimal foraging assumes that the animals to which the theory pertains are omniscient (Dicke and Grostal 2001), the infochemicalls emanating from the most important resources for an arthropod, such as food resources, predators, mates and competitors, are not impossible to be sensed and "known". In the light of the available evidence, food webs are always governed by infochemical webs. Moreover, feeding specialization and enemy free space can be seen as manifestations of optimal foraging theory. For instance, the finding of enemy free space drastically decreases the mortality imposed by natural enemies. Also, specialists make decisions in less time than generalists and this can have 'important consequences for fitness' (Dicke and Grostal, 2001).

6.14 Concluding remarks

The role of predation in shaping communities and influencing diversity has been known for many decades, at least as far as insects are concerned. However, in recent years the application of new methods on the vast accumulating data sets has provided more detailed insights into the functioning of predation. For example, new mathematical tools and the increase in computing power have given researchers the opportunity to further exploit the data from experimental studies and to move from abstract notions to testable hypotheses.

Furthermore, studies on predation have received a great boost from the change of attitude towards controlling harmful invertebrates. The shift from the use of chemical pesticides towards biological and integrated control has increased the importance of the use of predators as controlling agents and, therefore, has provided further data for the study of the role of predation.

This shift is also evident in the study of the role of chemical interactions in insect predation systems. The increase of precision of chemical analysis methods has assisted researchers to identify new natural compounds and to suggest roles for them.

Biodiversity is a common word nowadays and a lot of attention has been paid to its conservation and sustainable use. Accordingly, research on biodiversity issues has increased dramatically and has led to a reappraisal of the effect of predation on the diversity and structure of communities. What was up to now treated with generalizations has now become more specific.

Predation, especially in insects, is playing a very important role but only now are we becoming aware of the many specific effects that it may have.

References

Ahlering MA, Carrel JE (2001) Predators are rare even when they are small. Oikos 95: 471-475

Allan JD (1982) The effects of reduction in trout density on the invertebrate community of a mountain stream. Ecology 63: 1444-1455

Alphen J.J.M van, Jervis MA (1996) Foraging behaviour. In Jervis MA, Kidd NAC (eds) Insect Natural Enemies: Practical Approaches to their Study and Evaluation. Chapman & Hall, London, 1-62

Arbogast RT (1979) Cannibalism in *Xylocoris flavipes* (Hemiptera: Anthocoridae), a predator of stored-product insects. Entomologia Experimentalis et Applicata 25: 128-135

Arditi R, Ginzburg LR, Akcakaya HR (1991a) Variation in plankton densities among lakes: a case of ratio-dependent models. American Naturalist 138:1287-96

Arditi R, Perrin N, Saiah H (1991b) Functional responses and heterogeneities: an experimental test with cladocerans. Oikos 60:69-75

Arditi R, Saiah H (1992) Empirical evidence of the role of heterogeneity in ratio-dependent consumption. Ecology 73: 1544-1551

Arnqvist G, Henriksson S (1997) Sexual cannibalism in the fishing spider and a model for the evolution of sexual cannibalism based on genetic constraints. Evolutionary Ecology 11: 253-271

Bonner JT (1988) The Evolution of Complexity. Princeton University Press, Princeton

Brandt M, Mahsberg D (2002) Bugs with a backpack: the function of nymphal camouflage in the west African assassin bugs *Paredocla* and *Acanthaspis* spp. Animal Behaviour 63: 277-284

Brooks DR, McLennan DA (1991) Phylogeny Ecology and Behavior: A Research Program in Comparative Biology. University of Chicago Press, Chicago

Cantrell RS, Cosner C, Hutson V (1996) Spatially explicit models for the population dynamics of a species colonizing island. Mathematical Biosciences 136: 65-107

Caroll CR, Janzen DH (1973) Ecology of foraging ants. Annual Review of Ecology and Systematics 4: 231-257

Carrasco M, Montoya P, Cruz-Lopez L, Rojas JC (2005) Response of the fruit fly parasitoid *Diachasmimorpha longicaudata* (Hymenoptera:: Braconidae) to mano fruit volatiles. Environmental Entomology 34: 576-583

Claessen D, van Oss C, de Roos AM, Persson L (2002) The impact o size-dependent predation on population dynamics and individual life history. Ecology 83: 1660-1675

Connell JH (1980) Diversity and the coevolution of competitors, or the ghost of competition past. Oikos 35: 131-138

Dempster JP (1983) The natural control of populations of butterflies and moths. Biological Review 58: 461-481

Dennis B, Patil GP (1979) Species abundance, diversity and environmental predictability. In: Grassle JF, Patil GP, Smith W, Taillie C (eds) Ecological Diversity in Theory and Practice. International Cooperative Publishing House, Fairland 93-131

Dicke M, Grostal P (2001) Chemical detection of natural enemies by arthropods: an ecological perspective. Annual Review of Ecology and Systematics 32: 1-23

Dicke M, Sabelis MW (1988) Infochemical terminology: based on cost-benefit analysis than origin of compounds? Functional Ecology 2: 131-139

Dostalkova I, Kindlmann P, Dixon AFG (2002) Are classical predator-prey models relevant to the real world? Journal of Theoretical Biology 218: 323-330

Dukas R (2002) Behavioural and ecological consequences of limited attention. Philosophical Transactions of the Royal Society of London B Biological Sciences 357: 1539-1547

Eggert A-K, Sakaluk SK (1994) Sexual cannibalism and its relation to male mating success in sagebrush crickets, *Cyphodermis strepitans* (Haglidae: Orthoptera). Animal Behaviour 47: 1171-1177

Eisner T, Jones TH, Aneshansley DJ, Tschinkel WR, Silberglied RE, Meinwald J (1977) Chemistry of defensive secretions of bombardier beetles (Branchini, Metriini, Ozaenini, Paussini) Journal of Insect Physiology 23: 1383-1386

Elkinton JS, Carde RT (1984) Odor dispersion and chemo-orientation mechanisms. In Bell WJ, Carde RT (eds) Chemical Ecology of Insects. Chapman & Hall, London 73-92

Emmerson MC, Raffaelli D (2004) Predator-prey body size, interaction strength and the stability of a real food web. Journal of Animal Ecology 73: 399-409

Endler JA (1991) Interactions between predator and prey. In Krebs JR, Davies NB (eds) Behavioural Ecology, Blackwell, Boston 169-196

Fagan WF, Lewis MA, Neubert MG, van den Driessche P (2002) Invasion theory and biological control. Ecology Letters 5: 148-157

Forsyth DJ (1972) The structure of the pygidial defense glands of Carabidae (Coleoptera),,Transactions of the Zoological Society London,32: 249-309,

Fox LR (1975) Cannibalism in natural populations. Annual Review of Ecology and Systematics 6: 87-106

Giller PS (1984) Community Structure and the Niche. Chapman and Hall, London

Godfray HCJ (1994) Parasitoids. Behavioral and Evolutionary Ecology. Princeton University Press, Princeton

Greany PD, Tumlinson JH, Chambers DL, Boush GM (1977) Chemical mediated host finding by *Biosteres (Opius) longicaudatus*, a parasitoid of tephritid fruit fly larvae. Journal of Chemical Ecology 3: 189-195

Groner E, Ayal Y (2001) The interaction between bird predation and plant cover in determining habitat occupancy of darkling beetles. Oikos 93: 22-31

Gutierrez AP (1992) The physiological basis of ratio-dependent predator-prey theory: a metabolic pool model of Nicholson's blowflies as an example. Ecology 73: 552-1563

Hanski I (1991) The functional response of predator: worries about scale. Trends in Ecology and Evolution 6: 141-142

Hanski I, Gilpin M (1991) Metapopulation dynamics: Brief history and conceptual domain. Biological Journal of the Linnean Society 42: 3-16

Hespenheide HA (1975) Prey characteristics and predator niche width. In Cody ML, Diamond JM (eds) Ecology and Evolution of Communities,The Belknap Press, Harvard, Cambridge 158-180

Holt RD (1977) Predation, apparent competition, and the structure of prey communities. Theoretical Population Biology 12: 197-229
Holt RD, Lawton JH (1988) The ecological consequences of shared natural enemies. Annual Review of Ecology and Systematics 25: 495-520
Holt RD, Lawton JH (1994) The ecological consequences of shared natural enemies. Annual Review of Ecology and Systematics 25: 495-520
Huang X, Zu Y (2001) The LES population model: essentials and relationship to the Lotka-Volterra model. Ecological Modelling 143: 215-225
Inouye DW (1980) The terminology of floral larceny. Ecology 61:1251-1253
Jablonski D (1996) Body size and macroevolution. In Jablonski D, Erwin DH, Lipps JH (eds) Evolutionary Paleobiology, The University of Chicago Press, Chicago 89-122
Jackson JBC, Budd AF, Pandolfi JM (1996) The shifting balance of natural communities? In Jablonski D, Erwin DH, Lipps JH (eds) Evolutionary Paleobiology, The University of Chicago Press, Chicago 89-122
Janzen DH (1971) Seed predation by animals. Annual Review of Ecology and Systematics 2: 465-492
Jervis MA, Kidd NAC (eds) (1996) Insect Natural Enemies: Practical Approaches to their Study and Evaluation. Chapman & Hall, London
Johns PM, Maxwell MR (1997) Sexual cannibalism: who benefits? Trends in Ecology and Evolution 12: 127-128
Johnson DM (1991) Behavioral ecology of larval dragonflies and damselflies. Trends in Ecology and Evolution 6: 8-13
Jost C, Arino O, Arditi R (1999) About deterministic extinction in ratio-dependent predator-prey models. Bulletin of Mathematical Biology 61: 19-23
Kareiva PM (1983) Local movement in herbivorous insects: applying a passive diffusion model to mark-recapture field experiments. Oecologia 57: 322-327
Kats LB, Dill LM (1998) The sent of death: chemosensory assessment of predation risk by prey animals. Ecoscience 5: 361-394
Keddy PA (1989) Competition. Chapman and Hall, London
Kot M, Lewis MA, van den Driessche P (1996) Dispersal data and the spread of invading organisms. Ecology 77: 2027-2042
Krebs JR, Davies NB (eds) (1987) An Introduction to Behavioural Ecology. Blackwell, Oxford
Kreiter NA, Wise DH (2001) Prey availability limits fecundity and influences the movement pattern of female fishing spiders. Oecologia 127: 417-424
Kuang Y, Beretta E (1998) Global qualitative analysis of a ratio-dependent predator-prey models. Journal of Mathematical Biology 36: 389-406
Lafferty KD, Kuris AM (2002) Trophic strategies, animal diversity and body size. Trends in Ecology & Evolution 17: 507-513
Lawton JH (1984) Non-competitive populations, non-convergent communities, and vacant niches: The herbivores of bracken. In Strong DR, Simberloff D, Abele LG, Thistle AB (eds) Ecological Communities: Conceptual Issues and the Evidence, Princeton University Press, Princeton 67-100
Mappes J, Marples N, Endler JA (2005) The complex business of survival by aposematism. Trends in Ecology and Evolution 20: 588-603

Maynard Smith J (1974) Models in Ecology. Cambridge University Press, Cambridge

Menge BA, Sutherland JP (1976) Species diversity gradients: synthesis of the role of predation, competition and temporal heterogeneity. American Naturalist 110: 351-369

Miles J (1979) Vegetation Dynamics. Chapman & Hall, London

Miller DR, Borden JH, Lindgren BS (2005) Dose-dependent pheromone responses of *Ips pini*, *Orthotomicus latidens* (Coleoptera: Scolytidae), and associates in stands of lodgpole pine. Environmental Entomology 34: 591-597

Moore BP (1979) Chemical defence of Carabidae and its bearing on phylogeny. In Erwin TL, Ball GE, Whitehead DR, Halpern AL (eds) Carabid Beetles: Their Evolution, Natural History and Classification, Junk, Hague, 193-203

Moretti M, Barbalat S (2004) The effects of wildfires on wood-eating beetles in deciduous forests on the southern slope of Swiss Alps. Forest Ecology and Management 187: 85-103

Morin PJ, Lawler SP, Johnson EA (1988) Competition between aquatic insects and vertebrates: interaction strength and higher order interactions. Ecology 69: 1401-1409

Murray JD (2001) Mathematical Biology. I. An Introduction. Springer, New York

Ode PJ (2006) Plant chemistry and natural enemy fitness: effects on herbivore and natural enemy interactions. Annual Review of Entomology 51: 163-187

Odenbaugh J (2005) Idealized, inaccurate but successful: A pragmatic approach to evaluating models in theoretical ecology. Biology and Philosophy 20: 231-255

Parker AH (1965) The predatory behaviour and life history of *Pusilus tipuliformis* Fabricius (Hemiptera: Reduviidae). Entomologia Experimentalis et Applicata 8: 1-12

Pearson DL (1988) Biology of tiger beetles. Annual Review of Ecology and Systematics 33: 123-147

Pearson DL, Blum MS, Jones TH, Fales HM, Gonda E, Witte BR (1988) Historical perspective and the interpretation of ecological patterns: defensive compounds of tiger beetles (Coleoptera: Cicindellidae). American Naturalist 132: 404-416

Persons MH, Uetz GW (2005) Sexual cannibalism and mate choice decisions in wolf spiders: influence of male size and secondary sexual characters. Animal Behaviour 69: 83-94

Peschke K, Eisner T (1987) Defensive secretion of the tenebrionid beetle *Blaps mucronata* physical and chemical determinants of effectiveness. Journal of Comparative Physiology A Sensory Neural and Behavioral Physiology 161: 377-388

Petrakis PV (1989) A multivariate approach to the analysis of biotope structure with special reference to their avifauna in Prespa region, north-Western Greece, Biologia Gallo-Hellenica 16: 66-107

Petrakis PV (1990) Plant Heteroptera Associations in an East Mediterranean Ecosystem: an Analysis of Structure, Specificity and Dynamics, PhD Dissertation, College of Cardiff, University of Wales

Petrovskii SV, Morozov AY, Venturino E (2002) Allee effect makes possible patchy invasion in a predator-prey system. Ecology Letters 5: 345-352

Pettersson EM (2001) Volatile attractants for three Pteromalid parasitoids attacking concealed spruce bark beetles. Chemoecology 11: 89-95

Pianka ER (1978) Evolutionary Ecology. Harper & Row, New York

Polis GA (1981) The evolution and dynamics of intraspecific predation. Annual Review of Ecology and Systematics 12: 225-251

Polis GA, Myers CA, Holt RD (1989) The ecology and evolution of intraguild predation: Potential competitors that eat each other. Annual Review of Ecology and Systematics 20: 297-330

Raffa KF (2001) Mixed messages across multiple trophic levels: the ecology of bark beetle chemical communication systems. Chemoecology 11: 49-65

Raffa KF, Dahlsten DL (1995) Differential responses among natural enemies and prey to bark beetle pheromones. Oecologia 102: 17-23

Ranius T, Aguado LO, Antonsson K, Audisio P, Ballerio A, Carpaneto GM, Chobot K, Gjurašin B, Hanssen O, Huijbregts H, Lakatos F, Martin O, Neculiseanu Z, Nikitsky NB, Paill W, Pirnat A, Rizun V, Ruiconescu A, Stegner J, Süda I, Szwalko P, Tamutis V, Telnov D, Tsinkevich V, Versteirt V, Vignon V, Vögeli M, Zach P (2005) *Osmoderma eremita* (Coleoptera, Scarabaeidae, Cetoniinae) in Europe. Animal Biodiversity and Conservation 28: 1-44

Schlyter F, Svensson M, Zhang Q-H, Knizek M, Krokene P, Ivarsson P, Birgersson G (2001) A model for peak and width of signaling windows: *Ips duplicatus* and *Chilo partellus* pheromone component proportions - Does response have a greater window than production? Journal of Chemical Ecology 27: 1481-1511

Schoener TW (1989) Food webs from the small to the large. Ecology 70: 1559-1589

Sih A, Crowley P, McPeek M, Petranka J, Strohmeier K (1985) Predation, competition, and prey communities: A review of field experiments. Annual Review of Ecology and Systematics 16: 269-311

Sih A, Englund G, Wooster D (1998) Emergent impacts of multiple predators on prey. Trends in Ecology and Evolution 13: 350-355

Soluk DA, Collins NC (1988) Synergistic interactions between fish and stoneflies: facilitation and interference among stream predators. Oikos 52: 94-100

Southwood TRE, Henderson PA (2000) Ecological Methods. Blackwell Science, Oxford

Southwood TRE, Moran YD, Kennedy CEJ (1982) The assessment of arborial insect fauna: comparisons of knockdown sampling and faunal lists. Ecological Entomology 7: 331-340

Spencer M (2000) Are predators rare? Oikos 89: 115-122

Stearns SC (1992) The Evolution of Life Histories. Oxford University Press, Oxford

Stephens DW, Krebs JR (1986) Foraging Theory Princeton University Press, Princeton, New Jersey

Stewart AJA (1996) Interspecific competition reinstated as an important force structuring insect communities. Trends in Ecology & Evolution 11: 233-234

Strong DR (1984) Exorcising the ghost of competition past: Phytophagous insect communities. In Strong DR, Simberloff, Abele LG, Thistle AB (eds) Ecological Communities: Conceptual Issues and the Evidence, Princeton University Press, Princeton, 28-41

Strong DR, Lawton JH, Southwood TRE (1984) Insects on plants: community patterns and mechanisms. Blackwell, Oxford

Suarez AV, Case TJ (2002) Bottom-up effects on persistence of a specialist predator: ant invasions and horned lizards. Ecological Applications 12: 291-298

Takahara Y (2000) Individual base model of predator-prey system shows predator dominant dynamics. Biosystems 57: 173-185

Trautner J, Geigenmuler K (1987) Tiger Beetles, Ground Beetles, Illustrated Key to the Cincindelidae and Carabidae of Europe. Josef Margraf Publ., Aichtal, Germany

Tscharntke T (1997) Vertebrate effects on plant-herbivore food webs. In Gange AC, Brown VK (eds) Multitrophic Interactions in Terrestrial Systems. Blackwell, Oxford, 277-297

Vamosi SM (2005) On the role of enemies in divergence and diversification of prey: a review and synthesis. Canadian Journal of Zoology 83: 894-910

Veen FJF van, Morris RJ, Godfray HCJ (2006) Apparent competition, quantitative food webs, and the structure of phytophagous insect communities. Annual Review of Entomology 51: 187-208

Vet LEM, Dicke M (1992) Ecology of infochemicals used by natural enemies in a tritrophic context. Annual Review of Entomology 37: 141-172

Wallace AR (1879) The protective colours of animals. In Brown R ([ed) Science for All, Cassell, Petter, Galpin and Co, London, 128-137

Way MJ (1963) Mutualism between ants and honeydew producing Homoptera. Annual Review of Entomology 8: 307-344

Weiss MR (2006) Defecation behaviour and ecology of insects. Annual Review of Entomology 51: 635-661

Whittaker RH (1975) Communities and Ecosystems. MacMillan, New York

Will KW, Attygale AB, Herath K (2000) New defensive chemical data for ground beetles (Coleoptera: Carabidae): interpretations in a phylogenetic framework. Biological Journal of the Linnean Society 71: 459-481

Wise DH (2006) Cannibalism, food limitation, intraspecifc competition, and the regulation of spider populations. Annual Review of Entomology 51: 441-466

Xiao D, Ruan S (2001) Global dynamics of a ratio-dependent predator-prey system. Journal of Mathematical Biology 43:268-290

Yodzis P (1989) Introduction to Theoretical Ecology. Harper and Row, New York

7 Biological control of mosquito populations: An applied aspect of pest control by means of natural enemies.

Anna Samanidou–Voyadjoglou[1], Vassilios Roussis[2] and Panos V. Petrakis[3]

[1]National School of Public Health, Department of Parasitology, Entomology and Tropical Diseases, 196 Alexandras Ave, 11521 Athens, Greece, [2]University of Athens, Department of Pharmacy, Division of Pharmacognosy and Chemistry of Natural Products, Panepistemiopolis Zografou, 15771 Athens, Greece, roussis@pharm.uoa.gr, [3]National Agricultural Research Foundation, Institute of Mediterranean Forest Ecosystem Research, Lab. of Entomology, Terma Alkmanos, 1152 Ilissia, Athens, Greece, pvpetrakis@fria.gr

7.1 Abstract

Mosquitoes were recognized as a health and nuisance problem only in the last century. Since mosquito oviposition sites were initially associated with the expansion of human settlements, which as a rule is done by building actions of poor economies, the biological control became a necessity. Even with the development of chemical industry, the preparation of synthetic formulations of insecticides is environmentally hostile and ecologically unsafe since the main side effect of the application is the extinction of natural enemies of mosquitoes such as odonates, beetles, fishes and hemipterans in water pools apart from the induced resistance in a short time. The advancement of biological knowledge made available many new controlling methods of mosquito populations though substantially more expensive than synthetic insecticides. The most important of them is the set of semiochemicals (natural products) associated with the classical biological control by means of predators. Many investigations have proven that predators are able to control mosquitoes in ecosystems of variable size, nutrients and prey densities. Semiochemicals are employed in many aspects predation. Prey detection, oviposition site selection, chemical crypsis, kairomonal confusion is among the ways that predators and their prey mosquitoes are using to affect the outcome of predation. Microbials as killing agents, are extensively used in projects for mosquito control in order to minimize the environmental side effects of insecticides. However, they fail to eradicate pest populations. In the present report is described an integrated system for the control of mosquitoes from experience gained in Mediterranean areas of application. The control system is able to incorporate any future developments in mosquito population management such as new introductions or releases from rearing of predators, new genetic methods or inexpensive repellents and oviposition deterrents.

Keywords: Mosquitoes, Culicidae, predators, Odonata, Coleoptera, Hemiptera, population management, mosquito control, biological control, Mediterranean.

7.2 Introduction

Mosquitoes belong to the large family of Dipteran insects, Culicidae. They comprise the largest group of blood-sucking insects, which attack not only humans, but also many kinds of vertebrates, including mammals, birds, fish and reptiles and transmit pathogens to all these groups. More than 3200 different mosquito species are spread all over the world, with the exception of Antarctic, which is continuously covered by ice. Mosquitoes have been recorded at elevations of 4.300 m, in Kashmir, as well as 1160 m below sea level, in gold mines of South India (Harwood and James 1979).

The majority of mosquito species are nocturnal in habits. There are however several others active during daytime. Enormous mosquito populations are possible to be produced in any kind of water collections, large or small, temporary or permanent, stagnant or running, clear or polluted, fresh or brackish. The eggs of some species are capable to withstand dryness for several months and hatch normally when the sites where they were deposited flood again. These floodwater species cause severe nuisance problems in coastal areas and influence negatively the establishment for tourists or the fieldwork in rural areas.

The mosquitoes' ancestors appeared in the earth as early as Jurassic period, according to fossils trapped in amber. Since then, they evolved and adapted to every type of local environments with different water quality, different climate, temperature, and food. All these differences in adaptations resulted in the presently occurring species and subspecies. In spite of their long existence on the earth, it is only the end of 19th century that mosquitoes attracted the interest of scientists. That period coincides with the discovery that serious diseases such as malaria, filariasis or yellow fever, are transmitted to humans by certain mosquito species. It was realized then, that suppressing the mosquito vectors could attain successful control of these diseases. Although the accurate description of malaria's symptoms by Hippocrates in the 4rth century B.C. indicates that the disease was well known in Greece at that time, Alphonse Laveran discovered its causative agent, *Plasmodium* sp., in 1880 (Jones 1909). Intensive research followed by several scientists and in 1897 Ronald Ross pointed out the correlation between *Plasmodium* transmission and mosquito bites. Meanwhile, Manson, in 1896, proved the role of mosquitoes in the transmission of human filariae and early 1900's it became clear by W. Reed that mosquitoes also transmit yellow fever (Harrison 1978). All these diseases were often a disaster for human societies, leading to death millions of people and leaving the rest weak and unable to improve their living status. Scientists realized that diminishing the vector populations could succeed the control of these diseases. Therefore, intensive entomological research started, particularly in countries facing serious public health problems due to mosquitoes, with the aim to gain information necessary in the application of effective control measures. Several aspects dealing with mosquito systematics, their biology, life cycles, ecology, behavior and vector capacity were elucidated in these early days.

Systematic efforts for mosquito control actually started soon after the discovery of Ronald Ross that mosquitoes transmit malaria, although several natural substances with insecticidal properties were used before that time, indicating the extent of nuisance caused by mosquitoes. Records of using different substances against insects exist already in the ancient Greek literature (Panagiotakopulu 2000). Early methods in mosquito control included also drainage of large marshy areas, besides spraying with the existing natural insecticides. The discovery of the first synthetic insecticides, e.g. DDT and its derivatives, marks the beginning of a new era in the control of insects. As a consequence of the resistance mosquitoes developed to those organochlorine chemicals, further research directed towards the discovery of new insecticides with different mode of action (Ware 1983). Mosquito control relied thoroughly on the use of chemical insecticides for many years. No doubt, the chemical insecticides contributed to the extinction of many serious diseases transmitted by mosquitoes and other insects. The extensive use and the lack of adequate knowledge however, had a tremendously destructive impact to the environment and the wild life, including fish, birds, arthropod predators, insect pollinators and soil microorganisms. Rachel Carson (1962) presents dramatically the adverse effects of insecticides.

When all these side effects became obvious and awareness of environmental safety increased, alternative control measures were investigated and a revolution to biological control started in mid 1960 (Becker et al. 2003). The concept of using living organisms in controlling mosquitoes dates back to 19th century, when the first attempts were made to introduce dragonflies as predators in mosquito breeding places (Lamborn 1890). Meanwhile, it was noticed that several other organisms, aquatic or terrestrial, could consume mosquitoes as food. Very few, however, are effective and have been considered as possible biological control agents.

The oldest and better-known organisms successfully used in mosquito control are fish. The characteristic preference of some fish species in mosquito feeding had been already observed since the middle of the 19th century but their use in larviciding started after the discovery that mosquitoes transmit malaria. It is stated that in 1902 some species of fish were used in mosquito control in India for first time (Livadas and Sphangos 1940). However, systematic studies for large-scale application were conducted in the U.S.A., about the same period. It was proved then that most effective were fish of the genus *Gambusia* (Legner 1995).

In Europe this fish was imported in 1921 for first time in Spain and later in Italy. In Greece the use of *Gambusia* in antimalarial program first took place in 1927 when fish were transported from Italy and distributed in natural breeding sites in North Greece (Macedonia). Next year the Greek Red Cross transported *Gambusia* from Marseille for use, while the same year, the King of Italy donated a population of *Gambusia* to the Greek Government, which were distributed mainly in Attiki District.

Systematic distribution of larvivorous fish in the whole Country started in 1936 by the former Athens School of Hygiene (now National School of Public Health). The effort stopped when the application of the first synthetic insecticides started. Nevertheless, several populations of these fish still exist in several areas of the Country. Such an area is the marsh of Schinias in East Attiki (Livadas and Sphangos1940 and AS personal observation).

The use of *Gambusia* is no longer recommended in large scale mosquito control programs, because of the harmful impact of the fish on indigenous fish species (Service 1983). However, local research is required to identify indigenous species, which are adapted in the local geomorphology and climate, are most suitable for predation.

The possibility in using predators in mosquito control, either in the adult or in the larval stage, has been revised recently by Becker et al. (2003), who have included also a considerable number of related publications.

7.3 The basic suppression agents of mosquitoes in natural and anthropogenic ecosystems

Mosquitoes are recognized as a problem only in the last century. Perhaps at those time emerged the necessity to inhabit coastal marshes and inland wet biotopes that normally give rise to dense populations of mosquitoes. These human establishments together with global climate change are expected to raise in unprecedented levels the mosquito-borne or in general the arthropod-borne diseases –e.g. ARBO-viruses– (Epstein et al. 1998; Gubler, 1998; Colwell et al. 2006). In human prehistory the geomorphological processes in the coastal areas promote the existence of sand dunes and natural barriers to the inland waters that formed coastal swamps of brakish and fresh waters (examples are at Lerna, Argos, Greece in Jahns 1993; Schinias, Marathon, Greece in Petrakis 1990). The appropriate environment for the proliferation of mosquitoes rarely resulted in population outbreaks because of the action of natural enemies and the use of repellents and insecticides made of natural products, usually extracts of plant origin. On the other hand it must be remembered that the insecticides and repellents were used primarily for protection of the stored products and occasionally were used for air borne insects (Panagiotakopulu 2000).

Usually mosquito populations are controlled by means of variable agents, which apart from the commonly used synthetic chemical insecticides (Rodriguez et al. 2001; Kelly-Hope et al. 2005) include microbial insecticides (Park et al. 2001; Darboux et al. 2001; Wirth et al. 2005), fungal insecticides (Salgado 1997, 1998; Saunders and Bret 1997 Scholte et al. 2004;, nematodes (Perez-Pacheco et al. 2005), invertebrates such as damselflies and dragonflies (Insecta, Odonata) (Fincke et al. 1997; Singh et al. 2003), aquatic beetles (Insects, Coleoptera several familes e.g. Lundkvist et al. 2003), and backswimmers (*Notonecta* spp., Hemiptera, Heteroptera, Notonectidae), (Chesson 1984; Blaustein et al. 1995; Blaustein 1998; Eitam and Blaustein 2004; Eitam et al. 2002) with particularly adequate life history characteristics and metapopulation structure (Vepsäläinen 1974; Briers and Warren 2000). For the control of mosquitoes are also used vertebrate groups such as fishes, bats and birds to behaviour modifying agents. Important research includes guppy fish (*Poecilia reticulata*, Poeciliidae) in Elias et al. 1995; toads predated by odonates in Blaustein and Margalit 1996; Gambusia in Topal et al. 1994; and an application of *Gambusia patruelis* importation in Livadas and Sphangos 1941. The behaviour modifying agents are either synthetic or natural chemicals; often called 'semiochemicals', as they can change the behaviour of organisms even in trace amounts. Such semiochemicals are [1] synthetic repellents such as DEET (Ware 1983; Cranston et al. 1987); [2] alarm cues that induce defence compounds (Schoeppner and Relyea 2005); [3] sex, oviposition and aggregation pheromones (Millar et al. 1994; Mboera et al. 1999; Michelakis et al. 2005) that help in mate finding in the sexes of a species and the subsequent, after successful mating, oviposition; [4] kairomones (Schoeppner and Relyea 2005) that act for the benefit of the sensing organism; [5] cannibalism and ovicide (Sherratt & Church 1994); [6] natural repellents –mainly botanical- (Shaalan et al. 2005) such as the methanol extract of *Foeniculum vulgare*(Apiaceae) and Citronella oil; the last is an extract derived from *Andropogon nardus* (Poaceae) an indigenous plant of Sri Lanka and Sumatra and as all essential oil blends, citronella is not restricted only to the repellence of insect pests but it is also used as a antiseptic, deodorant, fungicide, insecticide, tonic and stimulant; [7] larvicidal natural products such as the leptostachyol acetate isolated from the roots of *Phryma leptostachya* var. *asiatica* (Park et al. 2005), vegetable tannins (Rey et al. 1999a & b) and the cell-wall fraction derived from decomposing alder leaves (David et al. 2000).

The mode of action of natural enemies varies with diel time, seasonal time and life stage of attacked mosquitoes. Bats are usually eating in flight by 'scooping' the adult mosquitoes and all the co-occurring insects. The web between the tail and the legs, the so-called '*inter-femoral pouch*', is used as a means to store the captured prey when in flight (Burton 1971; Burton 1980) and direct it to the mouth for eating with the aid of the third and fourth fingers.

Birds are not considered important consumers of mosquitoes mainly due to the non-overlap of the activities of the two groups. Many bird stomach analyses have shown that normally mosquitoes represent a small proportion of the daily diet. It is known that birds are opportunistic feeders and do not rely on one group of organisms for food intake (Cody 1981). The majority of mosquito species are nocturnal in habit, and the majority of birds feed in daytime, therefore it is not expected a bird species to restrict its diet to mosquitoes. However, when there is a dense population of mosquitoes these insects represent a substantial amount of their diet. Regardless to the population density, water fowl (Aves, Anatidae) as a rule, includes a large proportion of aquatic larvae in their diet. The only obstacle of these birds in controlling mosquito population is that their feeding is not selective and usually includes mosquito predators.

Since predators -especially invertebrate ones and the associated kairomones- do not induce any sort of resistance in the mosquitoes, they constitute the preferred control method. Especially in regions where either the public awareness for the environment is high or the funds for the suppression of mosquitoes are not enough for many adequate spray applications required in such anti-mosquito strategies. However, the control measures involving predators –either in the classical sense or their semiochemicals only- demand for a thorough knowledge of the existing ecosystems and many aspects have to be taken into account. Since predation is considered as the strongest selection pressure in natural ecosystems it is expected that many organisms such as mosquitoes in water ecosystems have evolved a variety of adaptive ways to avoid predation such as crypsis, chemical defences (for *Culex* predated by *Gambusia affinis* Angelon and Petranka 2002; for tadpoles predated by odonates Petranka and Hayes 1998) and behavioural strategies (Chivers et al. 1996), while the development of mechanical armour, in the existing evolutionary route, does not seem possible for these group of insects. Crypsis is used many times to indicate escape from predation in a more extended time frame. The most important escape from predation is usually done through oviposition, which constitutes the principal contribution of the mother to the antipredator defences of the offspring.

Unlike the oviposition behaviour elicited to gravid females in the Neotropical mosquito *Trichoprosopon digitatum* it was found that females preferred to oviposit in pools where other females have already oviposited and preferentially guarded their clutch. Also oviposition was more readily done in food rich pools (Sherratt and Church 1994). Cannibalism was practised by fourth instar larvae on conspecific larvae at a higher rate in food rich pools than in food poor pools, but no effect was detected when the amount of food changed. *T. digitatum* is selected for this type of cannibalism since the egg rafts are floating in the surface and for this they leave no space for more eggs. The crowding is reduced when cannibalism is exerted on the 2-day old larvae by larvae in the fourth instar, obviously because of the appropriate size relation. However, cannibalism is not always for the benefit of the practising organism. In Florida, the occurrence of the tiger mosquito *Aedes albopictus* varies according to the existence of a native mosquito inhabiting the bromeliaceous plant *Billbergia pyramidalis* (Bromeliaceae) (Lunibos et al. 2003). After a closer investigation by Lounibos et al. (2003) it was found that the fourth instar of the native mosquito *Wyeomyia* spp. and not the first instar deterred the oviposition of the tiger mosquito. Perhaps, this is the reason why *A. albopictus* is more numerous in northern Florida where the two native *Wyeomyia* spp. species are absent. The tiger mosquito has recently expanded its range as it is a tank inhabiting mosquito susceptible to human transportation of portable water habitats such fisherlings, potted plants for planting and already used tires for disposal. Used tires are a perfect means for the transportation of tiger mosquito eggs since there is no any arrangement capable in pouring the rain water trapped in their interior. On the other hand most mosquito eggs are capable of withstanding desiccation for many months.

When applying the classical biological control on mosquitoes the investigator has to take into account many strategies, which may be used by mosquitoes to escape predation together with many useful, or protected members of the local fauna. An example of the adverse effects caused by superficial introduction of natural mosquito enemies is the introduction in Greece of the mosquito fish *Gambusia* spp. by Metallinos in 1927 and Moutousis in 1928 with fisherlings taken from Italy and Marseille respectively (Livadas and Sphangos 1941). The fisherlings were used to enrich water bodies in northern Greece and Marathon Lake in Attica. At least in one case we are able to know the adverse effects on the indigenous fish fauna of antilarval predators. In Marathon area there is an endemic fish species, *Pseudophoxinellus marathonicus*, which is competitively inferior to *Gambusia* and it becomes extinct from water pools because the later either eats the eggs, hatchlings and fisherlings of *P. marathonicus* or performs better than the local endemic since it is readily acclimatised in a variety of environments (Pecl1995). After many introductions of the mosquito fish with speculative only damaging effects to local faunas it is now obvious that careful field trials must precede any introduction as Elias et al. (1995) suggests.

To assess the efficacy of predators several experiments have been carried out in tree holes –i.e. phytotelmata– as a rule in neotropical ecosystems (Fincke et al. 1997; Yanoviak 1999; 2001; Juliano and Gravel, 2002). The research group of Ola Fincke at Oklahoma State University working in tree holes of Barro Colorado Island, Panama observed that mosquitoes are very rare in the presence of odonate predators. In Barro Colorado there were four dominant odonate species (*Gynacantha membranalis* [Aeshnidae], *Megalopterus coerulatus, Mecistigaster linearis, M. ornata* [Pseudostigmatidae]) that predated on mosquito larvae and also on tadpoles of *Dendrobates auratus*, the predatory mosquito *Toxorynchites theobaldi* and more rarely on *Trichosporogon digitatum*. The researchers used artificial tree holes in order to keep track of the water parameters in each hole. In holes poor in food the mosquitoes were suppressed (Fig. 1). However, when the hole was richer –i.e. increased mosquito abundance– predation was not significant when odonate predators were small. In nutrient rich environments the larvae grew fast and the odonates ate them suppressing their numbers even in holes having excessively high mosquito populations. This indicates that local faunas are important in controlling mosquito populations without regardless of the densities.

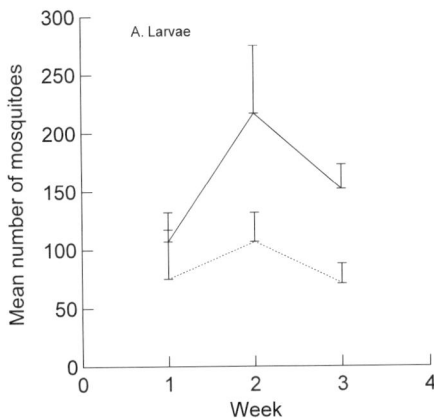

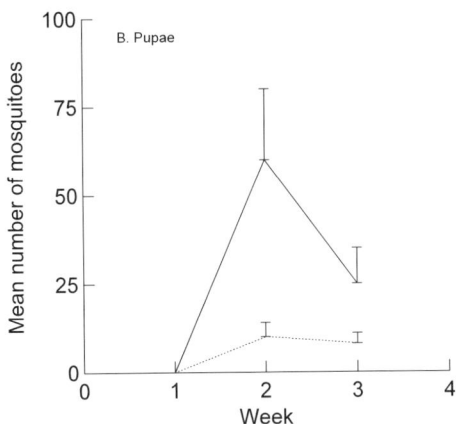

Fig. 1. Population densities in "tree holes" poor in nutrients. The continuous line is the control treatment while the dotted line corresponds to pools with odonates. Bar length is 1 SE (after Fincke et al. 1997)

When the enrichment of a biotope is planned in a project a feature of the site to be enriched is spatial complexity. If complexity is coupled with the biology of the various life stages of the predators then many predators can coexist in spite of the intraguild predation that happened in other ecosystems. Indeed, Hampton (2004) performed a series of mesocosm experiments where two notonectid predators, *Notonecta* and *Buenoa,* that were observed in the field to interact strongly since *Buenoa* has a size well in the range of prey sizes of *Notonecta* were studied. When the two species were offered two types of habitat, i.e. open water and vegetated water; *Buenoa* used at night both habitats while *Notonecta* during the day used vegetated water, as an ambush predator, while at night it moved freely among the two types of habitats.

The problem of habitat size in configuring the encounter rate between mosquitoes and predators was investigated in Japan in the period June-September (Sunahara et al. 2002). The mosquitoes were allocated to three groups according to the preferred habitat size. The first group comprised *Aedes (Stegomyia)* spp. and *Tripteroides bambusa,* which occurred mainly in small ($<0.1m^2$) habitats. In the second group there were *Oc. japonicus* and *Cx. kyotoensis*, which were found in larger habitats. The third group included *Cx. tritaeniorynchus* and *An. sinensis*, which preferred large habitats. The studied predators, i.e. notonectids, odonates and *Chaoborus*, showed a clear preference for large habitats and for this *Aedes (Stegomyia)* spp. and *Tripteroides bambus*a have never met these predators. As a rule they escape generalist-predators by staying in small water collections and the studied predators cannot control their populations.

When the release of mosquito predators is planned one trait of some mosquitoes that has to be taken into account is the capability in producing oviposition shifts as a part of the escape strategy from the predator. The mechanism responsible for this behaviour is primarily chemical. In the vicinity of Massachusetts the hypothesis of shift has been tested on the tree-hole mosquito *Aedes triseriatus* (Edgerly et al. 1998). The females of this species do not scatter their eggs in order to spread the risk associated with offspring mortality, a fact indicating that the mosquito does not exhibit opportunistic behaviour. When another mosquito with predatory larvae, *An. barberi,* was added in the tree-holes the females of *Ae. triseriatus* did not change the oviposition behaviour or the clutch size. It was also found that females are not able to detect (chemically?) the presence of the predator in the containers or the density of the predator was under the detectability threshold. However, the density of the predator cannot be very high since one predatory larva can consume daily some hundreds of *Ae. triseriatus*. The most striking behaviour of this mosquito is the positive response to co-existing conspecific eggs although larvae are not attracted to eggs. Early in the season females avoid tree-holes crowded with eggs as a fear for offspring intense competition or inevitable induced diapause. So, they shift oviposition to more advanced season even at the expense of competition from eggs, evidently not immediate. The quintessence of the egg number in the habitat signifies the 'water-holding capacity of the treehole' as the authors state (Edgerly et al. 1998).

7.4 The problem posed by synthetic chemical treatments and some toxins from biological preparations

The main problem posed by synthetic insecticides is the undesirable side effects such as the local extinction of aquatic taxa. Usually these control actions are not reported in the literature either from the side of the implementers or the environmental impact assessment team. Nevertheless, it exists as evidence in the notes of many scientists and usually it is communicated verbally in relevant discussions such as the scarcity of insects in entomological collections from Barcelona area in Catalonia, Spain verbally reported from Prof. Reinhart Remane (Petrakis 1986). Local authorities sprayed the broader area to remove mosquitoes as they are the main cause of nuisance to tourists visiting Barcelona. Since 1980, Petrakis sampled many lakes and other stagnant waters in Greece for aquatic and semiaquatic Heteroptera, and found not only a decrease of population densities but also local extinctions of several heteropteran species. While the main cause is the application of insecticides other causes certainly cannot be excluded such as the restriction of water bodies, urbanization or building of establishments for facilitation of visitors.

A more sophisticated example comes from the delta of river Po (Emilia-Romagna, Italy) (Veronesi et al. 2004). The periodical floods create temporary water habitats, which are inhabited by the salt marsh mosquito*es Oc.* caspius *and Oc. detritus.* Local nature conservation bodies criticized the practice of mosquito control in the area where many birds typically nest. The method chosen by local authorities is the enhancement of the already existing Mediteranean toothcarp fish *Aphanius fasciatus* (Cyprinodontidae). The enhancement included the construction of a network of ditches called runnels, to facilitate the fish in predating larvae of the mosquitoes. An additional action based on the chemical ecology of mosquitoes, was also anticipated. The extensive network of runnels is expected to prevent the oviposition of gravid female mosquitoes, or at least to decrease the attractiveness, in the local marshes (Veronesi et al. 2004). The expectations were based on the fact that several mosquito species minimise the risk of egg predation on their brood by chemically detecting the existence of predators (e.g. Blaustein et al. 2004). The kairomonal blend emanating from a fish predator is possibly a promising future line of research.

In another program in the low Canavese Valley (Piedmont, Italy) started in 1997 for the control of mosquito populations (Di Gia et al. 2004). The research team applied the traditional microbial preparations containing larvicidal toxins derived from *Bacillus thuringiensis* var. *israelensis* (Bti). Since these toxins are not selective and kill many natural enemies of mosquitoes the researchers surveyed the rice fields of the area in order to evaluate the status of the local fauna of mosquito predators. It was found that the most abundant Odonata were *Anax imperator, Orthetrum albistylum, O. cancellatum, Libellula depressa, Sympetrum pedemontanum* and *Crocothemis erythraea* while the fish fauna of the area consisted of the species *Leuciscus suffia, L. cephalus, Gobio gobio, Esox lucius, Padogobius matensi, Lepomis gibbosus* and the most efficient mosquito eater *Gambusia affinis*. All these species predate on four mosquito species, namely *Oc. caspius, Culex pipiens, Cx. modestus* and *Anopheles maculipennis*. According to Di Gia et al. (2004) the species of predators are indicative of the functionality of local biodiversity. Unfortunately, the effect of the Bti application on the odonate larvae is not known for Canavese Valley.

The natural products used in controlling mosquito populations as insecticides are usually of plant or fungal origin (Shaalan et al. 2005). Some of these products are widely sold as patented brands (Salgado 1997, 1998; Scott and Kaushik 2000; Cetin et al. 2005). Other natural products such as tannins are derived from live (seeds in Yang et al. 2003; roots in Park et al. 2005) or decayed (Rey et al. 1999a & b; David et al. 2000) plant material. The advantage of these products over the synthetic chemical insecticides is the reduced resistance to the active compounds, the quick degradability of the insecticides so as not to pollute the groundwater table, the trophic (Saunders and Bret 2000) and the suite of characters that make these compounds environmentally safe, in the sense that they are not biologically active to non target organisms. Almost in all tropical countries for many years the farmers used the botanical extracts, either crude or refined of the neem tree *Azadirachta indica* (Meliaceae) with only a few environmental problems (Ishman et al. 1991; Sundaram et al. 1997). However, the short life time of the active ingredient azadirachtin, which ranges from 24 h to 6.85 days of the commercial formulation Margosan-O © necessitates the repetition of the applications in order to cover a sufficiently large time period. In bioassays transacted by Scott and Kaushik (2000) in specifically designed microcosms it was found that neem oil (the formulation Margosan-O) did not affect the filter feeding planctonic organisms *Daphnia* sp. and *Culex* sp. In the same experiment the benthic ecosystems of the artificial microcosms were destroyed since the multiple applications of the neem oil as a result of the short life of the active ingredient killed the fundamental invertebrate species *Chironomus riparius* (Diptera, Chironomidae).

The problem of the resistance developed in the insects is the matter of many debates such as the one hosted by the journal *American Entomologist* (2005, vol. 51) (Alves et al. 2005; Cook et al. 2005; Hamilton et al. 2005; Kalcar et al. 2005; MacKay et al. 2005; Ragson et al 2005; Roberts et al. 2005; Watson et al. 205; Wooley et al. 2005). Among the many points of the debate was the efficiency of the traditional (= synthetic chemical treatments such as those employing DDT or organophosphates) over the natural biological methods or the newer –cf. molecular- methods such as the development of transgenic methods of vector management. Old chemical insecticides such as DDT have been abandoned world wide for a number of reasons the most important being the very slow degradability and incorporation into trophic webs (Curtis and Lines2000). The most important feature of the long debate is the renewed interest for old and environmentally unsafe chemicals. Actually, the most disappointing is the inability of the scientists to produce an efficient, cheap and quick ecologically compatible method.

The resistance of the insects to the various chemical insecticides is possibly the most adverse effect rendering them inapplicable in spraying programs. For instance Rodriguez et al. (2001) have detected resistance of Cuban and Vanezuelan *Aedes aegypti* (Diptera: Culicidae) population to organophosphates, pyrethroids and synergistic compounds. The involvement of esterases and monoxygenases in the resistance was also assessed. It was found that the insect is highly resistant to organophosphates, which according to the authors, represents a serious threat to control operations. Importantly the mosquito did not show resistance to pyrethroids except for a Cuban population to cypermethrin. Kelly-Hope et al. (2005) have also studied resistance of *Anopheles culifacies* and *A. subpictus* in Sri Lanka to a range of organophosphates and pyrethroid insecticides. They found that there is a striking inland-coastal pattern of spatial resistance with more resistance exhibited to malathion. This chemical has been accused for non-selectivity and pollution of the water table. For this reason EPA (2000) has proposed a revision of the risk assessment procedures for the registration of the respective brand. Importantly, the mechanism of resistance is always very specific (Hemmingway and Ranson 2000) and the spread of resistant populations, i.e. the respective genes, is very quick because it is based on the significantly increased selection of resistant strains over susceptible ones. In this respect genetic invasions play a primary role (Chevillon et al.

1995). Cross-resistance is also a problem that is always engaged in control operations.

On the basis of the developed resistance and the way it is incorporated in the ecological webs of a particular ecosystem, many human communities prefer the classical biological control. Probably this occurs because any biological control program necessitates the rehabilitation of the natural environment and the ceasing of any chemical application. In cases where the environment was not taken into account many species of the local fauna of natural enemies went locally extinct among other adverse effects (for predatory heteropterans see in Petrakis 1990; Petrakis and Roussis 2001a & b). The odonate predators of mosquitoes are now very sparse in Schnias, evidently as a result of the restriction or disappearance of many water bodies; the scarcity of suitable substrates made the odonate larvae actually absent from Schinias area, Attica, Greece. Nevertheless adults can be found, as they are strong flyers especially in early spring when the summer visitors have not arrived in the area and the applications of insecticides have not yet started (personal observation).

In the context of the alleviation of killing non-target organisms and damaging the existing ecosystems several attempts have been made. Schwartz et al. (2003) devised a controlled-release system (CRS) to deliver an insect growth regulator (IGR) to the surface of a water hole in order to kill the existing larvae of mosquitoes. IGR' s are selected as insecticide agents that mimick the moulting hormone of insects (or juvenile hormone) or interfere with the biosynthesis of chitin. This action typically results to the death of the larval or pupal stage of the insect while it keeps the minimal impact to mammals, birds, fishes and bees. In addition it has insecticide results even in cases of resistance to conventional insecticides. The IGR chosen for CRS experiments is cyromazine under the commercial name Larvadex Trigard (Novartis Crop Protection). This compound is environmentally degraded by a multitude of processes giving melamine. Melamine undergoes several transformations, all environmentally induced, and results in a range of compounds that are hazardous to the environment (Lim 1990; Yokley et al. 2000). The developed CRS is responsible for the increase in duration of the contact of the targeted mosquito larvae and pupae increasing substantially the efficacy of cyromazine. Simultaneously, it minimizes the quantities used and in effect the impact to the environment. In this way it is expected to balance the increased production coast induced by the CRS process and materials.

7.5 The chemical basis of predation on mosquitoes

The best systems to study the chemical ecology of organisms are the aquatic ecosystems. They are not only well defined and manageable but they present a wealth of interactions (see for instance Moore et al. 1996; Jeffries 2002; Stoks 1998). Paradise (2000) investigated the role of resource and abiotic factors in a series of simulated treeholes as model ecosystems detritus-based and found that the biotic interactions are influenced by pH and acidification determines the outcome of biotic interactions. The predator-prey interaction presents also a wealth of modes and actions (e.g. Slusarczyk 1995) that are directly applicable in mosquito control programs.

It is a common sense that the oviposition of mosquitoes is greatly influenced by chemical compounds. The messages concerning the quality of the substrate, the crowding of the water substrate and its future and the existence of egg and larval predators are all chemical compounds conveyed to gravid female mosquitoes. Female mosquitoes respond to these stimuli according to the strategy adopted by the species. Although these situations are very complex involving behavioural alteration they are worth mentioning and important in integrated control systems for mosquitoes. In the tree hole mosquito *Aedes triseriatus* predated by the mosquito *Toxorynchites rutilus* it was found that the presence, either actual or perceived, of the predator resulted in shifts of metamorphosis time, which in turn results in reduced weight at pupation and reduced eclosion rates (Hechtel and Juliano 1997). All these assume that the prey mosquito has the necessary plasticity to manipulate metamorphosis time. The prey mosquito perceives the predator through chemical cues left by the activity of the predator such as feces, eaten or half-eaten prey, kairomones of the predator emanating from the cuticle or gland and evacuated into the water.

Traditionally, oviposition is the activity that directly affects the reproductive output of an organism. For this reason many investigators have concentrated on this activity of gravid females. Kiflawi et al. (2003) investigating the oviposition strategy of the mosquito *Culiseta longiareolata* found that there is a trade-off between the fitness costs of predation risk imposed by the existence of *Notonecta maculata* larval predator and the density of the emerged larvae. Since mosquitoes suffer from extreme larval predation from notonectids (Blaustein 1999; Eitam et al. 2002; Eitam and Blaustein 2004) a strong aversion for pools containing the predator is expected. These authors have never observed a behavioural sequence for the detection of the quality or the heterogeneity of the oviposition substrate. Blaustein et al. (2004) working with the same predator and the same mosquito has added the midge *Chironomus riparius* and demonstrated that the mosquito avoided the high predation risk by *N. maculata* by using chemical cues since mosquitoes already have receptors for the chemicals, chemical detection is common in aquatic environments and finally all the other ways of the detection of a predator are not reliable in the water. The unidentified kairomone responsible for this particular chemical detection is also found in experiments conducted by Wisenden et al. (1995) involving the fathead minnow as a fish predator.

In other type of experiments Schoeppner and Relyea (2005) working with tadpoles of *Hyla versicolor* as predators, were able to separate the effect of predator pheromones and the alarm pheromones released by the crushing of ten different prey species by the predator. The ten prey species were tadpoles of the species *Rana sylvatica* and *R. pipiens*, the salamander *Ambystoma maculatum*, the damselfly *Lestes* spp., the dragonfly *Sympetrum* spp. and the two snail species *Physa acuta* and *Stagnicola elodes*. Importantly, they also discovered that the response of the prey is phylogenetically related to the emitter of the chemical cue. We were not able to find a similar study involving mosquitoes at least in the alternative crushed prey species and possibly it shows an issue, which demands further investigation.

An important aspect of oviposition behaviour comes from the fact that mosquitoes, at least the investigated ones, avoid visiting pools with the "odour" of an already met predator. Whether this avoidance has a side effect the oviposition of less egg rafts by means of less time devoted to oviposition remains to be discovered. In an important experiment involving the oviposition behaviour of the mosquitoes *Culiseta lineolata* and *Culex paticinctus* in pools containing a range of *Notonecta maculata* predators (Eitam and Blaustein, 2004) it was discovered that gravid mosquitoes oviposited in predator free artificial pools. However, the number of predators did not change the oviposition behaviour of mosquitoes. The authors state that the threshold of kairomones emanating from *Notonecta maculata* remains to be discovered since this type of interaction could be used in mosquito control programs. Since kairomones are expensive chemical compounds and their synthesis is a difficult task, the detection of such a threshold is important before application. On the other hand it is important to study this kairomonal blend in the field since it is expected to affect the behaviour of other potential prey species. Especially, in importation of predators in mosquito control programs that aim to enhance the predatory efficacy of the local fauna of natural enemies, the effect of a pheromone blend may disrupt other ecosystem processes.

7.6 Towards an integrated system of mosquito control

Vector control programs for many years were based almost entirely upon the use of chemicals. The serious adverse effects on the environment, the non-target organisms, the natural enemies, human health etc, were recognized some decades ago and a shift in the policy of pest control shifted gradually towards a revision of the up to now strategies. Emphasis is given to the safety of the ecosystem and the humans, by applying an integrated vector control system involving environmental management and biological methods, with reduced use of chemicals. The main advantage in the use of alternative methods is that it delays the development of pesticide resistant mosquito strains.

Integrated vector control, in terms of WHO (1983) can be consider as "the utilization of all appropriate technological and management techniques to bring about an effective degree of vector suppression in a cost-effective manner". It must now be added that managing effectiveness must not precede environmental and ecological safety. Approaches vary with different vectors, ecosystem types and contribution degree of the vectors in public health (e.g. disease transmission or nuisance).

Becker and coworkers (2003) made a thorough survey of the different mosquito control methods including biological ones and present examples of routine mosquito control programs applied in different European countries. In the same report is also discussed in detail the prerequisites for the successful implementation of an integrated mosquito control program.

The basic steps that must be followed in such a program are:

1. Monitoring the seasonal population densities in both the adult and the larval stages and the compilation of detailed entomological and ecological data. Several techniques are available to accomplish this step, e.g. human bait catches, CO_2 baited traps and oviposition traps for monitoring and estimating adult densities, dipping for monitoring and estimating larval densities. This step is also necessary to evaluate the effectiveness of the different control measures applied.

2. Mapping, numbering and characterization of all potential breeding sites

according to mosquito seasonal densities, plant association, occurrence of predators and other non target organisms. Geographical Information Systems (GIS) have become in recent years very useful tools for collection, storage and analysis of data dealing with entomological and ecological information (Burrough et al 1998).

3. Decision upon the type of insecticidal agent and selection of appropriate equipment and application techniques. Methods that can be used in mosquito control are: chemical, biochemical, biological, genetic and physical.

Currently the use of chemical insecticides is mainly restricted against the adult stage and only if there is a need for urgent reaction. Most commonly used adulticides include organophosphates, carbamates, pyrethroids and botanicals, e.g. pyrethrum. WHO (1997) classifies pesticides suitable to be used either against adults or larvae, according to their oral toxicity. Although these non-selective substances are effective in controlling mosquitoes, they also affect non-target organisms and should be avoided as the sole method when environmental or health concern is increased.

Biochemical agents, such as insect growth regulators (IGR), include either chitin synthesis inhibitors (e.g. diflubenzuron) or juvenile hormone analogues (e.g. methoprene). It is advisable IGRs to be used as larvicides instead of the conventional chemical ones, since they are declared to combine selectivity, lower vertebrate acute toxicity and prolonged action. Contrary to this, recent investigations have revealed that the active ingredients of IGR blend seriously alter the natural enemy complex by killing important predators such as *Chrysoperla carnea* (Celli et al. 1997). In addition the branch of EPA at California has included one such active ingredient such as fenoxycarb (CAS #72490-01-8), in the list of carcinogenic compounds (OEHHA 2000).

Biological control is generally defined as the use of natural enemies including pathogens, parasites and predators in reducing pest populations in natural habitats. The incorporation of biological measures in an Integrated Mosquito Control Program requires a careful selection of the antagonistic organism, so that the human protection is achieved without affecting the biodiversity and without inducing ecological problems. Experimental studies over the last century revealed a great diversity of living organisms, including microbes, fungi, protozoa, nematodes, invertebrate and vertebrate predators, as promising mosquito control agents. Weiser (1991) discusses extensively all groups of organisms that have been tested as potential biological control agents, including aspects on field trials, handling, transporting and laboratory activity testing of the isolates on different groups of vectors. Woodring and Davidson (1996) give an overview of the research on different organisms as biological mosquito control agents and comment on the possibilities of using them effectively under field conditions. According to these authors, several parameters dealing with environmental conditions and bioecological properties of both mosquitoes and the control agents should be taken into consideration in order to optimize their control activity and minimize their adverse effect on other natural enemies. Davidson and Becker (1996) give also a review of the pathogens affecting mosquitoes and emphasize on the two types of bacteria, *Bacillus thuringiensis* var *israelensis* and *Bacillus sphaericus*, the toxins of which are successfully and widely used the latest years in mosquito control programs.

In summary, the following groups of organisms have been studied as potential mosquito control agents but few of them gave promising results and even fewer gained commercial interest for mass production and application:

Pathogens. They include viruses, bacteria, protozoa, and fungi. Two microbial insecticides, based on the toxins produced by *Bacillus sphaericus* and *Bacillus thuringiensis israelensis* are already in use worldwide since many years. From the pathogenic fungi that attack mosquitoes, most promising for use in mosquito control is the larval pathogen *Lagenidium giganteum*. However, several problems dealing with the persistence in the habitat and the storage of the infectious stage of the fungus still remain unsolved. The best studied group from the protozoa is microsporidia. However, their complex life cycle and usually low pathogenicity are factors that prevent their mass production and therefore their use in control programs.

Parasites. Research with nematodes has shown that the newly hatched larvae of some *Romanomermis* species (Mermithidae). penetrate the integument of mosquito larvae and develop inside their body. Although mass rearing was achieved and commercial use started in 1970' s, problems with transportation and survival under some environmental conditions hindered its application (Becker et al. 2003).

Invertebrate predators. Hydrozoa (*Hydra* sp.), flatworms (*Planaria* and *Turbelaria*), insect predators (beetles, caddisflies, dragonflies, several Hemiptera, e.g. backswimmers, and *Toxorhynchites* larvae. All of these have not proved to be applicable in large-scale programs at present. It is worthy to mention that dragonflies were the first recognized mosquito predators and attempts for their use were made as early as the end of 19th century (Lamborn 1890).

Vertebrate predators. Fish, birds, bats and tadpoles. Some species of fish, e.g. *Gambusia* have been successfully used since the beginning of 1900's as biological control agents against mosquito larvae. Until today their use is recommended in several types of breeding sites. Birds, bats and tadpoles have been proven poor predators, because of lack of selectivity in mosquito larvae consumption.

Genetic control of insects can be achieved by several methods. The general concept of these methods is the gradual replacement of the target population after the release of a genotype with desirable properties. Efforts for mosquito control by means of genetic methods started as early as the decade of 1960. The sterile male technique (SIT), successfully used at that time in the control of the screwworm fly (Knipling 1959), was tried on several mosquito species (Karamjit 1996). The high cost and the selectivity of the method is a limiting factor for including it in an integrated mosquito control program at present. Other promising genetic methods include hybrid sterility, cytoplasmic incompatibility, competitive displacement and chromosomal translocations. The progress in molecular genetics in recent years promises future achievements in the application of genetic control methods. Current efforts focus in the introduction of disease refractory genes into vector populations (Karamjit 1996).

Physical methods in mosquito control include mainly environmental management and personal protection from mosquito bites.

4. Training of personnel for field or laboratory work. It is obvious that a successful integrated control program demands thorough knowledge of both the target organisms and the biological control agent in terms of their biology, ecology, population dynamics and degree of efficiency. On the other hand, several environmental parameters of the ecosystem where the application is going to take place should be examined. All these require well-trained staff and cooperation of scientists from all implicated areas.

5. Public awareness. Appropriate information of the inhabitants about the significance of mosquito control and the techniques they are used is necessary for the acceptance of the treatments especially in the urban environment. This is achieved by lectures, film shows, popular science programs and spots on TV, Internet, leaflet circulation or any other related means.

6. Community participation in the control of mosquitoes, especially of species that breed in the human environment. In recent years the development of the community participation in mosquito control, after appropriate training, has gained significant consideration and consists part of the primary health care (WHO 1983).

7.7 Acknowledgements

The funding of the Hellenic Secretariat for Research and Development to VR and PVP is also greatly acknowledged. Many people helped in the collection of literature, discussions and suggestions for this paper. The reviewers of the manuscript are greatly acknowledged for their suggestions in the discipline of entomology and natural product chemistry. The fact that many works were not included in the manuscript is due to the limited space and resources.

References

Alves AP, Campbell LA, Macedo PA, Smith JD (2005) Traditional vector control research should receive higher priority than transgenic efforts to control human and animal disease: proposition. American Entomologist 51: 107-108

Angelon KA, Petranka JW (2002) Chemicals of predatory mosquitofish (*Gambusia affinis*) influence selection of oviposition site by *Culex* mosquitoes. Journal of Chemical Ecology 28:797-806

Becker N, Petric D, Zgomba M, Boase C, Dahl C, Lane J, Kaiser A (2003) Mosquitoes and their control. Kluwer Academic / Plenum Publ., New York

Blaustein L (1998) Influence of the predatory backswimmer, *Notonecta maculata*, on invertebrate community structure. Ecological Entomology 23: 246-252

Blaustein L (1999) Oviposition site selection in response to risk of predation: evidence from aquatic habitats and consequences for population dynamics and community structure. In Wasser SP (ed) Evolutionary Theory and Processes. Kluwer Academic Publishers, Dordrecht, The Netherlands 441-456

Blaustein, L, Margalit J (1996) Priority effects in temporary pools: nature and outcome of mosquito larva-toad tadpole interactions depend on order of entrance. Journal of Animal Ecology 65: 77-84

Blaustein L, Kotler BP, Ward D (1995) Direct and indirect effects of a predatory backswimmer (*Notonecta maculata*) on community structure of desert temporary pools. Ecological Entomology 20: 311-318

Blaustein L, Kiflawi M, Eitam A, Mangel M, Cohen JE (2004) Oviposition habitat selection in response to risk of predation in temporary pools: mode of detection and consistency across an experimental venue. Oecologia 138: 300-305

Briers RA, Warren PH (2000) Population turnover and habitat dynamics in *Notonecta* (Hemiptera: Notonectidae) metapopulations. Oecologia 123: 216-222

Burrough P, McDonnel A, Rachael A (1998) Principles of Geographical Information Systems. Oxford University Press

Burton M (1971) Wild Animals. F.Warne, London

Burton JA (1980) Wild Animals. Collins, London

Carson R (1962) Silent Spring. Fawcett Publ. Greenwich, Connecticut

Cetin H, Yanikoglu A, Cilek JE (2005) Evaluation of the naturally-derived insecticide spinosad against *Culex pipiens* L. (Diptera: Culicidae) larvae in septic tank water in Antalya, Turkey. Journal of Vector Ecology 30:151-154

Celli GL, Bortolotti C, Nanni C, Porrini, Sbrenna G (1997) Effects of the IGR fenoxycarb on eggs and larvae of *Chrysoperla carnea* (Neuroptera: Chrysopidae). Laboratory test.. In Haskell PT, McEwen PK, (eds) New studies in Ecotoxicology. The Welsh Pest Management in Forum, Cardiff, 15-18

Chase JM, Knight TM (2003) Drought-induced mosquito outbreaks in wetlands. Ecology Letters 6:1017-1024

Chesson J (1984) Effect of notonectids (Hemiptera: Notonectidae) on mosquitoes (Diptera: Culicidae): predation or selective oviposition? Environmental Entomology 13: 531-538

Chevillon C, Addis G, Raymond M, Marchi A (1995) Population structure in Mediterranean islands and risk of genetic invasion in *Culex pipiens* L. (Diptera: Culicidae). Biological Journal of the Linnean Society 55:329-343

Chivers DP, Wisenden BD, Smith RJF (1996) Damselfly larvae learn to recognize predators from chemical cues in the predator's diet. Animal Behaviour 52: 315-320

Cody ML (1981) Habitat selection in birds: the roles of vegetation structure, competitors and productivity. BioScience 31: 107-113

Colwell R, Epstein P, Gubler D, Hall M, Reiter P, Shukla J, Sprigg W, Takafuji E, Trtanj J (2006), Global Climate Change and Infectious Diseases,, Emerging Infectious Diseases (electronic journal), 4(3)

Cook K, Kent L, Lundgren J, Marlow E (2005) The federal government should support the use of pesticides previously banned in the United States to fight vector-borne diseases in developing countries: con position. American Entomologist 51: 112

Cranston PS, Ramsdale CD, Snow KR, White GB (1987) Adults, larvae and pupae of british mosquitoes (Culicidae) - with notes on their ecology and medical importance. Freshwater Biological Association, Scientific Publication No 48

Curtis CF, Lines JD (2000) Should DDT be banned by international treaty? Parasitology Today 16: 119-121

Darboux I, Nielsen-LeRoux C, Charles J-F, Pauron D (2001) The receptor of *Bacillus sphaericus* binarytoxin in *Culex pipiens* (Diptera: Culicidae) midgut: moleclar cloning and expression. Insect Biochemistry and Molecular Biology 31:981-990

David J-P, Rey D, Marigo G, Meyran J-C (2000) Larvicidal effect of a cell-wall fraction isolated from alder decaying leaves. Journal of Chemical Ecology 26: 901-913

Davidson EW, Becker N (1996) Microbial control of vectors. In Beaty BJ, Marquardt WC (eds) The Biology of Disease Vectors. University Press of Colorado, USA, 549-563

Di Gia I, Vietti Niclot MM, Eusebio Bergo P, Mosca A (2004) First investigations on predators in mosquito breeding sites The 3rd European Mosquito Control Association Worhshop, Osijek, Croatia pp. 36

Edgerly JS, McFarland, Morgan P, Livdahl T (1998) A seasonal shift in egg-laying behaviour in response to cues of future competition in a treehole mosquito. Journal of Animal Ecology 67: 805-818

Eitam A, Blaustein L, Mangel M (2002) Effects of *Anisops sardea* (Hemiptera: Notonectidae) on oviposition habitat selection of mosquitoes and other dipterans and on community structure in artificial pools. Hydrobiologia 485: 183-189

Eitam A, Blaustein L (2004) Oviposition habitat selection by mosquitoes in response to predator (*Notonecta maculata*) density. Physiological Entomology 29: 188-191

Elias M, Islam MS, Kabir MH, Rahman MK (1995) Biological control of mosquito larvae by Guppy fish. Bangladesh Med Res Counc Bull. 21: 81-86

EPA (U.S. Environmental Protection Agency) (2000) Memorandum. Malathion: revisions to the preliminary risk assessment for the registration eligibility decision (RED) document. Chemical 057701, Case 0248, Barcode D265482

Epstein PR, Diaz HF, Elias S, Grabherr G, Graham NE, Martens WJM (1998) Biological and physical signs of climate change: focus on mosquito-borne disease. Bulletin of the American Meteorological Society 78: 409-417

Fincke OM, Yanoviak SP, Hanschu RD (1997) Predation by odonates depresses mosquito abundance in water-filled tree holes in Panama. Oecologia 112: 244-253

Gubler DJ (1998) Resurgent vector-borne diseases as a global health problem. Emerging Infectious Diseases (electronic journal), 4(3)

Hamilton RM, Rebek EJ, Saltzman KD (2005) Eradication of insect vectors of disease should receive priority over vector management: pro position. American Entomologist 51: 105-106

Hampton S (2004) Habitat overlap of enemies: temporal patterns and the role of spatial complexity. Oecologia 138: 475-484

Harrison G (1978) Mosquitoes, Malaria and Man: A History of the Hostilities Since 1880. Clarke, Irwin & Co, New York

Harwood RF, James MT (1979) Entomology in Human and Animal Health. MacMillan Publ., NewYork

Hechtel LJ, Juliano SA (1997) Effects of a predator on prey metamorphosis: Plastic responses by prey or selective mortality? Ecology 78: 838-881

Hemmingway J, Ranson H (2000) Insecticide resistance in insect vectors of human disease. Annual Review of Entomology 45: 371-391

Ishman MB, Koul O, Arnason JT, Stewart J, Salloum GS (1991) Developing a neem-based insecticide for Canada. Memoirs of the Entomological Society of Canada 159: 39-47

Jahns S (1993) On the Holocene vegetation history of the Argive plain (Peloponnese, southern Greece). Vegetation History and Archaeobotany 2: 187-204

Jeffries MJ (2002) Evidence for individualistic species assembly creating convergent predator: prey ratios among pond invertebrate communities. Journal of Animal Ecology 71:173-184

Jones WHS (1909) Malaria and Greek History. Manchester University Press, Manchester

Juliano SA, Gravel ME (2002) Predation and the evolution of prey behavior: an experiment with tree hole mosquitoes. Behavioral Ecology 13: 301-311

Kalcar OE, Nauman J, Paysen E, Reeves W, Staeben J (2005) Publicly funded mosquito control efforts in urban areas should take precedence over private concerns regarding pesticide exposure: con position. American Entomologist 51: 110

Karamjit SR (1996) Genetic control of vectors. In Beaty BJ, Marquardt WC (eds) The Biology of Disease Vectors. University Press of Colorado, USA, 564-574

Kelly-Hope LA, Yapabandara AMGM, Wickramasinghe MB, Perera MDB, Siyambalagoda RRMLR, Herath PRJ, Galappaththy GNL, Hemingway J (2005) Spatiotemporal distribution of insecticide resistance in *Anopheles culicifacies* and *Anopheles subpictus* in Sri Lanka. Transactions of the Royal Society of Tropical Medicine and Hygiene 99: 751-761

Kiflawi M, Blaustein L, Mangel M (2003) Oviposition habitat selection by the mosquito *Culiseta longiareolata* in response to risk of predation and conspecific larval density. Ecological Entomology 28: 168-173

Knipling EF (1959) Sterile-male method of population control. Science 130:902-904

Lamborn RH (1890) Dragonflies vs. mosquitoes. Can the mosquito pest be mitigated? Studies in the life history of irritating insects, their natural enemies and artificial checks by working entomologists. D. Appleton Co., New York

Legner EF (1995) Biological control of Diptera of medical and veterinary importance. Journal of Vector Ecology 20: 59-120

Lim LO (1990) Disposition of cyromazine in plants under environmental conditions. Journal of Agricultural and Food Chemistry 38: 860-864

Livadas GA, Sphangos JC (1941) Malaria in Greece 1930-1940. Pyrgos Press, Athens

Lounibos LP, O' Meara GF, Nishimura, Escher RL (2003) Interactions with native mosquito larvae regulate the production of *Aedes albopictus* from bromeliads in Florida. Ecological Entomology 28: 551-558

Lundkvist E, Landin J, Jackson M, Svensson C (2003) Diving beetles (Dytiscidae) as predators of mosquito larvae (Culicidae) in field experiments and in laboratory tests of prey preference. Bulletin of Entomological Research 93: 219-226

MacKay AJ, Oremus GR, Watson EJ (2005) Publicly funded mosquito control efforts in urban areas should take precedence over private concerns regarding pesticide exposure: introduction to debate. American Entomologist 51: 108-110

Mboera LEG, Mdira KY, Salum FM, Takken W, Pickett JA (1999) Influence of synthetic oviposition pheromone and volatiles from soakage pits and grass infusions upon oviposition site selection of *Culex* mosquitoes in Tanzania. Journal of Chemical Ecology 25: 1855-1865

Michelakis A, Mihou A, Couladouros EA, Zounos AK, Koliopoulos G (2005) Oviposition responses of *Culex pipiens* to a synthetic racemic *Culex quinquefasciatus* oviposition aggregation pheromone. Journal of Agricultural and Food Chemistry 53: 5225-5229

Millar JG, Chaney JD, Beehler JW, Mulla MS (1994) Interaction of the *Culex quinquefasciatus* egg raft pheromone with a natural chemical associated with oviposition sites. J. Am. Mosq. Control Assoc. 10: 374-379

Moore RD, Newton B, Sih A (1996) Delayed hatching as a response of streamside salamander eggs to chemical cues from predatory sunfish. Oikos 77: 331-335

OEHHA (2000) Office of Environmental Health Hazard Assessment: Chemicals Listed Effective June 2, 2000 as Known to the State to Cause Cancer: Chloroprene, Cobalt sulfate heptahydrate, and Fenoxycarb, under Proposal 65, California, Environmental Protection Agency

Panagiotakopulu E (2000) Archaeology and Entomology in the Eastern Mediterranean: Research into the history of insect synanthropy in Greece and Egypt. British Archaeological Reports, International Series 836: 1-146

Paradise C (2000) Effects of pH and resources on a processing chain interaction in simulated treeholes. Journal of Animal Ecology 69: 651-658

Park HW, Delecluse A, Federici BA (2001) Constraction and characterization of a recombinant *Bacillus thuringiensis* subsp. *israelensis* strain that produces Cry 11B. Journal of Invertebrate Pathology 78: 37-44

Park I-K, Shin S-C, Kim C-S, Lee H-J, Choi W-S, Ahn Y-J (2005) Larvicidal activity of lignans identified in *Phryma leptostachya* var. *asiatica* roots against three mosquito species. Journal of Agricultural and Food Chemistry 53: 969-972

Pecl K (1995) Fishes of Lakes and Rivers,,Magna Books, Leicester, UK

Perez-Pacheco R, Rodriguez-Hernandez C, Lara-Reyna J, Montes-Belmont R, Ruiz-Vega J, 2005, Control of the mosquito Anopheles pseudopunctipennis (Diptera: Culicidae) with Romanomermis iyengari (Nematoda: Mermithidae) in Oaxaca, Mexico,,Biological Control,32:137-142

Petrakis PV (1986) Round table on management of nature reserves in Prespa: A summary. In Drosopoulos S (ed) Proceedings of the Second International Congress Concerning the Rhynchota Fauna of Balkan and Adjacent Regions. Mikrolimni Prespa, Greece pp 69-70

Petrakis PV (1990) Plant Heteroptera Associations in an East Mediterranean Ecosystem: an Analysis of Structure, Specificity and Dynamics. PhD Dissertation, University of Wales, College of Cardiff

Petrakis PV, Roussis V (2001a) Bioindication value of Hellenic aquatic Heteroptera: An algorithmic approach. In Management and Sustainable Development of Water Ecosystems. Proceedings of the 10th PanHellenic Ichthyological Congress, Chania, Crete

Petrakis PV, Roussis V (2001b) A general framework for multidimensional Hutchinsonian niche analysis of aquatic organisms for use in bioindication of water ecosystems. In Management and Sustainable Development of Water Ecosystems. Proceedings of the 10th PanHellenic Ichthyological Congress, Chania, Crete

Petranka J, Hayes L (1998) Chemically mediated avoidance of a predatory odonate (*Anax junius*) by American toad (*Bufo americanus*) and wood frog (*Rana sylvatica*) tadpoles. Behavioral Ecology and Sociobiology 42: 263-271

Ragson JL, Styer LM, Minnick SL (2005) Traditional vector control research should receive higher priority than transgenic efforts to control human and animal disease: con position. American Entomologist 51: 108

Rey D, Cuany A, Pautou M-P, Meyran J-C (1999a) Differential sensitivity of mosquito taxa to vegetable tannins. Journal of Chemical Ecology 25: 537-548

Rey D, Pautu MP, Meyran JC (1999b) Histopathological effects of tannins on the midgut epithelium of aquatic Diptera larvae. Journal of Invertebrate Pathology 73: 173-181

Roberts M, Hartzer K, McCoy C (2005) The federal government should support the use of pesticides previously banned in the United States to fight vector-borne diseases in developing countries: pro position. American Entomologist 51: 111-112

Rodriguez MM, Bisset J, de Fernandez DM, Lauzan L, Soca A (2001) Detection of insecticide resistence I *Aedes aegypti* (Diptera: Culicidae) from Cuba and Venezuela. Journal of Medical Entomology 38: 623-628

Salgado VL (1997) The mode of action of spinosad and other insect control products. Down to Earth 52:35-44

Salgado VL (1998) Studies on the mode of action of spinosad: Insect symptoms and physiology correlates. Pesticide Biochemistry and Physiology 60: 91-102

Saunders DG, Bret BI (1997) Fate of spinosad in the environment. Down to Earth 52: 14-20

Schoeppner NM, Relyea RA (2005) Damage, digestion, and defence: the roles of alarm cues and kairomones for inducing prey defences. Ecology Letters 8: 505-512

Scholte E-J, Knols BGJ, Samson RA, Takken W (2004) Entomopathogenic fungi for mosquito control: A review. Journal of Insect Science, www.insectscience/4.19

Schwartz L, Wolf D, Markus A, Wybraniec S, Wiesman Z (2003) Controlled-release systems for the delivery of cyromazine into water surface. Journal of Agricultural and Food Chemistry 51: 5972-5976

Scott IM, Kaushik NK (2000) The toxicity of a neem insticide to populations of culicidae and other aquatic invertebrates as assessed in in situ microcosms. Archives of Environmental Contamination and Toxicology 39: 329-336

Service MW (1983) Biological control of mosquitoes-has a future? Mosquito News 43: 113-120

Shaalan EA-S, Canyon D, Younes MWF, Abdel-Wahab H, Mansour A-H (2005) A review of botanical phytochemicals with mosquitocidal potential. Environment Iternational 31: 1149-1166

Sherratt TN, Church SC (1994) Ovipositional preferences and larval cannibalism in the Neotropical mosquito *Trichoprosopon digitatum* (Diptera: Culicidae). Animal Behaviour 48: 645-652

Singh RK, Dhiman RC, Singh SP (2003) Laboratory studies on the predatory potential of dragon-fly nymphs on mosquito larvae. Commun Dis. 35: 96-101

Slusarczyk M (1995) Predator-induced diapause in *Daphnia*. Ecology 76: 1008-1013

Stoks R (1998) Effect of lamellae autotomy on survival and foraging success of the damselfly *Lestes sponsa* (Odonata: Lestidae). Oecologia 117: 443-448

Sunahara T, Ishizaka L, Mogi M (2002) Habitat size: a factor determining the opportunity for encounters between mosquito larvae and aquatic predators. Journal of Vector Ecology 27: 8-20

Sundaram KMS, Sundaram A, Curry J, Sloane L (1997) Formulation selection, and investigation of azadiractin - a persistence in some terrestrial and aquatic components of a forest environment. Pesticide Science 51: 74-90
Topal J, Miklosi A, Csanyi V (1994) Culicidae - *Gambusia* Abstracts. Acta Biologia Hungarica 45: 87-99
Vepsäläinen K (1974) The life cycles and wing lengths of finnish *Gerris* Fabr. species (Heteroptera, Gerridae) Acta Zoologica Fennica 141: 1-73
Veronesi R, Pandolfi N, Alberani A, Bellini R (2004) Enhancing th emosquito predation activity of *Aphanius fasciatus* (Cyprinodontiformes: Cyprinodontidae) by runneling in the Po river delta area. In The 3rd European Mosquito Control Association Worhshop, Osijek, Croatia 27
Ware GW (1983) Pesticides. Theory and Application. W.H. Freeman, New York
Watson EJ, MacKay AJ, Oremus GR (2005) Eradication of insect vectors of disease should receive priority over vector management: con position. American Entomologist 51: 106
Weiser J (1991) Biological Control of Vectors. Wiley, West Sussex, UK
WHO (1983) Technical report series 688. Geneva, 72
WHO (1997) Pesticide evaluation scheme. Chemical methods for the control of vectors and pests of public health importance, (Chavasse DC, Yap HH (eds.) 129
Wirth MC, Jiannino JA, Federici BA, Walton WE (2005) Evolution of resistance toward *Bacillus sphaericus* or a mixture of *B. sphaericus*+Cyt1A from *Bacillus thuringiensis*, in the mosquito, *Culex quinquefasciatus* (Diptera: Culicidae). Journal of Invertebrate Pathology 88: 154-162
Wisenden BD, Chivers DP, Brown GE, Smith RJF (1995) The role of experience in risk assessment: avoidance of areas chemically labelled with fathead minnow alarm pheromone by conspecifics and heterospecifics. Ecoscience 2: 116-122
Woodring J, Davidson EW (1996) Biological control of mosquitoes, In Beaty BJ, Marquardt WC (eds) The Biology of Disease Vectors. University Press of Colorado, USA, 530-548
Wooley S (2005) Publicly funded mosquito control efforts in urban areas should take precedence over private concerns regarding pesticide exposure: pro position. American Entomologist 51: 109-110
Yang Y-C, Lim MY, Lee H-S (2003) Emodin isolated from *Cassia obtusifolia* (Leguminosae) seed shows larvicidal activity against three mosquito species. Journal of Agricultural and Food Chemistry 51: 7629-7631
Yanoviak S (1999) Effects of *Megistogaster* spp. (Odonata: Pseudostigmatidae) and *Culex mollis* (Diptera: Culicidae) on litter decomposition in neotropical treehole microcosms. Florida Entomologist 82: 462-468
Yanoviak SP (2001) Predation, resource availability, and community structure in Neotropical water-filled tree holes. Oecologia 126: 125-133
Yokley RA, Mayer LC, Rezaaiyan R, Manuli ME, Cheung MW (2000) Analytical method for the determination of cyromazine and melamine residues in soil sampling. Journal of Agricultural and Food Chemistry 48: 3352-3358

8 A case for cannibalism: Confamilial and conspecific predation by naticid gastropods, Cretaceous through Pleistocene of the United States Coastal Plain

Patricia H. Kelley[1] and Thor A. Hansen[2]

[1]Department of Earth Sciences, University of North Carolina Wilmington, Wilmington, North Carolina, 28403-5944 USA, kelleyp@uncw.edu;
[2]Department of Geology, Western Washington University, Bellingham, Washington, 98225, USA, thorenet@cc.wwu.edu

8.1 Abstract

Cannibalism is a common phenomenon in animals, but some previous authors have concluded that cannibalism by shell-drilling naticid gastropods was caused by predator ineptitude, especially early in the evolution of naticids. The suggestion that naticids were less efficient predators earlier in their history may be considered consistent with the hypothesis of escalation. According to the hypothesis of escalation, biological hazards, such as predation, have increased through geologic time. If naticids were less efficient early in their history, and if cannibalism is an indicator of predator ineptitude, then the incidence of cannibalism should be greatest early in the history of the naticid predator-prey system. Based on this hypothesis, we predicted a temporal decrease in the frequency of cannibalism. We tested this hypothesis by determining the incidence of predation by naticid gastropods on naticid gastropods, including intraspecific cannibalism, from the Cretaceous through the Pleistocene. Drilling frequencies (percent of naticid specimens with complete naticid drillholes) were determined for samples of naticids from twenty-three stratigraphic levels in the U.S. Atlantic and Gulf Coastal Plain (~3,400 naticid specimens).

Contrary to this hypothesis, drilling on naticids (including intraspecific cannibalism) increased through time. Drilling frequency for all Cretaceous samples combined was 0.02; comparable results were 0.11 and 0.26 for the

Paleogene and post-Paleogene respectively. Spearman rank correlation coefficients between drilling frequency and stratigraphic position were statistically significant for both confamilial and conspecific naticid predation. Drilling frequencies were not correlated with the abundance of naticids in the fauna, nor were they correlated with drilling frequencies on the gastropod fauna as a whole. Most attempts to drill naticid prey were successful, and no trends in frequency of incomplete drillholes (prey effectiveness) occurred between the Cretaceous and Pleistocene. These results may indicate that naticids were not less efficient early in their history. More plausibly, cannibalism may be an inappropriate measure of predator ineptitude. In that case, the increase in cannibalism may indicate increasing naticid predatory capabilities through time; because naticids are highly mobile prey, cannibalism may require greater predator efficiency than do non-cannibalistic predation events. Predation on naticids is energetically profitable and typically successful once drilling is initiated, suggesting that cannibalism may be an attractive alternative for an efficient predator rather than a hallmark of ineptitude.

Keywords: Naticid gastropod, drilling predation, cannibalism, Cretaceous, Paleogene, Neogene, Pleistocene.

8.2 Introduction

In the movie *Charlie and the Chocolate Factory*, based on the book of the same name by Roald Dahl, Willy Wonka invites five lucky children to tour his fanciful chocolate factory. As they enter the first room filled with delectable sweets, Johnny Depp, as Willy Wonka, declares, "Everything in this room is eatable. Even I'm eatable. But that is called cannibalism, my dear children, and is in fact frowned upon in most societies." Humans are both fascinated and repelled by cannibalism, and we often attach our own anthropocentric interpretations to cannibalism when observed in other organisms.

Nevertheless, cannibalism is relatively common among many vertebrate and invertebrate animals (Polis 1981). Cannibalism has been reported from time to time for predatory gastropods, both in the field (e.g., Basedow 1996, for the muricid gastropod *Murex trunculus*) and laboratory (e.g., the fasciolariid gastropod *Triplofusus giganteus*; Dietl 2003). Cannibalism also has been reported for naticid gastropods. Naticids are primarily infaunal predators that drill through the shells of their victims in order to consume their prey (Ziegelmeier 1954). Naticids prey on a variety of taxa;

juvenile naticids attack ostracods (Reyment 1999; Reyment and Elewa 2003), and possibly foraminifers (Reyment 1966, 1967; Livan 1937; Arnold et al. 1985), whereas adult naticids occasionally drill scaphopods (Yochelson et al. 1983). The primary prey of naticid gastropods, however, are bivalves and gastropods, including other naticids.

Naticid shells bearing naticid drillholes date to the Cretaceous Period (Kitchell et al. 1986; Kelley and Hansen in press). In some cases, the predator and prey may have represented different species of naticids. Nevertheless, examples of actual intraspecific cannibalism are known. For example, the Miocene St. Marys Formation of Maryland includes two common species of naticid, *Euspira heros* and *Neverita duplicata*, but individual assemblages are dominated by single species. Kelley (1991) argued that naticid drilling of St. Marys naticids represented actual intraspecific cannibalism. Cannibalism has also been observed in the laboratory within extant naticid species (Kitchell et al. 1981).

Perhaps due to our own aversion to cannibalism, conspecific predation by naticids has been viewed as anomalous by some authors and attributed either to unavailability of alternative prey or to predator ineptitude (Paine 1963, Stanton and Nelson 1980, Hoffman et al. 1974). For example, Stanton and Nelson (1980), studying the Eocene Stone City Formation of Texas, and Hoffman et al. (1974), for the Miocene Korytnica Clays of Poland, claimed that cannibalism and multiply bored shells indicate that naticids bore any shell they find, including already empty shells and other naticids. In particular, Stanton and Nelson (1980, p. 128) stated that cannibalism indicated that, "in the early Cenozoic the naticids had not yet quite gotten the knack of the new, boring mode of predation." Similarly, Kojumdjieva (1974) doubted the validity of aquarium observations of naticid cannibalism and attributed such occurrences to the artificiality of laboratory conditions. These interpretations contrast with the conclusion (Kitchell et al. 1981; Kelley 1991; Dietl and Alexander 1995) that cannibalism by naticids is a predictable result of selective predation and therefore not anomalous.

The view that Eocene naticids were less adept as predators might be seen as consistent with Vermeij's (1987, 1994) hypothesis of escalation. This hypothesis states that biological hazards such as predation have increased through the Phanerozoic. Vermeij (1987) cited drilling by predatory gastropods as evidence for escalation, suggesting that an increase in gastropod drilling frequencies occurred between the Cretaceous and sometime in the Eocene, when drilling achieved modern levels. If the hazard of naticid predation has increased through time, then perhaps naticids are more efficient predators today than earlier in their history.

If naticids were less efficient predators early in their history, and if cannibalism is an indicator of ineptitude on the part of the predator, then the incidence of cannibalism should be greatest early in the history of the naticid predator-prey system. Based on this hypothesis, we predicted a decrease in the frequency of cannibalism through time. This study tests this hypothesis with a database on naticid drilling from the Cretaceous through Pleistocene in the U.S. Gulf and Atlantic Coastal Plains.

8.3 Materials and methods

Data on naticid cannibalism were extracted from a database on naticid predation we have compiled to test Vermeij's hypothesis of escalation (Kelley and Hansen 1993, 1996a, 2003, and 2006). The database has been described in detail, including sampling levels and methods used, by Kelley and Hansen (1993, 1996a, and 2006). In brief, with the exception of the Cretaceous data (which were obtained from the Norman Sohl collection of the U.S. Geological Survey at Reston, Virginia), all data come from bulk samples. The samples were wet sieved using 1-mm screens, hand picked to extract all shell material, and all gastropods and bivalves were identified to species level. Drilling data were collected for all whole or nearly whole bivalve and gastropod specimens. The complete database includes approximately 148,000 specimens from 28 formations of Cretaceous through Pleistocene age. Included in this tally are data from Graham (1999), Huntoon (1999), and Jones (1999). In addition, data on naticid drilling presented by Melland (1996), which were not previously included in the database, were added for this study.

We extracted from this database information on all naticid specimens within the samples (3,364 specimens). Twenty-three stratigraphic levels are represented (Table 1). Not all formations in the original database were included in the present study; for instance, the Corsicana and Kincaid had inadequate preservation to collect drilling data on gastropods (Kelley and Hansen 1993). In addition, the Yazoo Formation was excluded from study; it was deposited in an outer neritic environment (Smith and Zumwalt 1987), in contrast with the inner to middle shelf environments represented by the remainder of the formations. We also excluded the Eastover Formation because only seven naticid specimens were present in our samples.

Table 1. Stratigraphic units studied, including age and occurrence of naticid gastropod species. Fm, Formation; Mbr, Member

Fm or Mbr	Age	One species	One species dominant	Multiple species
Ripley Fm	Late Cretaceous	X		
Providence Fm	Late Cretaceous			X
Brightseat Fm	Early Paleocene	X		
Matthews Landing Mbr/ Naheola Fm	Middle Paleocene	X		
Bells Landing Mbr/Tuscahoma Fm	Late Paleocene			X
Bashi Mbr/Hatchetigbee Fm	Early Eocene			X
Cook Mountain Fm	Late middle Eocene			X
Gosport Fm	Late middle Eocene			X
Moodys Branch Fm	Late Eocene			X
Red Bluff Fm	Early Oligocene		X	
Mint Spring Fm	Early Oligocene		X	
Byram Fm	Early Oligocene	X		
Belgrade Fm	Late Oligocene		X	
Calvert Fm	Middle Miocene		X	
Drumcliff Mbr/Choptank Fm	Middle Miocene		X	
Boston Cliffs Mbr/Choptank Fm	Middle Miocene		X	
Little Cove Point Mbr/St. Marys Fm	Middle/Late Miocene		X	
Windmill Point Mbr/St. Marys Fm	Late Miocene		X	
Rushmere Mbr/Yorktown Fm	Early Pliocene			X
Moore House Mbr/Yorktown Fm	Late Pliocene		X	
Chowan River Fm	Late Pliocene		X	
James City and Waccamaw Fms	Early Pleistocene		X	
Flanner Beach and Neuse Fms	Late Pleistocene		X	

In previous work, we had calculated drilling frequencies for individual formations. However, different members within a formation may differ in composition of the naticid assemblage. For example, within the Yorktown Formation, our samples of the Rushmere Member were dominated by *Euspira heros*, whereas those of the Moore House Member were dominated by *Neverita duplicata*. Wherever possible, we therefore treated samples from different members of a formation separately. Six stratigraphic levels had two or more species present in approximately even numbers (Table 1); these samples are referred to as containing "multiple naticid species." For those formations, it was impossible to verify that naticid predation on naticids represented true intraspecific cannibalism. However, samples from four stratigraphic levels contained only one naticid species: the Ripley Formation, Brightseat Formation, Matthews Landing Member of the Naheola Formation, and Byram Formation. In 12 additional formations or members the naticid faunas were dominated by a single species. Thus for 16 of the 23 stratigraphic levels studied (more than two-thirds of the samples), actual intraspecific cannibalism most likely occurred. (The Rushmere Member also was dominated by a single species, *Euspira heros*, but the one drilled specimen was a *Tectonatica pusilla*; thus we did not count this unit as exhibiting intraspecific cannibalism.)

As is the case for the non-naticid species in the database, data include the total number of specimens of each naticid species within each stratigraphic level and the occurrence of complete and incomplete naticid drillholes. Naticid drillholes are recognized by their typically parabolic shape in cross section (designated by the ichnogenus *Oichnus* Bromley 1981; see Kelley and Hansen 2003 for a comparison of drillholes produced by different families of drilling gastropods). Complete drillholes are those that penetrate to the interior of the shell and are presumed to have caused mortality of the individual. Incomplete holes do not penetrate to the interior of the shell and usually indicate failed predation, although on rare occasions incomplete holes may result from successful attacks in which predators abandon drilling and consume prey that suffocated during drilling (Ansell and Morton 1987). Drilling frequencies (DF) were analyzed for each stratigraphic level by dividing the number of specimens with complete naticid drillholes by the total number of specimens. In addition, we calculated "prey effectiveness" (Vermeij 1987; Kelley and Hansen 1993, 1996a, 2003, 2006; Kelley et al. 2001) as the ratio of incomplete drillholes to total attempted (complete plus incomplete) holes.

Several analyses were conducted using different subsets of the naticid data: 1) all 23 stratigraphic levels; 2) only samples with at least 40 naticid specimens; 3) all 16 samples thought to represent actual intraspecific cannibalism, regardless of sample size; 4) samples representing intraspecific cannibalism that contained at least 40 naticid specimens. (Note: the 40 specimen cut-off is more rigorous than that employed in some earlier studies. For example, Vermeij 1987 reported drilling frequencies on "abundant species," which he defined as including 10 or more individuals.) We explored the temporal pattern of cannibalism using Spearman's rank correlation coefficient to determine if a significant relationship occurred between drilling frequency and stratigraphic position. We also calculated the product-moment correlation coefficient between cannibalism and naticid abundance to determine whether drilling frequency could be an artifact of abundance of predators. Temporal patterns in cannibalism were compared to drilling on the gastropod fauna as a whole using product-moment correlation coefficients. (We recognize the potential for autocorrelation of these two datasets, because the naticid data are a subset of the data for the gastropod fauna; however, lack of correlation between the two datasets, despite the potential for autocorrelation, may be particularly informative. See Kelley and Hansen 2006 for further discussion.)

8.4 Results

Drilling frequency for naticids varied widely among stratigraphic levels (Table 2) from <0.01 in the Cretaceous Ripley Formation to 0.36 in the Miocene Windmill Point Member of the St. Marys Formation.

Table 2. Drilling data and relative abundance of naticid gastropods for formations (Fm) and members (Mbr) studied. N, total number of naticid specimens; CD, number of naticid specimens with one or more complete naticid drillholes; Inc., number of drillholes that were incomplete; RA, relative abundance of naticids as a proportion of the total gastropod fauna; DF, drilling frequency (given for both naticids and the total gastropod fauna

Fm or Mbr	N	CD	Inc.	RA	Naticid DF	Gastropod DF
Ripley	167	1	0	0.069	0.006	0.037
Providence	137	4	0	0.266	0.029	0.060
Brightseat	18	3	0	0.043	0.167	0.377
Matthews Landing	15	3	0	0.051	0.200	0.465
Bells Landing	46	3	0	0.061	0.065	0.352
Bashi	74	5	0	0.273	0.068	0.203
Cook Mountain	228	32	9	0.083	0.140	0.162
Gosport	10	2	0	0.05	0.200	0.075
Moodys Branch	108	14	1	0.057	0.130	0.098
Red Bluff	92	1	0	0.176	0.011	0.135
Mint Spring	229	26	0	0.310	0.114	0.145
Byram	41	7	0	0.230	0.171	0.174
Belgrade	40	6	0	0.115	0.15	0.118
Calvert	215	40	1	0.362	0.186	0.202
Drumcliff	184	65	1	0.095	0.353	0.256
Boston Cliffs	25	2	0	0.064	0.080	0.352
Little Cove Point	563	150	7	0.094	0.266	0.385
Windmill Point	415	149	3	0.156	0.359	0.371
Rushmere	43	1	0	0.102	0.020	0.165
Moore House	241	62	2	0.165	0.258	0.135
Chowan River	161	26	0	0.050	0.161	0.145
James City	297	56	1	0.119	0.189	0.091
Flanner Beach	15	4	0	0.025	0.267	0.060

When samples with fewer than 40 naticid specimens or those with multiple naticid species were excluded from consideration, drilling frequencies still ranged from 0 to 0.36.

Drilling frequencies tabulated by series (using all samples) exhibited a general pattern of increase through time (Table 3). The frequency of drilling on naticids was 0.02 in the Cretaceous (5 of 304 naticid specimens had complete naticid drillholes). Drilling frequencies ranged between 0.10 and 0.13 for Paleogene series; overall drilling on naticids for the Paleogene was 0.113 (102 of 901 naticid specimens drilled). In contrast, drilling during the Neogene through Pleistocene was significantly greater, ranging at the series level from 0.19 to 0.29. Overall drilling for all post-Paleogene samples was 0.257 (555 of 2159 specimens drilled). Drilling in the Paleogene was significantly greater than in the Cretaceous ($\chi^2 = 26.302$, $p = 0.0000$) and significantly less than in the post-Paleogene ($\chi^2 = 78.025$, $p = 0.0000$).

Table 3. Drilling frequency and ratio of incomplete to total attempted drillholes by series

Series	Drilling Frequency	Incomplete:Total holes
Upper Cretaceous	0.016	0
Paleocene	0.114	0
Eocene	0.126	0.159
Oligocene	0.100	0
Miocene	0.290	0.028
Pliocene	0.200	0.023
Pleistocene	0.192	0.016

The apparent increase in cannibalism through time is supported by rank correlations between stratigraphic position and drilling frequency (Table 4). Spearman's rank correlation coefficient was statistically significant for each of the four analyses (all samples, large samples only, samples that likely represent true cannibalism, large samples representing true cannibalism).

Table 4. Spearman rank correlation coefficients for drilling frequency (DF) vs. stratigraphic position of samples, and product moment correlation coefficients of naticid DF with relative abundance (RA) of naticids and with DF for the total gastropod fauna

Analysis	N	DF vs stratigraphic position	Naticid DF vs RA	Naticid DF vs gastropod DF
All samples	23	0.5012*	-0.1470	0.3061
Large samples	18	0.6718**	-0.0758	0.5399*
Actual cannibalism - all samples	16	0.5559*	-0.0771	0.3486
Actual cannibalism - large samples	12	0.6923*	-0.0878	0.7469*

Fig. 1 illustrates the increase in actual cannibalism in samples with >40 specimens.

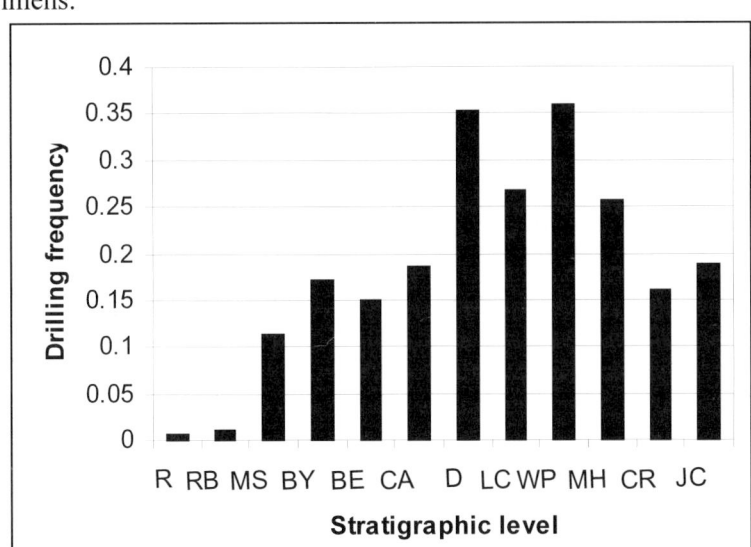

Fig. 1. Drilling frequency through time for samples larger than 40 specimens that represent actual cannibalism. Abbreviations for stratigraphic levels: R, Ripley; RB, Red Bluff; MS, Mint Spring, BY, Byram; BE, Belgrade; CA, Calvert; D, Drumcliff; LC, Little Cove Point; WP, Windmill Point; MH, Moore House; CR, Chowan River; JC, James City/Waccamaw

Naticids were important constituents of Cretaceous and Cenozoic gastropod faunas (Table 2), representing an average of 11% of the gastropods in these assemblages (3364 of 31,061 total gastropods in all samples combined). The proportion of naticids varied among individual assemblages from 2 to 36% of the gastropods present. Correlation coefficients between the frequency of drilling on naticids and relative abundance of naticids were nonsignificant for all four analyses (Table 4, Fig. 2).

When all samples were included, the correlation coefficient between drilling frequencies on naticids and on the gastropod fauna as a whole was nonsignificant (Table 4, Fig. 3). Likewise, when only samples inferred as representing actual cannibalism are included, the correlation between naticid DF and DF for the total gastropod fauna is nonsignificant. However, when samples with fewer than 40 specimens were excluded from analysis, correlation between drilling frequencies on naticids and on the gastropod fauna as a whole were statistically significant ($r = 0.5359$ for all samples and $r = 0.7469$ for conspecific predation; Table 4).

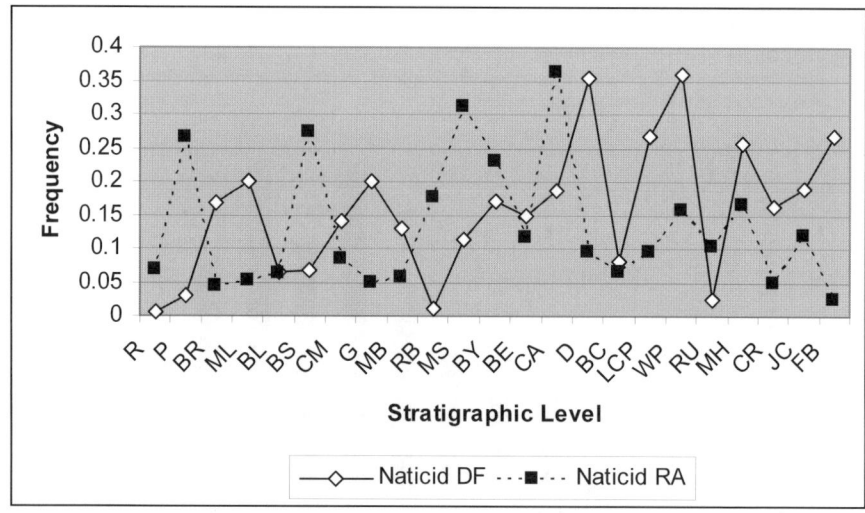

Fig. 2. Frequency of drilling on naticid gastropods (naticid DF) and frequency of naticids in the total gastropod sample (naticid RA) through time. Abbreviations for stratigraphic levels: R, Ripley; P, Providence; BR, Brightseat; ML, Matthews Landing; BL, Bells Landing; BS, Bashi; CM, Cook Mountain; G, Gosport; MB, Moodys Branch; RB, Red Bluff; MS, Mint Spring, BY, Byram; BE, Belgrade; CA, Calvert; D, Drumcliff; LCP, Little Cove Point; WP, Windmill Point; RU, Rushmere; MH, Moore House; CR, Chowan River; JC, James City/Waccamaw; FB, Flanner Beach/Neuse

Incomplete drilling of naticids was rare (Table 2). Most stratigraphic levels included no naticid specimens that had been drilled unsuccessfully. Only one species, *Neverita limula* of the Cook Mountain Formation, exhibited substantial incomplete drilling. This species included 7 incomplete drillholes and 11 complete drillholes. As a result, the ratio of incomplete to total attempted holes (prey effectiveness of Vermeij 1987) in the Eocene was much higher than for any other series (0.159; Table 3).

No general temporal pattern of incomplete drilling was apparent in the data; nearly all drilling attempts by naticids on naticids were successful. In addition, none of the naticid specimens in the Cretaceous or Paleogene bore more than one drillhole. Seven Miocene naticid specimens had more than one complete naticid drillhole (out of a total of 414 complete holes in the Miocene), as did two specimens from the Moore House Member of the Yorktown Formation (out of a total of 64 complete drillholes in the Moore House). No other cases of multiple drilling were present.

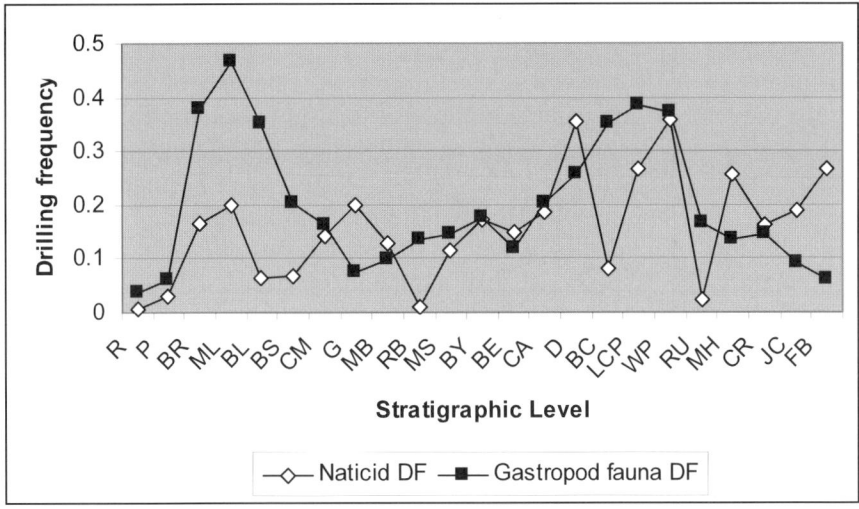

Fig. 3. Drilling frequency (DF) for naticid gastropods compared to the total gastropod fauna through time. Stratigraphic abbreviations as in Fig. 2

8.5 Discussion

Results of this study are similar to those of most previous studies that reported drilling frequencies on naticid prey for individual formations. For example, Taylor et al. (1983) reported that 3% of naticaceans from the Blackdown Greensand (Albian of England) exhibited naticid drillholes, a result consistent with our drilling frequency of 0.02 for the Cretaceous. (However, more recently Kase and Ishikawa 2003 have argued that the Albian holes were not drilled by naticids.) Previously reported Eocene drilling frequencies on naticids range from about 0.10 to 0.20 (Taylor 1970, Paris Basin; Adegoke and Tevesz 1974, Nigeria; Stanton et al. 1981, Texas). The range of Eocene drilling frequencies reported in the present study is similar (0.07-0.20). Published frequencies of drilling on Miocene naticids (0.03-0.22, reported by Colbath 1985, Hoffman et al. 1974, Kowalewski 1990, Dietl and Alexander 2000, and Złotnik 2001) are mostly less than in our study (0.08-0.36). However, our results for the Pliocene (DF = 0.20) and Pleistocene (DF = 0.19) are consistent with most published results: Pliocene of Emporda, Spain, 0.21 (Hoffman and Martinell 1974); Pleistocene of Fiji, 0.22 (Kabat and Kohn 1986) and 0.19 (Kohn and Arua 1999); Pleistocene of North Carolina, USA, 0.17 (Dietl and Alexander 2000). An exception is a drilling frequency of 0.08 reported by Boekschoten 1967 for *Polinices* sp. from the Pliocene of Belgium. Walker 2001 reported a high frequency of cannibalism (~0.5) on 39 specimens of *Natica scethra* from the upper Pliocene of Ecuador, but results may not be comparable to those of this study because they are from a deep-water environment. (Hansen and Kelley 1995 found significantly greater drilling frequencies in the deeper water Yazoo Formation than in the nearly contemporaneous Moodys Branch Formation in the Eocene).

The history of predation by naticid gastropods on naticids therefore does not support the hypothesis that cannibalism decreased through time. Temporal patterns in confamilial (and in many cases conspecific) predation indicate that, despite fluctuations, an overall increase occurred between the Cretaceous and Paleogene and again in the Neogene. The frequency of such predation in the Cretaceous was less than 0.03, whereas comparable results for the Paleogene and post-Paleogene are 0.11 and 0.26 respectively, results that are consistent with those of most previously published studies. In addition, despite the fluctuating pattern of drilling frequencies in our database, rank correlation coefficients between drilling frequency and stratigraphic position were statistically significant for all analyses. Thus predation by naticids on naticids, including actual intraspecific cannibalism, increased through time, contrary to the hypothesis tested.

Rejection of this hypothesis may indicate that naticid predators were not less efficient early in their history. Some evidence, however, supports the hypothesis that naticid predatory efficiency increased through time. Kelley and Hansen (1996b) examined selectivity of prey size and prey species for four Eocene bivalve assemblages. Species selectivity was analyzed using cost-benefit analysis, in which cost of predation was measured by prey thickness (proportional to drilling time) and benefit was determined by the internal volume of the prey item. Predators were predicted to select the prey item with the lowest cost:benefit ratio in the size range that could be handled by the predator. Although intraspecific prey size selectivity occurred for most Eocene bivalve prey species studied, interspecific prey selectivity was weaker than in Neogene through Recent assemblages. We concluded that prey selectivity and thus predator efficiency appeared to be less developed in the Paleogene compared to the post-Paleogene, as we predicted from the hypothesis of escalation (Kelley and Hansen 1996b).

An alternative explanation for these contradictory results is that cannibalism is not indicative of predator ineptitude. Previous work appears to support this alternative. Kelley (1991) found that cannibalism by Miocene naticids from the St. Marys Formation of Maryland followed the same "rules" of prey selectivity as did predation on bivalve prey (Kelley 1988). A high degree of drillhole site selectivity indicated a capacity for precise manipulation of prey, and prey choice was consistent with selective predation to maximize energy gain as indicated by cost-benefit analysis. Similar results were reported by Dietl and Alexander (1995), who examined Recent collections of *Euspira heros* and *Neverita duplicata*. The results of Kelley (1991) and Dietl and Alexander (1995) were interpreted to support the conclusion, based on laboratory experiments (Kitchell et al. 1981), that cannibalism was not anomalous. If cannibalism is not a result of predator ineptitude, the incidence of cannibalism may not be an appropriate test of the hypothesis that efficiency of naticid predators increased through time.

What then controls the degree of cannibalism within a fossil assemblage? What accounts for the marked increase in drilling on naticids between the Cretaceous and Neogene? The pattern of temporal increase in cannibalism is not an artifact of naticid abundance. Although fluctuations in naticid abundance and drilling on naticids appear to be somewhat in phase for the Neogene (Fig. 2), if anything the two are inversely correlated through the Paleogene. In fact, all correlation coefficients calculated for naticid abundance and drilling frequency are negative, though nonsignificant. Thus the increase in cannibalism is not a simple consequence of increasing representation of naticids in the assemblages.

Nor does the temporal pattern of cannibalism mimic closely the history of drilling on the gastropod faunas as a whole. Kelley and Hansen (1993, 1996a, 2003, 2006) documented a fluctuating pattern of drilling frequencies characteristic of the mollusc fauna as a whole and to an extent by the bivalve and gastropod components of the fauna. Drilling frequencies on the total gastropod fauna (Table 2, Fig. 3) were low to moderate in the Cretaceous and then increased dramatically after the Cretaceous-Tertiary boundary. Drilling frequencies on all gastropods declined through the Eocene but increased above the Eocene-Oligocene boundary. A late Oligocene decrease was followed by a significant rise in drilling after the middle Miocene extinction and by another decline in the Plio-Pleistocene. Kelley and Hansen (1996a) hypothesized that drilling increased after mass extinctions because extinctions selectively eliminated the highly armored (escalated) species of a fauna, producing a recovery fauna that is vulnerable to naticid predation. Further analysis of the susceptibility of armored species to extinction, however, has not supported this hypothesis (Hansen et al. 1999). The question of whether species that were physiologically escalated (rather than morphologically escalated) were preferentially eliminated by mass extinctions remains open, however (Dietl et al. 2002).

Kelley and Hansen (2006) compared temporal patterns in assemblage-level drilling frequencies with those exhibited by lower taxa (individual families, genera and species of bivalves and gastropods). Of two gastropod families (turritellids and naticids) and six bivalve families examined, only the poorly represented noetiid bivalves and the naticid gastropods yielded patterns inconsistent with those exhibited at the class level. The present study, which employed finer stratigraphic resolution than the study by Kelley and Hansen (2006), upholds the conclusions of the previous work. Because the naticid database is not independent of the database for all gastropods, the potential exists for autocorrelation of the two data sets (Kelley and Hansen, 2006). Despite this potential for autocorrelation, the correlation between DF for all gastropods and for naticids is not statistically significant when all samples are included. Nor is it significant when only samples representing true cannibalism are included. However, elimination of smaller samples yielded significant correlations between total gastropod drilling frequency and that for naticid gastropods. Elimination of samples with few naticids increases the chance for autocorrelation of the two datasets because naticids contribute more substantially to the total gastropod fauna. We suggest that the factors (as yet unresolved) that control the fluctuations in drilling frequency for the gastropod fauna as a whole may not be sufficient to explain the temporal changes in cannibalism.

Paine (1963) and Taylor (1970) suggested that cannibalism by naticids may result from the lack of alternative prey. In many organisms, cannibalism is associated with a decline in alternative prey (Polis 1981). Under such conditions of food limitation, cannibalism may increase because hunger increases foraging activities, diets expand, and food-deprived conspecific prey may be more vulnerable to attack. Dietl (2003) attributed cannibalism by the fasciolariid gastropod *Triplofusus giganteus* to hunger induced by lack of alternative resources. In the present case, however, assemblages containing cannibalized naticids include abundant drilled bivalves and other groups of gastropods, suggesting that cannibalism was not a "last resort" of starving naticid predators.

A possible deterrent to cannibalism would occur if the cannibalistic interaction proved dangerous to the predator. Dietl and Alexander (2000) have argued that confamilial naticid predation can be considered dangerous because the "hunted" individual may turn on its predator and become the "hunter." However, if a size differential exists between conspecifics during attempted cannibalism, the smaller individual will be killed (Kitchell et al. 1981) without endangering the larger individual. In such cases, smaller naticid prey would not be considered dangerous to larger conspecific predators, obviating this possible deterrent to cannibalism.

As prey, naticids are more difficult to subdue than most bivalve prey, because of their greater degree of mobility (Kitchell et al. 1981; Kelley 1991; Dietl and Alexander 1995; Złotnik 2001). Nevertheless, the relatively thin naticid shell can be penetrated more rapidly, once prey are subdued. The rarity of incomplete boreholes observed in this study, with the exception of one Eocene species, indicates that predation on naticids was typically successful once drilling commenced. Although Taylor (1970) also reported a high frequency of failed drilling on Eocene naticids, previous studies of drilling on Miocene and Pleistocene naticids also reported the absence or rarity of incomplete drilling (Kowalewski 1990, Kelley 1991, Kabat and Kohn 1986, Kohn and Arua 1999, Dietl and Alexander 2000). The high frequency of success in drilling a subjugated naticid, coupled with a low cost-benefit ratio of naticid prey items relative to their thicker bivalve counterparts (Złotnik, 2001), indicates that cannibalism may be more profitable than predation on bivalves or non-naticid gastropods.

Cannibalism may be beneficial for other reasons as well. Conspecifics may be an important resource in terms of biomass (Polis 1981), but the benefit to the predator goes beyond simple energetics. From the perspective of the individual predator, cannibalism also eliminates potential enemies, both competitors and predators (Kelley, 1991). Competition for both mates and food resources can be reduced by cannibalism; as argued by Polis (1981), cannibals contribute relatively more genes to the population both through the energetic benefits of cannibalism and the decreased genetic contribution of competitors. Cannibalism may also serve as a means of population regulation (Polis 1981, Kitchell, 1986). For instance, Basedow (1995) reported that cannibalism in extant *Lunatia alderi* (Atlantic coast of Portugal) appears to be density dependent and suggested it damps population fluctuations. Cannibalism may increase resistance to extinction because populations thus tend to be more stable and also because populations honed by cannibalism may contain more vigorous individuals (Polis 1981). Obviously, limits to cannibalism must occur; in the extreme it may produce local extinctions of the predator (Polis 1981). Indeed, instances of cannibalism in some organisms may be maladaptive (the result of stress, accidents, or non-natural conditions), but in many cases cannibalism should be favored by natural selection (Polis 1981).

In contrast to the interpretation of Stanton and Nelson (1980) and Hoffman et al. (1974), we thus argue that naticid cannibalism may be an advantageous interaction that also requires greater predator efficiency than do non-cannibalistic predation events. The temporal increase in naticid cannibalism may therefore be interpreted as the result of an increase in predator efficiency between the Paleogene and Neogene, perhaps in part due to natural selection that favored cannibalism. Such an increase in the post-Paleogene fossil record is also supported by the occurrence of very low levels of incomplete and multiple drilling in the total fauna during the Neogene and Pleistocene (Kelley et al. 2001), indicating increased predator capabilities relative to prey defenses among the mollusc fauna as a whole. Further testing of our hypotheses concerning evolution of predatory traits awaits the availability of valid hypotheses of naticid phylogenies.

Interestingly, Dietl and Alexander (2000) reported a relatively high frequency of incomplete drilling on the Recent naticids *Neverita duplicata* and *Euspira heros* (representing a sevenfold and threefold increase in prey effectiveness of the two species, respectively, since the Miocene). They linked this increase in failed drilling to increases in naticid shell thickness and a shift of preferred drilling site towards the (thicker) umbilical area of the naticid shell since the Miocene. Repositioning of the drillhole site was considered advantageous in preventing egress of the prey's foot during drilling (Dietl and Alexander 2000). They also observed a decrease in the slope of cost-benefit curves between the Miocene and Recent for both *Euspira heros* and *Neverita duplicata*, suggesting that predation on naticids has become increasingly profitable since the Miocene. Thus since the Miocene, increases in naticid predatory capabilities apparently were accompanied by coevolution of defenses by increasingly attractive naticid prey (Dietl and Alexander 2000). Confamilial predation, including true cannibalism, therefore represents a delicate balance between the capabilities of (highly efficient) naticid predators and prey, rather than a default alternative resulting from predator inefficiency.

8.6 Conclusions

1. Predation by naticid gastropods on naticids, including intraspecific cannibalism, increased from the Cretaceous through Pleistocene of the Gulf and Atlantic Coastal Plain. These results are consistent with previously published data on drilling of naticids.
2. Drilling frequencies are not an artifact of naticid abundance, nor is the temporal pattern of cannibalism consistent with drilling on the gastropod fauna as a whole.
3. Most attempts to drill naticid prey were successful, and no trends in frequency of incomplete drillholes (prey effectiveness) occurred between the Cretaceous and Pleistocene.

4. Although previous authors have suggested that cannibalism is a result of predator inefficiency, we argue that predation on highly mobile naticids requires a high degree of predator efficiency. Thus the increase in cannibalism from the Cretaceous through Neogene may indicate an increase in predator capabilities through time.

8.7 Acknowledgements

This study was based on a database collected with support of National Science Foundation grants EAR 8915725 to Kelley and Hansen and EAR 9405104 to Kelley and EAR 9406479 to Hansen. The following individuals provided assistance in the field and laboratory: D. Dockery, D. Haasl, L. Ward, N. Gilinsky, S. Templeton, C. Hampton, A. Tobias, V. Melland, R. Sickler, S. Graham, A. Huntoon, M. Jones, B. Farrell, E. Akins, K. Bradbury, and G. Kays. We thank R. Reyment and an anonymous reviewer for their comments.

References

Adegoke OS, Tevesz MJS (1974) Gastropod predation patterns in the Eocene of Nigeria. Lethaia 7: 17-24

Ansell AD, Morton B (1987) Alternative predation tactics of a tropical naticid gastropod. J Exper Mar Biol Ecol 111: 109-120

Arnold AJ, D'Escrivan F, Parker WC (1985) Predation and avoidance responses in the Foraminifera of the Galapagos hydrothermal mounds. J Foram Res 15: 38-42

Basedow T (1995) Analysis of shells on beaches and rocky coasts of Europe led to new findings on nutritional and population ecology of predatory marine gastropods (Muricidae and Naticidae). Zoologische Beiträge 36(1): 29-48

Basedow T (1996) Contribution to the nutritional and population ecology of *Murex trunculus* (L. 1767) (Gastropoda, Muricidae), derived from findings at the north coast of Crete (Eastern Mediterranean). With remarks on two species of Naticidae (Gastropoda). Zoologische Beiträge 37(2): 157-170

Boekschoten GJ (1967) Palaeoecology of some Mollusca from the Tielrode sands (Pliocene, Belgium). Palaeogeogr Palaeoclimat Palaeoecol 3: 311-362

Bromley RG (1981) Concepts in ichnotaxonomy illustrated by small round holes in shells. Acta Geol Hisp 16: 55-64

Colbath SL (1985), Gastropod predation and depositional environments of two molluscan communities from the Miocene Astoria Formation at Beverly Beach State Park, Oregon. J Paleontol 59: 849-869

Dietl GP (2003) First report of cannibalism in *Triplofusus giganteus* (Gastropoda: Fasciolariidae). Bull Mar Sci 73: 757-761

Dietl GP, Alexander RR (1995) Borehole site and prey size stereotypy in naticid predation on *Euspira (Lunatia) heros* Say and *Neverita (Polinices) duplicata* Say from the southern New Jersey coast. J Shellfish Res 14: 307-314

Dietl GP, Alexander RR (2000) Post-Miocene shift in stereotypic naticid predation on confamilial prey from the mid-Atlantic shelf: coevolution with dangerous prey. Palaios 15: 414-429

Dietl GP, Kelley PH, Barrick R, Showers W (2002) Escalation and extinction selectivity: morphology versus isotopic reconstruction of bivalve metabolism. Evolution 56: 284-291

Graham SE (1999) A test of the Kelley-Hansen escalation model using Neogene fossil assemblages from the Atlantic Coastal Plain of North America. Unpubl MS Thesis, Western Washington Univ

Hansen TA, Kelley PH (1995) Spatial variation of naticid gastropod predation in the Eocene of North America. Palaios 10: 268-278

Hansen TA, Kelley PH, Melland VD, Graham SE (1999) Effect of climate-related mass extinctions on escalation in molluscs. Geology 27: 1139-1142

Hoffman A, Martinell, J (1984) Prey selection by naticid gastropods in the Pliocene of Emporda (Northeast Spain). Neues Jahrb Geol Paläontol, Monatschefte 1984: 393-399

Hoffman A, Pisera A, Ryszkiewicz M (1974) Predation by muricid and naticid gastropods on the Lower Tortonian mollusks from the Korytnica clays. Acta Geol Polon 24: 249-260

Huntoon AG (1999) Naticid gastropod predation in the Pleistocene of North Carolina. Unpubl MS Thesis, Univ North Carolina at Wilmington

Jones MA (1999) Cost-benefit analysis of naticid gastropod predation across the Plio-Pleistocene boundary from North Carolina and Virginia. Unpubl MS Thesis, Univ North Carolina at Wilmington

Kabat AR, Kohn AJ (1986) Predation on early Pleistocene naticid gastropods in Fiji. Palaeogeogr Palaeoclimat Palaeoecol 53: 255-269

Kase T, Ishikawa M (2003) Mystery of naticid predation history solved: Evidence from a "living fossil" species. Geology 31:403-406

Kelley PH (1988) Predation by Miocene gastropods of the Chesapeake Group: stereotyped and predictable. Palaios 3: 436-448

Kelley PH (1991) Cannibalism by Chesapeake Group naticid gastropods: a predictable result of stereotyped predation. J Paleontol 65: 5-79

Kelley PH, Hansen TA (1993) Evolution of the naticid gastropod predator-prey system: an evaluation of the hypothesis of escalation. Palaios 8: 358-375

Kelley PH, Hansen TA (1996a) Recovery of the naticid gastropod predator-prey system from the Cretaceous-Tertiary and Eocene-Oligocene extinctions. Geol Soc Spec Publ 102: 373-386

Kelley PH, Hansen TA (1996b) Naticid gastropod prey selectivity through time and the hypothesis of escalation. Palaios 1: 437-445

Kelley PH, Hansen TA, Graham SE, Huntoon AG (2001) Temporal patterns in the efficiency of naticid gastropod predators during the Cretaceous and

Cenozoic of the United States Coastal Plain. Palaeogeogr Palaeoclimat Palaeoecol 166(1/2): 165-176

Kelley PH, Hansen TA (2003) The fossil record of drilling predation on bivalves and gastropods. In Kelley PH, Kowalewski M, Hansen TA (eds). Predator-Prey Interactions in the Fossil Record, Kluwer Academic/Plenum Press 113-139

Kelley PH, Hansen TA (2006) Comparisons of class- and lower taxon-level patterns in naticid gastropod predation, Cretaceous to Pleistocene of the U.S. Coastal Plain. Palaeogeogr Palaeoclimat Palaeoecol

Kitchell JA (1986) The evolution of predator-prey behavior: naticid gastropods and their molluscan prey. In Nitecki M, Kitchell JA (eds). Evolution of Animal Behavior: Paleontological and Field Approaches, Oxford Univ Press, 88-110

Kitchell JA, Boggs CH, Kitchell JF, Rice JA (1981) Prey selection by naticid gastropods: experimental tests and application to the fossil record. Paleobiology 7: 533-552

Kitchell JA, Boggs CH, Rice JA, Kitchell JF (1986) Anomalies in naticid predator behavior: a critique and experimental observations. Malacologia 27: 291-298

Kohn AJ, Arua I (1999) An early Pleistocene molluscan assemblage from Fiji: gastropod faunal composition, paleoecology and biogeography. Palaeogeogr Palaeoclimat Palaeoecol 146: 99-145

Kojumdjieva E (1974) Les gasteropodes perceurs et leurs victimes du Miocene de Bulgarie du Nord-Ouest. Bulgarian Acad Sci Bull Geol Inst Ser Paleontolo 23: 5-24

Kowalewski M (1990) A hermeneutic analysis of the shell-drilling gastropod predation on mollusks in the Korytnica Clays (Middle Miocene; Holy Cross Mountains, Central Poland). Acta Geol Polon 40: 183-213

Livan M (1937) Über Bohr-locher an rezenten und fossilen Invertebraten, Senckenbergiana 19: 138-150

Melland VD (1996) The effects of mass extinction on escalation of the naticid gastropod predator-prey system at the Eocene-Oligocene boundary. Unpubl MS Thesis, Univ North Dakota

Paine RT (1963) Trophic relationships of 8 sympatric gastropods. Ecology 44: 63-73

Polis GA (1981) The evolution and dynamics of intraspecific predation. Ann Rev Ecol Sys 12: 225-251

Reyment R (1966) Preliminary observations on gastropod predation in the western Niger Delta. Palaeogeogr Palaeoclimat Palaeoecol 2: 81-102

Reyment R (1967) Paleoethology and fossil drilling gastropods. Trans Kansas Acad Sci 70: 33-50

Reyment R (1999) Drilling gastropods. In Savazzi E (ed.) Functional Morphology of the Invertebrate Skeleton, John Wiley & Sons, Ltd 197-204

Reyment RA, Elewa AMT (2003) Predation by drills on Ostracoda. In Kelley PH, Kowalewski M, Hansen TA (eds.) Predator-Prey Interactions in the Fossil Record, Kluwer Academic/Plenum Press 93-112

Smith SM, Zumwalt GS (1987) Gravity flow introduction of shallow water microfauna into deep water depositional environments. Mississippi Geology 8: 1-7.

Stanton RJ, Nelson PC (1980) Reconstruction of the trophic web in paleontology: community structure in the Stone City Formation (middle Eocene, Texas). J Paleontol 54: 118-135

Stanton RJ, Powell EN, Nelson PC (1981) The role of carnivorous gastropods in the trophic analysis of a fossil community. Malacologia 20: 451-469

Taylor JD (1970) Feeding habits of predatory gastropods in a Tertiary (Eocene) molluscan assemblage from the Paris basin. Palaeontology 13: 254-260

Taylor JD, Cleevely RJ, Morris NJ (1983) Predatory gastropods and their activities in the Blackdown Greensand (Albian) of England. Palaeontology 26: 521-553

Vermeij GJ (1987) Evolution and Escalation: An Ecological History of Life. Princeton Univ Press, 527 p

Vermeij GJ (1994) The evolutionary interaction among species: selection, escalation, and coevolution. Ann Rev Ecol Syst 25: 219-236

Walker SE (2001) Paleoecology of gastropods preserved in turbiditic slope deposits from the Upper Pliocene of Ecuador. Palaeogeogr Palaeoclimat Palaeoecol 166: 141-163

Yochelson EL, Dockery D, Wolf H (1983) Predation of sub-Holocene scaphopod mollusks from Southern Louisiana. US Geol Survey, Prof Paper 1282, 13 p

Ziegelmeier E (1954) Beobachtungen über den Nahrungserwerb bei der Naticide *Lunatia nitida* Donovan (Gastropoda Prosobranchia). Helgolander Wissenschaftliche Meeresforschung 5:1-33

Złotnik M (2001) Size-related changes in predatory behaviour on naticid gastropods from the middle Miocene Korytnica Clays, Poland. Acta Palaeontol Polon 46: 87-97

9 On models for the dynamics of predator-prey interaction

Richard A. Reyment

Naturhistoriska riksmuseet, Stockholm, Sweden, richard.reyment@nrm.se

9.1 Abstract

A brief account of attempts at modelling the dynamics of predator-prey systems is presented. The original deterministic Lotka-Volterra model is not biologically realistic. Later workers have attempted to modify the basic features of that model to include biologically relevant parameters and to introduce the stochastic aspect into the computations. Mathematical models cannot solve a problem; only point to factors needing biologically oriented attention. The problem of formulating multi-tiered predation models is discussed in relation to myticulture.

Keywords: predator-prey systems, mathematical models, stochastic processes.

Little did the Ukrainian Alfred Lotka (1925) nor the Italian Vito Volterra (1926) imagine that when they, independently of each other, introduced the predator-prey biomathematical model into Science that they were laying the foundation to a rewarding area of mathematical research, and one that today extends far beyond the original biological confines of the problem as originally conceived. This is known as the subject of quasipolynomial systems. As an example of modern developments, we can take Hernandez-Bermejo and Fairén (2001) on Volterra-Lapunov stability for n-dimensional conservative systems with non-linearities of arbitrary degrees.

In its original formulation, the Lotka-Volterra model describes interactions between two species in an ecosystem, a predator and a prey in wholly deterministic terms. The underlying concept is not fundamentally biological in that the model is borrowed from physical chemistry for rates of reaction where molecules in solution interact by randomly colliding (Swift, 2002, p. 58). The adaptation of the chemical situation involves two equations, one that describes how the prey-population changes in numbers and the second that expresses how the number of predators (i.e. individuals) changes. The system is expressed by two differential equations, to wit:

$$dH/dt = (a_1 - b_1 P)H \tag{9.1}$$

$$dP/dt = (-a_2 + b_2 H)P \tag{9.2}$$

All constants > 0. The subscript 1 denotes prey, subscript 2 denotes predator.

Here,

$$H = \text{density of prey individuals} \tag{9.3}$$

P = density of predators
a_1 = intrinsic rate of increase in the prey-population in the absence of predation
b_1 = coefficient of the rate of predation which expresses the diminution in a_1 by an amount $b_1 P$ when P prey-animals are present.
b_2 = the reproduction rate of predators in relation to each prey individual consumed
a_2 = the mortality rate of the predators, which perforce dwindles to nothing in the absence of prey.

The solution of the quadratic differential equations yields

$$a_2 \ln H - b_2 H + a_1 \ln P - b_1 P = \text{constant} \tag{9.4}$$

This solution represents a family of closed curves in which each member corresponds to a different value of the constant, determined by the initial values assigned to P and H. The original Lotka-Volterra model represents a system with neutral stability, which implies that prey and predator undergo oscillations with amplitudes that bear no relation to the biology of the two species involved, and which depend solely on the arbitrary values assigned to P and H. This deterministic system, and its stochastic counterpart, have been discussed in detail by Bartlett (1957) who showed that there is no damping effect towards the point of equilibrium, a_2/b_2, a_1/b_1, where $dH/dt = dP/dt = 0$ and that the deterministic path of small cycles around this point is given by an ellipse. Bartlett also showed that in the stochastic model, the effect of random drift, in the absence of damping, eventually results either in the extinction of the predator, or the extinction, first of the prey, and then of the predator due to starvation. Leslie and Gower (1960) reasoned that the stationary state is in a sense unstable. However, a more elaborate definition of the Lotka-

Volterra model, which includes a damping term, can be expected to lead to greater stability and fluctuation around an equilibrium state with only a slight chance of random extinction of one of the species (Leslie 1948). Swift (2002) used a birth-death formulation to develop a stochastic version of the Lotka-Volterra model. Of recent date, attempts have been made to introduce a quantitative evolutionary aspect into predational population dynamics (Drossel et al., 2000). A relatively early attempt at updating the Lotka-Volterra model to accommodate the effects of a fluctuating environment is due to Moran (1953).

In the general form of the theory of predator-prey interaction it is assumed, albeit in some cases tacitly, that the predator (or parasite) depends for subsistence on a single species of prey (or host) and cannot turn to an alternative source of nourishment. (This is a coarse generalization inasmuch as predaceous gastropods, for example, can change diet from one species to another when the prey animal has been annihilated at a locality (Fischer-Piette, 1935; LeBreton and Lubet, 1992; Moore, 1958).) The case for two prey-species being hunted by a predator species was considered by Lotka (1925, pp. 88-97) in a tripartite deterministic model.

A biologically more reasonable model, that results in damped oscillations towards a stable equilibrium level in both populations, is due to Leslie and Gower (1960). The basic problem was stated as being concerned with the behaviour of predator-prey systems in the region of the stationary state and the chances of random extinction for one or the other of two species. Two situations may be envisaged here.

Situation 1:
a) There is an ample supply of food for the prey

b) The environment is spatially limited.

c) All members of the prey-population are exposed to the same risk of attack by the predator.

Situation 2:
Criterion c) above is relaxed so that only a fraction of the prey is exposed to risk.

The Leslie-Gower predator-prey model results in damped oscillations. Pielou (1976, p.91) summarizes the significance of the pertinent equations in terms of two models: all of the constants > 0.

Model 1: Does not allow for density dependent regulation

$$dH/dt = (a_1 - c_1P)H \qquad (9.5)$$

$$dP/dt = (a_2 - c_2P/H)P \qquad (9.6)$$

Model 2: Allows for density-dependent regulation

$$dH/dt = (a_1 - b_1H - c_1P)H \qquad (9.7)$$

$$dP/dt = (a_2 - c_2P/H)P \qquad (9.8)$$

The greater the ratio P/H, the smaller is the number of prey individuals per predation event and thence the slower the increase of the population of predators. The visible effect of both of these models is that each species population undergoes a damping of oscillations over time towards a state of equilibrium.

In the stochastic model of Leslie and Gower (1960) a variety of possible birth-rate and death-rate functions for predator and prey can be introduced. Leslie and Gower (1960, p. 227) also considered the situation taken up by the Russian zoologist G. F. Gause (1934) where only a part of the prey is exposed to risk owing to some environmental or behavioural factor. Gause (1934) pointed out the importance of a refuge in enabling the prey to survive in his experiments with various species of protozoans, even to the extent of the local population of predators dying off. Slight elaborations of the model of Leslie and Gower have been proposed between which it is not easy to discriminate (cf. Pielou, 1976, pp. 92-95; 107). Pielou points out that the operationally necessary simplifications involved in many attempts at modelling ecological situations are often lacking in credibility. She lists as examples of this the requirement that the environment be spatially homogeneous, the predator-prey system is closed and cannot receive immigrants or lose individuals due to emigration, that each population responds to changes in its own and the other's sizes instantly, variations in age-structure are disregarded, competitive abilities are independent of size of the populations, stochastic effects can be disregarded. A moment's reflection will bring to light that some of these situations are not feasible. For example, the loss or gain of individuals, age-structure, and the influence of stochastic variation in populations. Pielou (1976, pp. 108-110) reviews the effects of deviations from the ideal mathematical model. She concludes that mathematical models in theoretical ecology are not genuinely useful for answering questions, but have their use for raising them.

Chapter 11 in Emlen (1973) provides a fairly comprehensive coverage of deterministic model-building for predator-prey systems with emphasis on host-parasite interaction. Another useful textbook reference is Chapter 22 in Roughgarden (1979) in which a simple development of the Lotka-Volterra model and its inherent weaknesses is given.

An example of how complicated predator-prey interactions can be in nature is the excellent study of Le Breton and Lubet (1992) on *Mytilus edule*. In this study, which may be regarded as an extensive elaboration of the work of Fischer-Piette (1935), problems connected with the industrial farming of mussels are addressed (myticulture). The primary problem is parasitism by the trematode *Proctoeces maculatus* on *Mytilus*. This is a first order predator- prey (parasite) situation. The problem was found to go deeper than that, however, since it was not known at the time how the parasites came into contact with the mussels. The solution lay with the drill *Nucellus lapillus*, a muricid that drills barnacles by seeking out larval stages of *Balanus balanoides*, *Chthamalus stellatus* and *Eliminius modestus* (the "cyprid "stage). Infected *Nucellus* could be shown to be the vector introducing the parasite when forced to switch prey to mussels, as was first elucidated by Fischer-Piette (1935). An effective ecological model would need to take account of not only the doubly tiered predation condition, but also crowding, variation in balanid populations and prey-switching when balanids had been predated upon to near extinction, and, not least, the ability of the trematode to adjust its life-cycle.

Finally, an aspect of the predator-prey system that seems to have been overlooked in constructing mathematical models, as far as I am aware, concerns the expenditure of energy involved in hunting and subduing prey (the foraging factor) and when it is time to call off a costly attack. An introduction to the subject is the monograph by Stephens and Krebs (1986). Reyment (1988) applied foraging analysis to cephalopods with respect to feeding, including predation.

References

Bartlett MS (1957) On theoretical models for competitive and predatory biological systems. Biometrika 44: 27-42

Drossel B, Higgs PG, McKane AJ (2000) The influence of predator-prey population dynamics on the long-term evolution of food web structure. Arxiv:nlen, AO/0002032, 2000: 1-40

Emlen JM (1973) Ecology: an evolutionary approach. Addison-Wesley publishing Company 493 p

Gause GF (1934) The struggle for existence. Williams and Wilkens, Baltimore (Translated from the Russian)

Fischer-Piette E (1935) Histoire d'une moulière. Bull Biol France et Belgique 69: 153-177

Hernandez-Bermejo B, Fairen V (2001). Volterra-Lapunov stability of n-dimensional conservative systems with non-linearities of arbitrary degree. Journal of Mathematical Analysis and Applications 256: 242-256

LeBreton J, Lubet T (1992) Résultats d'une intervention sur une parasitose à *Protoeces maculatus* (Trematode, Diogenea) affectant la myticulture de l'Ouest Cotentin. Société Française de Malacologie "Les Mollusques marins, biologie et aquaculture", Ifremo, Actes de Colloques 14: 107-118

Leslie PH (1958) A stochastic model for studying the properties of certain biological systems by numerical methods. Biometrika 45: 16-31

Leslie PH, Gower JC (1958). The properties of a stochastic model for two competing species. Biometrika 45: 316-330

Lotka AJ (1925). Elements of Mathematical Biology. Dover Publications, New York 465 p

Moore HB (1958). Marine Ecology, Wiley and Sons 493 p

Moran PAP (1953). The statistical analysis of the Canadian lynx cycle. I. Structure and prediction. Australian Journal of Zoology 1: 163-173

Pielou EC (1976) Mathematical Ecology. Wiley and Sons, New York 385 p

Reyment RA (1988) A foraging model for shelled cephalopods. In Wiedmann J, Kullman J (eds) Cephalopods Present and Past (1988). Schweizerbart'sche Verlagsbuchhandlung, Stuttgart 687-203

Roughgarden J (1979).Theory of Population Genetics and Evolutionary Ecology: an Introduction. MacMillan, New York 634 p

Stephens DW, Krebs JR (1986) Foraging Theory. Princeton University Press, New Jersey 248 p

Swift RJ (2002) A stochastic predator-prey model. Irish Mathematical Society Bulletin 48: 57-63

Volterra V (1926) Variazioni e fluttuazioni del numero d'individui in specie animali conviventi. Mem R Accad Naz dei Lincei ser VI, vol. 2

10 Evolutionary consequences of predation: avoidance, escape, reproduction, and diversification

R. Brian Langerhans

Department of Organismic and Evolutionary Biology, 26 Oxford St., Harvard University, Cambridge, MA 02138, langerhans@oeb.harvard.edu

10.1 Abstract

One of the most important and obvious forces shaping organismal traits is predation. Prey have evolved diverse means of enhancing the probability of survival in the face of predation, and these means fall into two classes of antipredator strategies: (1) avoidance of predatory encounters, and (2) escaping after encountering a predator. A range of antipredator defenses—including behavioral, morphological, physiological, and chemical defenses—serve to either reduce the probability of detection by a predator or enhance the probability of surviving after detection by a predator. However, the recognition that reproductive strategies (e.g. offspring number, reproductive lifespan) are typically strongly influenced by mortality regimes induced by predators, highlights that most but not all "antipredator traits" fall into one of these two categories—that is, some life history traits influence only fecundity, not survival. Life history evolution has not traditionally been included in reviews of antipredator adaptations, however this chapter reveals that the conceptual link between life histories and predation broadens and refines our understanding of predation's role in phenotype evolution.

While ecologists have long recognized the importance of predation in population- and community-level dynamics, a varied history exists for the study of predation's role in influencing evolutionary change. Despite the wealth of antipredator adaptations present in organisms, research investigating the significance of predation in biological evolution has received considerably less attention than other ecological factors (e.g. competition, mate attraction). However, predation can generate divergent selection among prey populations in several different ways, and is predicted to represent a major source of evolutionary change. Recent empirical work supports this claim. This chapter reviews the varied forms of evolutionary strategies prey have evolved to mitigate malicious attempts of natural predators, and the potential importance of predators in driving phenotypic divergence and speciation.

Keywords: Antipredator adaptations, divergent natural selection, fitness tradeoffs, life history evolution, phenotypic plasticity, predator-prey interactions, reproductive isolation, speciation.

10.2 Introduction

This chapter primarily examines classical predation in animals (i.e. consumption of one animal by another), however predation can be broadly defined to include all transfers of energy from one organism to another, including herbivores, parasites, parasitoids, and pathogens. To illustrate the generality of several concepts, examples will be provided from this more general definition of predation. Rather than offer an extensive review of antipredator strategies, this chapter highlights conceptual points meant to enhance our investigation and understanding of the role of predation in the evolution of prey traits.

The transfer of energy among organisms has long received considerable attention from ecologists (e.g. Elton 1927; Lindeman 1942; Huffaker 1958; Holling 1959; Paine 1966; Addicott 1974; Pimm 1982; Kerfoot and Sih 1987; Lima and Dill 1990; Polis and Winemiller 1996; de Ruiter et al. 2005). The importance of consumptive interactions in the distribution and abundance of organisms can be easily observed by a cursory glimpse of some important ecological concepts and terms: keystone predation, food web ecology, food-chain length, top-down effects, trophic cascade. While ecological consequences of predation have been firmly established, research into the role of predation in driving evolutionary change has lagged far behind studies centering on other ecological factors, such as resource competition (reviewed in Vamosi 2005). This is puzzling considering the vast array of antipredator adaptations present in extinct and extant organisms, the early attention predation received in the developing field of evolutionary biology (e.g. Müller 1879; Wallace 1879; Poulton 1890; Beddard 1892; Thayer 1896), and the repeated propositions that predation was likely a significant force of evolutionary change in need of more focused attention (e.g. Cott 1940; Worthington 1940; Fryer 1959; Askew 1961; Ehrlich and Raven 1964; McPhail 1969; Stanley 1979; Vermeij 1987; Schluter 2000). For more than a century, most evolutionarily oriented research investigating predation either examined fossil evidence—where predation has long enjoyed substantial respect as a driving force of evolutionary change (e.g. see chapters 2-3, 7-9, 13-15)—or examined color patterns (i.e. crypsis, aposematism, mimicry) (see references in Komárek 1998; Ruxton et al. 2004). Outside of these two areas, evolutionary studies of predation have been comparatively minimal until a renewed focus began to emerge in the mid-1970s (e.g. Farr 1975; Gilbert 1975; Ricklefs and O'Rourke 1975; Holt 1977; Harvey and Greenwood 1978; Reznick and Endler 1982; reviewed in Vamosi 2005).

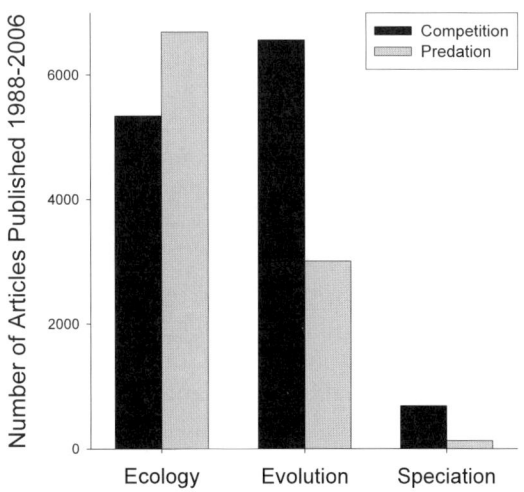

Fig. 1. Numbers of papers published in the general fields of ecology, evolution, and speciation involving the subjects of competition (black bars) and predation (grey bars). Data are from a search of the Institute for Scientific Information Science Citation Index conducted in March 2006. Searches included the following terms for each field of inquiry and subjects: Ecology = ecology, ecological, food web, population dynamic*, species distribution*, species abundance*, population distribution*, population abundance*; Evolution = evolution, evolutionary, diversification, divergence, differentiation; Speciation = speciation, reproductive isolation, reproductive isolating, sexual isolation, sexual isolating; Competition = competition (excluding the terms predation and predator), Predation = predation, predator (excluding the term competition). Results are similar if "herbivory," "herbivore," and "herbivores" are included in the search for Predation, if the term "food web" is excluded from the search for Ecology, and if searches for one subject are not exclusive of the other factor (i.e. if the term "competition" were allowed in searches involving "predation" and vice versa)

To illustrate the disparity in the academic attention received by competition and predation among ecologically and evolutionarily oriented studies, I conducted a literature search compiling the number of recently published scientific articles involving either predation or competition in the general fields of ecology, evolution, and speciation. While predation receives considerable attention in ecological studies, it is clearly overshadowed by competition in evolutionary studies (Fig. 1).

During the temporal span of the search, the numbers of papers of a primarily ecological nature involving these two subjects is relatively similar (≈25% more papers involving predation than those on competition); however, studies involving competition far outnumber studies involving predation for papers investigating evolution (more than twice as many) or speciation (more than 5 times as many). Thus, while it is widely recognized that predation is among the most important ecological factors structuring natural communities (e.g. Paine 1966; Addicott 1974; Jeffries and Lawton 1984; Kerfoot and Sih 1987; Holt and Lawton 1994; Wellborn et al. 1996; Jackson et al. 2001; Shurin and Allen 2001; Almany and Webster 2004), its potentially important role in phenotypic divergence, speciation, and diversification rates has only recently attracted significant attention (e.g. Reimchen 1994; Endler 1995; McPeek et al. 1996; Reznick et al. 1997; Blackledge et al. 2003; Langerhans et al. 2004; Nosil 2004; Vamosi and Schluter 2004; Vamosi 2005; Nosil and Crespi 2006b). Surprisingly, the extension of predation's importance in the distribution and abundance of species to its consequences for the distribution and abundance of phenotypes within and among species has been relatively slow in development. This chapter is meant to illustrate that when antipredator traits are reviewed with a broad, evolutionary perspective, the evidence overwhelmingly points to predation as a major force of evolutionary divergence across a wide range of phenotypes, and its role in speciation particularly demands future investigation.

10.3 Solving the problem of being eaten: avoidance and escape

Predation is a fundamental and pervasive component of ecosystems. Virtually every organism is perceived as a potential prey by some other organism. Obviously, an organism has little chance of reproducing and proliferating its lineage if it is severely injured or consumed by predators. Thus, selection has presumably strongly favored prey traits that increase the probability of survival and reproduction amidst predators. For the purposes of this chapter, "antipredator traits" are traits that predator-mediated natural selection has played a role in shaping, although they may have a different function in some current populations and may be influenced by other factors as well.

Previous authors have described various conceptual frameworks for understanding antipredator traits (e.g. Edmunds 1974; Vermeij 1982; Sih 1987; Lima and Dill 1990; Endler 1991; Caro 2005). I believe the most natural approach perceives predator-prey interactions from a prey's perspective, and categorizes antipredator traits based on the manner in which the traits influence prey fitness (i.e. individual viability, offspring viability, individual fecundity) and the chronological sequence of the possible components of predatory encounters—from pre-detection through recognition, attack, capture, and consumption (Table 1). Antipredator traits are highly diverse, spanning behavior, morphology, physiology, chemistry, and life history (reviewed in Edmunds 1974; Janzen 1981; Kerfoot and Sih 1987; Vermeij 1987; Greene 1988; Caro 2005). The approach taken in this chapter builds upon those of Sih (1987), Endler (1991), and Caro (2005), in an attempt to provide a more unified framework for understanding all possible types of antipredator traits.

To enhance survivorship in the face of predation, prey have evolved two types of strategies: (1) avoidance of predatory encounters, and (2) escaping after encountering a predator. Predator avoidance is defined as a reduction in the probability of detection by a predator. Predator escape is defined as a reduction in the probability of consumption after detection by a predator. Prey have evolved an astounding arsenal of defenses to avoid and escape predation (see Table 1). Importantly, antipredator traits do not necessarily enhance individual survivorship with predators, but might rather influence offspring survivorship or individual fecundity. Prey generally exhibit numerous antipredator traits, and as Table 1 illustrates, traits can influence both avoidance and escape, as well as affect different aspects of fitness. Further, traits can impact other traits through various means (e.g. physiological, architectonical), and trait correlations might take on a number of different forms (e.g. codependence: where one trait is determined by another; complementation: where specific combinations of trait values are required to achieve a particular function; see DeWitt et al. 1999; DeWitt and Langerhans 2003). Prey can additionally employ different traits in different ecological contexts and at different developmental stages. For example, different antipredator strategies might be utilized during different life stages or ages (e.g. employ crypsis when in larval form, rapid retreat when adult). Thus, effects of predation on phenotype evolution can be quite complex.

Table 1. Framework for understanding antipredator traits based on the manner in which traits directly influence prey fitness and the chronological stage of the predatory encounter in which they are employed. All prey traits are discussed in the text. Acronyms are as follows, ASF: avoiding a predator's sensory field, DSF: avoiding detection within a predator's sensory field, ATD: attack deterrence, CPD: capture deterrence, CND: consumption deterrence, Y: yes

	How the Antipredator Trait Directly Enhances Fitness				
	Increase Individual Survivorship		Increase Offspring Survivorship		
Prey Trait	Avoidance	Escape	Avoidance	Escape	Increase Individual Fecundity
Activity level	ASF				
Crypsis	DSF				
Development time	ASF				
Use predator-free habitat	ASF				
Active defense		CPD, CND			
Aposematism		ATD			
Attack diversion		CPD			
Autotomy		CPD			
Chemical defense		ATD, CPD, CND			
Death feigning		ATD			
Deimatic behavior		ATD, CPD			
Mimicry		ATD			
Protective morphologies		ATD, CPD, CND			
Rapid retreat, protean behavior		CPD			
Grouping	ASF, DSF	ATD, CPD, CND			
Use protective habitat	ASF, DSF	ATD, CPD, CND			
Vigilance	ASF, DSF	ATD, CPD, CND			
Reproductive timing			ASF, DSF	ATD, CPD, CND	Y
Reproductive effort:					
-Offspring size			ASF, DSF	ATD, CPD, CND	
-Parental care			ASF, DSF	ATD, CPD, CND	
-Offspring number			ASF, DSF	ATD, CPD, CND	Y
-Frequency of offspring production					Y
-Reproductive lifespan					Y

The most important aspect of complexity involving antipredator traits is their likelihood of exhibiting tradeoffs with other aspects of performance. That is, their utility in defense against one predator might come at the detriment of defense against another predator, or the production of offspring, or the ability to acquire resources or mates. Such tradeoffs should generate divergent selection across environments differing in predator regime, and thus should be important in evolutionary divergence (see 10.7).

10.4 Predator avoidance: winning without a fight

Prey have evolved numerous means of reducing the probability of detection by a predator (see Table 1). These means can be divided into two categories: 1) avoiding a predator's sensory field, or 2) avoiding detection within the predator's sensory field. Prey remove themselves (and their offspring, see 10.6) from a predator's sensory field primarily by using habitat with decreased probability of predator presence, by reducing their activity level, and by reducing time spent in vulnerable life stages. Prey avoid detection within a predator's sensory field primarily through crypsis.

10.4.1 Steering clear of a predator's realm: avoiding a predator's sensory field

Many organisms have evolved means of avoiding contact with predators through their habitat use. Some prey live their entire lives, or spend much of their time, in holes, fissures, crevices, and other sheltered refugia (termed anachoresis). Others use ephemeral habitats that are too temporary in nature or too frequently disturbed to be available to predators, while others use stressful habitats that are too physically severe to allow persistence of predators (e.g. Edmunds 1974; Sih 1987). Prey might use these low-predation habitats exclusively, or might exhibit temporal habitat shifts, thus remaining active, but in different habitats at different times. Utilizing low-predation environments reduces the probability of a prey entering a predator's sensory field, however the use of more general protective habitats might also influence other aspects of predator avoidance and escape (see 10.5.4).

A common antipredator response of many animals is reduced activity level, where probability of detection is reduced by limiting activity to relatively safe situations and/or minimizing time available for detection. For instance, many organisms restrict activity to particular times of the day (e.g. nocturnal and crepuscular activity) or seasons (e.g. diapause, dormancy, cyst formation) (reviewed in Stein 1979; Hairston 1987; Stemberger and Gilbert 1987).

Timing of developmental schedules is an important, but underappreciated form of predator avoidance. Some organisms exhibit rapid growth during vulnerable life stages to quickly reach a less vulnerable stage, or inhibit growth into a vulnerable stage until a relatively safe time period (Williams 1966; Istock 1967; Wilbur 1980; Roff 1992). To date, most empirical research on this topic has centered on the influence of predation on the timing of hatching and metamorphosis in amphibians (e.g. Werner 1986; Sih and Moore 1993; Warkentin 1995; Chivers et al. 2001; Saenz et al. 2003; Vonesh 2005). A related, but distinct, topic is reproductive timing, which describes schedules of reproductive events (rather than growth rates or life-stage changes) and primarily affects aspects of fitness other than individual viability. That topic is discussed in section 10.6.1.

10.4.2 Hiding in plain sight: avoiding detection within a predator's sensory field

Crypsis, the phenomenon where an organism resembles a random sample of relevant aspects of its environment, is a highly important form of predator avoidance, and is found in many disparate taxa (reviewed in Cott 1940; Norris and Lowe 1964; Edmunds 1974; Endler 1978; Caro 2005). Crypticity is meant to reduce detection within the sensory field of a predator, and includes reduction of visual, tactile, chemical, and electrical detection. For example, Queen parrotfish (*Scarus vetula*) surround themselves at night with a transparent mucus cocoon. This cocoon greatly reduces odors emanated from the fish, hiding its scent from predators. Effectiveness of crypsis often depends on a prey's ability to remain still, move subtly, utilize appropriate background environments, and coordinate behaviors with appropriate changes in body coloration.

Classic examples of crypsis include remarkable cases of background matching by prey organisms (e.g. flat-tailed geckos, sargassum fish, transparent zooplankton), but crypsis also includes more simple cases of blending in with the background, such as disruptive coloration, barred color patterns, general color matching, and countershading. These less spectacular forms of crypsis are extremely common and often highly effective. Countershading is ubiquitous among many taxa (e.g. fish, mammals, birds) and helps break up the body outline by exhibiting darker coloration dorsally and paler coloration ventrally (obscuring the ventral shadow formed by overhead lighting which reveals the body form) (Thayer 1896; Poulton 1902; Cott 1940; Kiltie 1988; Ruxton et al. 2004). The survival advantage of crypsis has now been demonstrated in a number of taxa (e.g. Cott 1940; Dice 1947; Cain and Sheppard 1954; Kettlewell 1956; Edmunds 1974; Caro 2005), and results of early experiments on body color in fish and grasshoppers are presented in Fig. 2.

Some authors consider particular types of mimicry as special cases of crypsis, however I distinguish crypsis from mimicry by the occurrence of detection. That is, cryptic organisms avoid detection by predators—predator avoidance—whereas organisms mimicking inedible organisms or objects allow detection, but avoid recognition as a palatable prey item—predator escape. Thus, mimicry is discussed below under the subject of predator escape.

10.5 Predator escape: prey fight back

Many prey traits enhance the probability of survival despite detection by a predator. These traits can be divided into three categories based on three chronological components of a predatory encounter: 1) attack deterrence, 2) capture deterrence, and 3) consumption deterrence.

10.5.1 Don't even think about it: attack deterrence

Antipredator traits that comprise attack deterrence are those that reduce the probability that a predator will actually attempt an attack once detection has occurred. These traits are meant to advertise detection or quality in an attempt to stimulate the predator to withdraw. That is, prey produce signals that they have detected the predator and/or are of poor quality (i.e. difficult to capture, handle, digest, or altogether inedible).

A common method of attack deterrence is deimatic behavior. Deimatic behaviors alert a predator that it has been detected and are meant to frighten, confuse, or intimidate the predator. These behaviors often involve alarm calls in social animals that additionally serve to alert other prey individuals that a predator has been detected. In deimatic behavior, animals often attempt to appear strong, healthy, and large, and sometimes emit sounds, display bright colors (e.g. color flashes, revealing hidden areas of the body), eyespots, or weapons, and adopt a stereotypic posture (Edmunds 1974). Some animals even eject body fluids during deimatic behaviors, such as the blood squirting behavior of horned lizards, the spraying of anal gland fluids in skunks, and the fluid jets and sprays of many arthropods (Eisner and Meinwald 1966).

Body fluids ejected during deimatic displays might affect a predator's senses (e.g. obscure vision), cause harm (e.g. fluids that are toxic to the predator), or simply confuse the predator. Examples of deimatic behaviors include the stotting and leaping of gazelles, flashing of colors in many cephalopods, erection of brightly colored or eyespotted wings in stick insects, chelae displays in crabs, crayfish and scorpions, and the striking eyespot displays of the Brazilian toad *Physalaemus nattereri*.

Aposematism is an important means of deterring an attack, and is common in both plants and animals (e.g. Edmunds 1974; Bowers 1993; Komárek 1998; Mallet and Joron 1999; Lev-Yadun et al. 2002; Härlin and Härlin 2003). Aposematic organisms have dangerous attributes (e.g. painful weapons, foul tastes), and advertise this fact via characteristic colors, structures, or other signals. Bright colors (often red, yellow, or orange) and spines are common signals for aposematic prey.

For aposematism to be advantageous, predators must either sample some prey and learn to avoid attacking those with particular warning signals, or possess an evolved avoidance response for prey with particular warning signals—both of these possibilities have been demonstrated in various taxa.

Mimicry is another way to deter an attack, and is often discussed as an offshoot of aposematism because many (but not all) cases of mimicry involve aposematic prey as the model. Müllerian mimicry describes the phenomenon where an unpalatable or otherwise dangerous prey species (the mimic) evolves a resemblance to another aposematic prey species (the model), benefiting from the fact that predators already avoid the model species (and thus, reciprocally benefiting the model species as well) (Müller 1879).

Examples of Müllerian mimics include several genera of wasps and bees, and many heliconiid butterflies. Batesian mimicry describes the same phenomenon, with the exception that the mimetic species is actually palatable (Bates 1862). Thus, Müllerian mimicry involves honest signals, while Batesian mimicry involves counterfeit signals. Examples of Batesian mimics include the beetle *Clytus arietis* and the hoverfly *Helophilus hybridus* which both resemble *Vespula* wasps (Mostler 1935).

Further, mimicry need not involve resemblance to aposematic organisms, but can involve resemblance to any dangerous or unpalatable object. For example, many organisms resemble inedible objects, such as leaves, twigs, and bird droppings, and still others resemble predators or competitors of their predator—and these represent cases of Batesian mimicry since the prey are truly edible (Cott 1940; Edmunds 1974; Bowers 1993; Brakefield et al. 1992; Wiklund and Tullberg 2004).

Prey appearing unpalatable due to their resemblance to particular organisms or objects avoid attack due to a recognition failure in the predator. That is, the predator can detect the prey, but it does not recognize the individual as a prey item.

A similar case involves detection by the predator, but a failure to initiate a killing response, despite the possibility of initial recognition as a prey item. This is typically accomplished by feigning death, termed thanatosis. This response is widespread in arthropods, reptiles, amphibians, birds, and mammals; classic examples include chickens, American opossums, and the African ground squirrel (e.g. Ewer 1966; Robinson 1969; Edmunds 1974; Greene 1988; Caro 2005).

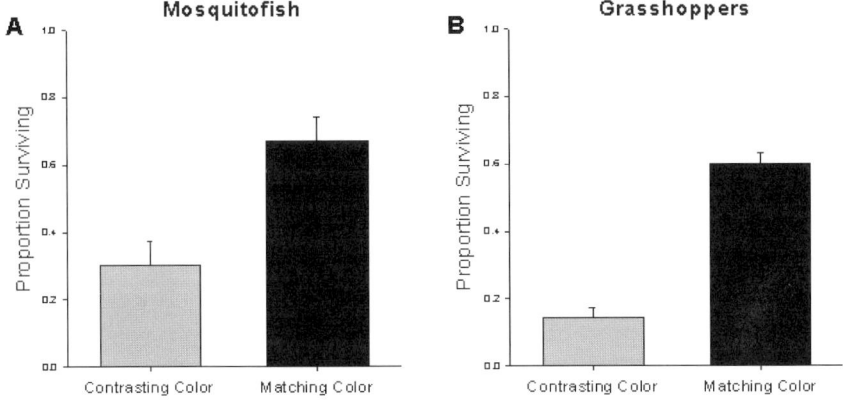

Fig. 2. Proportion of surviving prey individuals during experiments with predators exhibiting either a contrasting or matching body color with respect to its background environment. (A) Differential survival during predatory encounters with Galapagos penguins (*Spheniscus mendiculus*) for pale- and dark-bodied Western mosquitofish (*Gambusia affinis*) in pale and dark background environments (paired *t*-test, one-tailed P = 0.0006, eight experimental blocks, N = 1,046 fish; data from Sumner 1934). (B) Differential survival during predatory encounters with several bird species for several species of grasshoppers representing four color morphs in four corresponding background environments (paired *t*-test, one-tailed P < 0.0001, 25 experimental blocks, N = 758 grasshoppers; data from Isley 1938)

In these cases, predators often lose interest in the prey and move away, or relax attention temporarily providing the prey an opportunity to retreat. In some cases of death feigning, prey might actually deter consumption, rather than attack, by stiffening the body in a position that operationally increases its size and thus reduces a gape-limited predator's ability to consume the prey (Honma et al. 2006).

10.5.2 Catch me if you can: capture deterrence

Once a prey has been attacked, it can reduce the probability of successful capture in several ways. One of the most widespread and common forms of capture deterrence is rapid retreat, where the organism attempts to quickly flee from the oncoming attack. To rapidly displace themselves from the predator, prey can run, jump, swim, fly, drop, or slither away. In some cases, the retreat may seem quite slow to the human observer, however it may still be adequate to escape the relevant predator (e.g. rapid retreats in many snails); thus "rapid" retreat is a relative term in reference

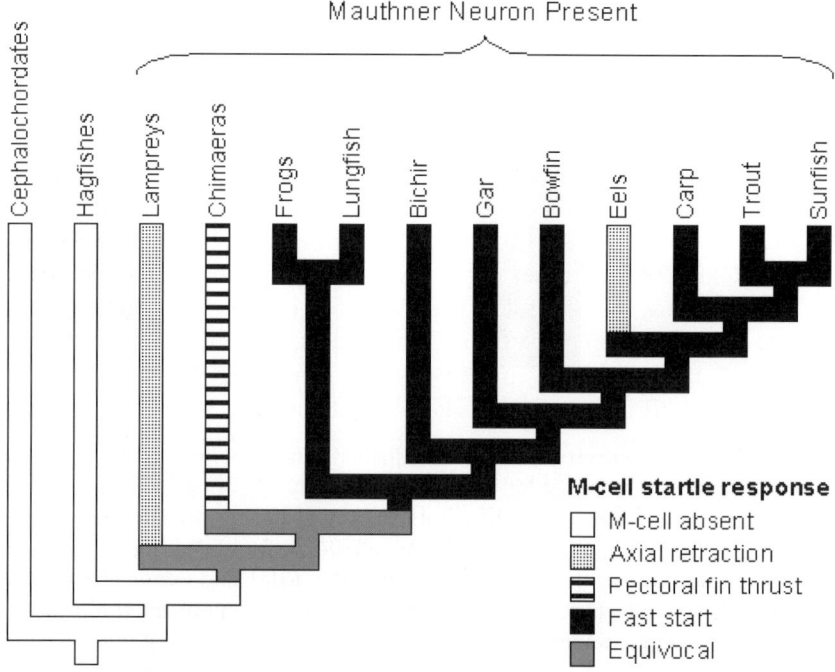

Fig. 3. Mauthner-cell initiated rapid retreats in chordates, illustrating the fundamental importance of predator escape responses. For frogs, Mauthner neurons produce a fast-start in tadpoles and elicit an escape jump in adults (Will 1986; Hoff and Wassersug 2000). Data from Hale et al. (2002)

to the speed of the oncoming attack. A classic example of rapid retreat is the Mauthner-cell initiated fast-start escape mechanism present in most anurans and fishes (Domenici and Blake 1997; Hale et al. 2002). This mode of rapid locomotion is highly important in surviving predatory strikes (Walker et al. 2005) and is conserved in general form across a wide range of organisms (Fig. 3). Rapid retreat is often combined with protean behavior, which describes irregular, unpredictable escape patterns and displays serving to confuse, disorient, and evade the predator (e.g. Humphries and Driver 1967; Humphries and Driver 1971; Edmunds 1974; Driver and Humphries 1988). For example, many prey flee from predators in an erratic, zigzagging, or bouncing fashion, or combine multiple behaviors in unpredictable manners. A highly studied predator-prey interaction involving rapid flight and strong protean behavior is the bat-moth system (e.g. Roeder 1962; Roeder 1965; Acharya and Fenton 1992; Rydell et al. 1995; Waters 2003). Efficacy of rapid retreat can also be enhanced by close proximity to a refuge, contrasting coloration, the flashing of colors during the retreat, grouping behaviors, or ejection of body fluids, such as the smoke screen effect of ink ejection in many octopus and luminescent clouds in some squids (e.g. Edmunds 1974; Helfman et al. 1997; Brooke 1998; Caro 2005; Palleroni et al. 2005).

Many prey have evolved an attack diversion to reduce the probability of capture during an attack. That is, many prey exhibit "predator lures" which divert the attack of the predator to less vulnerable objects or regions of the body (e.g. Edmunds 1974; Caldwell 1982; Riley and Loxdale 1988; Van Buskirk et al. 2003). For example, some prey exhibit defection marks which direct attacks to nonessential or distasteful parts of the body, such as the false eyes and antenna of some butterflies, the false heads of some snakes, and conspicuously colored tails of some lizards. Autotomy, the ability to break off a part of the body when attacked, is a common mode of attack diversion in many organisms, and is often combined with deflection marks centered in these expendable body regions. In these cases, predators are left with only a small part of the prey, while the prey survives another day (e.g. lizard tails, mollusc papillae, arthropod limbs). A remarkable example of attack diversion is found in the cuttlefish, which sometimes ejects a cloud of dark, viscous ink. This ink remains as a discrete unit for some time, resembling the cuttlefish in general size and color; predators often attack this cloud while the cuttlefish pales in color and flees (Holmes 1940; Boycott 1958).

10.5.3 Go ahead, try and eat me: consumption deterrence

Even after a prey has been detected, attacked, and captured, there is still a chance for it to avoid consumption—and a number of antipredator adaptations represent consumption deterrence. Many organisms prevent consumption, and sometimes capture, through active defense, whereby organisms behaviorally interact with the predator using weapons evolved specifically for defensive purposes (e.g. claws, spines), as well as structures more commonly employed during food capture (e.g. teeth/jaws, stingers), intraspecific interactions (e.g. antlers, horns), and a range of other activities (e.g. limbs, hooves) (e.g. Edmunds 1974; Vermeij 1982; Caro 2005). In each of these cases, behavioral actions are taken by prey to reduce the efficiency of a predator's capture, handling, and consumption. Included in this category are the electric shocks produced by some fishes when seized. In some cases, active defense is aided by expulsions or secretions of the prey that are not necessarily noxious, but rather mechanically interfere with consumption. For example, physid snails sometimes deposit egg jelly and eggs on the feeding parts of attacking crayfish. This typically results in the crayfish predator dropping the snail to scrape clean the mouthparts and limbs, while the snail crawls away (DeWitt 1996; DeWitt et al. 2000).

Numerous prey have evolved protective morphologies that reduce the mechanical efficiency of consumption without requiring active deployment. Examples of protective morphologies include unwieldy size and shape, and defensive armor and spines; such defenses are common in both plants and animals (Edmunds 1974; Zaret 1980; Jeffries and Lawton 1984; Myers and Bazely 1991). While many of these morphologies enhance consumption deterrence, some may also affect attack deterrence (e.g. armor or spines can provide signals for aposematic prey) or capture deterrence (e.g. difficult to seize small or large prey).

Another common means of reducing the post-capture probability of consumption is chemical defense. Many organisms possess noxious chemicals, sometimes associated with weapons or protective morphologies, which are emetic, produce bad tastes, foul odors, or painful experiences for the predator. Noxious chemicals can also be used in other modes of predator escape (i.e. attack deterrence, capture deterrence), however they are often employed as a method of last resort. For instance, many organisms have chemical defenses only released after capture or injury. One example of highly toxic prey are newts; some of which can actually be swallowed, their toxins eventually killing the predator, and then safely emerge from the dead predator's mouth (Brodie 1968; Hanifin et al. 1999). Chemical defenses are also very common in plants (e.g. Karban and Baldwin 1997; Agrawal et al. 1999; Tollrian and Harvell 1999). An alternative strategy to the inhibition of consumption is to actually pass safely through the predator's gut; this phenomenon typically involves chemical defenses and has been described in some plants and invertebrates (e.g. Vinyard 1979; Aarnio and Bonsdorff 1997).

10.5.4 Multitasking prey: all-purpose antipredator traits

Some antipredator traits have the potential to influence all chronological stages of predator-prey encounters. For example, predator detection (vigilance) can affect survivorship in a number of ways. Many prey, but not all, have evolved the ability to recognize their predators (reviewed in Caro 2005). Predator recognition is a fundamental prerequisite of vigilance. That is, only prey that possess the capability of distinguishing between predators and nonpredators can be vigilant, or employ activities to see, smell, hear, or chemically or electrically detect predators. Because this definition of vigilance is conceptual in focus (describing actions prey take to detect predators), it is broader than what is often operationally used in vigilance studies (e.g. when measuring vigilance it is often difficult to discern whether prey organisms are listening for predators while foraging). Many organisms utilize environments where predators occasionally occur, and use vigilance to increase the probability of detecting the predator before being detected themselves. Once a prey detects a predator, it can then perform a variety of subsequent actions to avoid or escape predation. For instance, vigilant prey often remain near a refuge so that when a predator is detected, they can quickly seek refuge until the predator threat has passed (all the while going unnoticed by the predator). One striking example of prey vigilance is the ability of many moths to the detect ultrasonic echolocation calls of bat predators—an ability that has apparently evolved at least six separate times and increases survival through early detection (e.g. Treat 1955; Roeder 1962; Hoy et al. 1989; Yack and Fullard 2000; Waters 2003). While vigilance can enhance survivorship, it can also suffer costs, such as lost time or energy that could have been used for other important activities (e.g. Brown 1999; Gauthier-Clerc et al. 1998; Caro 2005).

Many prey use protective habitats to avoid or escape predation. In these cases, prey utilize microhabitats that offer protection from predators in the form of a complex, structured setting, or an otherwise dangerous or impenetrable environment for the predator (e.g. shells of hermit crabs). Protective habitats can enhance prey survivorship in many ways. For instance, many aquatic organisms inhabit complex habitats such as macrophytic vegetation and coral reefs, and some plants and invertebrates live in close association with plants exhibiting high levels of antiherbivore defenses, such as cactus or chemically-defended seaweeds. These environments can hide prey, obscure prey, cause predators to halt their attacks, cause difficulty in negotiating prey capture, and reduce the ability of a predator to successfully consume and digest prey. Many animals temporarily utilize stressful environments that predators have difficulty penetrating, such as low-oxygen or high-temperature aquatic microhabitats. One striking example of the use of such protective habitats is the finding that many fish species that were presumed extinct subsequent to introduction of predatory Nile perch (*Lates niloticus*) in East African lakes have actually persisted in the swampy fringes of the lakes, where structural complexity and hypoxic conditions reduce detection and capture abilities of the predator (Chapman et al. 1996a; Chapman et al. 1996b; Chapman et al. 2002).

Conspecific or multispecies grouping behaviors (e.g. flocking, herding, schooling) can affect survivorship in a number of ways, and have received considerable research (e.g. Hamilton 1971; Vine 1971; Morgan and Godin

1985; reviewed in Caro 2005). In one type of grouping behavior, prey associate themselves with organisms that predators avoid (i.e. the enemy of my enemy is my friend), thus reducing the probability of entering a predator's sensory field. Many prey often aggregate with other organisms within a predator's sensory field, where individual prey within the group reduce their probability of being detected by a predator; although the group itself might be more easily detected than an individual. One advantage of grouping that has been long discussed is the increase in vigilance efficiency, as groups often detect predators earlier, even though each individual may spend less time being vigilant when in groups (e.g. Darwin 1871; Galton 1871; Miller 1922; Pulliam 1973; Kenward 1978; Trehorne and Foster 1980). Efficacy of deimatic behaviors can also sometimes be improved by grouping behaviors (Edmunds 1974; Humphries and Driver 1967). Further, grouping behaviors can reduce the probability of an attack through a dilution effect (e.g. Hamilton 1971; Bertram 1978; Viscido et al. 2001). By grouping, individuals reduce their probability of being attacked by effectively diverting attacks to other members of the group. Grouping can also increase the efficacy of rapid retreat (e.g. Miller 1922; Welty 1934), as well as active defense (e.g. Edmunds 1974; Pulliam and Caraco 1984).

10.6 Reproductive strategies: transcending predators through life history traits

Most antipredator traits increase fitness amidst predation by increasing individual survivorship (all those described above). However, predator-mediated selection can also favor traits that do not necessarily affect individual survivorship, but rather serve to increase offspring survivorship or individual reproductive output (Roff 2002). Increasing any of these three factors (i.e. individual survival, offspring survival, individual fecundity) can enhance fitness in a predatory environment by augmenting lineage proliferation. While many antipredator traits can indirectly increase fecundity by increasing viability—and thus, a prey's ability to produce offspring—there are some traits that actually influence fecundity directly. In this section, I discuss traits that directly influence offspring survival and/or individual fecundity, and refer to these as life history traits. Traits influencing offspring survivorship (i.e. reproductive timing, parental care, offspring size, offspring number) either enhance predator avoidance or escape, while traits only influencing fecundity (i.e. frequency of offspring production, reproductive lifespan) do not.

10.6.1 Know when to hold 'em, know when to fold 'em: reproductive timing

Predator-induced mortality regimes can favor particular reproductive schedules (e.g. early vs. late sexual maturity) that enhance individual fecundity (e.g. Stearns and Crandall 1981; Mangel and Clark 1988; Crowl and Covich 1990; Reznick et al. 1990; Roff 2002). Prey organisms can further enhance fitness in the face of predation by timing reproductive events (e.g. mating, fertilization, pregnancy, birthing) to correspond with time periods where offspring might enjoy relatively high survivorship. Such adjustments in reproductive timing can increase the offspring's probability of either avoiding or escaping predators. For example, prey might produce offspring at a time when the level of crypsis is elevated, when predators exhibit low densities or are absent, when predators are weak or otherwise more vulnerable to prey defenses, or when environmental conditions reduce a predator's prey capturing abilities. To date, few studies have examined shifts in reproductive timing in response to predator-induced mortality, although a number of examples are known for the effects of abiotic sources of mortality (e.g. desiccation, reduced energy sources due to winter; Thomson 1950; Lack 1954; Newman 1988; Brinkhof et al. 1993; Van Noordwijk et al. 1995; Reznick et al. 2006b). Reproductive timing can also influence individual viability by minimizing negative effects of offspring production (e.g. pregnancy, egg production) on capture deterrence. That is, prey are often more easily captured during reproductive events, and thus the timing of these events can be selected to avoid major individual fitness costs.

10.6.2 Putting all your eggs in one basket and flooding the market: reproductive effort

The reproductive effort of an organism describes the allocation of energy towards reproduction during its lifespan. This includes several important life history attributes that can be influenced by predator-induced mortality regimes. Because many prey populations experience unsaturated environments (i.e. few limiting resources) under high predation levels, they are generally believed to be r-selected (i.e. density-independent selection maximizing per capita population growth rate) in the classic r-K continuum of life history theory (MacArthur 1962; MacArthur and Wilson 1967). Thus, prey species often exhibit reproductive strategies reflecting a maximization of lineage growth rate. Here, I discuss how reproductive effort, and components thereof, might be shaped by predation.

To maximize lineage growth, life history theory generally predicts the evolution (or induction) of increased reproductive effort under high predation pressure (Kozlowski and Uchmanski 1987; Abrams and Rowe 1996; Roff 2002). Parental care can be thought of as synonymous with reproductive effort as defined here, and includes all supply of energy towards reproduction. Many prey organisms employ high levels of parental care to enhance fecundity amidst predators. For example, prey can protect and provide nutrients for embryos and juveniles through nest site selection, nest morphology, active defense of offspring, and production of yolk, endosperm, and a placenta. Parental care of juveniles increases offspring survivorship by enhancing predator avoidance and escape (e.g. hiding offspring from predators, diverting predator attention from offspring, active defense of offspring from predators).

While predation may generally favor increased reproductive effort, the manner in which this effort is allocated can vary. For instance, an individual can produce one large offspring or several small offspring with equivalent levels of reproductive effort. Let us now consider how predation might influence offspring number and size.

The number of offspring per reproductive bout, or clutch size, can be selected to maximize lineage growth in response to predation. Such an optimization of clutch size can increase fecundity without altering offspring survival. In this scenario, offspring survivorship can be held constant, and the clutch size leading to the greatest lineage growth rate will be selected. Clutch size can also enhance predator escape ability for offspring. First, a low clutch size might ensure little competition among progeny and thus result in good health (which might often increase escape ability). Second, a high clutch size can provide an indirect form of attack deterrence by flooding the environment with offspring. In this situation, higher numbers of juveniles can result in a lower per capita probability of an attack. A striking example of this phenomenon is reproductive synchrony (or emergence/metamorphosis synchrony), where high clutch sizes may be combined with a developmental timing strategy to produce very high densities of juveniles, effectively reducing individual attack probabilities, and possibly gaining other advantages of grouping (e.g. mast seeding in oaks, periodical cicadas; Darling 1938; Lloyd and Dybas 1966; Hamilton 1971; Janzen 1981; Gochfeld 1982; Ims 1990; DeVito et al. 1998).

Offspring size—including embryo size, egg size, birth size, emergence size, hatching size, seed size—can influence both predator avoidance and escape. Thus, many prey organisms have evolved offspring sizes that enhance their survival in the presence of predators (e.g. Lloyd 1987; Janzen et al. 2000; Moran and Emlet 2001). For example, small offspring are often difficult to detect, while large offspring are often difficult to capture or consume. Offspring size is often correlated with offspring number, and thus will often evolve in a coordinated fashion (see discussion of trait correlations below).

Two additional components of reproductive effort that might often reflect the selective influence of predation are reproductive lifespan and frequency of offspring production (Roff 2002). These two traits can serve to increase individual fecundity in the face of predation. For example, prey organisms might exhibit a long reproductive lifespan (i.e. early age at maturity, late onset of senescence) and/or frequent production of offspring. These strategies can maximize fecundity in an environment with high levels of mortality. For example, Reznick et al. (2006a) found that guppies (*Poecilia reticulata*) that evolved with predators exhibit a longer reproductive lifespan than guppies that evolved in the absence of major fish predators. Further, Hubbs (1996) demonstrated that mosquitofish species (genus *Gambusia*) inhabiting relatively low-predation environments exhibit longer interbrood intervals. While strategies such as exhibiting many, frequent bouts of reproduction can sometimes decrease individual survivorship (e.g. mating, pregnancy, and parental care might increase vulnerability to predation), these strategies may still be favored as they can produce a net increase in fitness.

10.7 Predators spawn phenotypic diversity of prey: plasticity, divergence, and speciation

Effects of predators on prey phenotypes can be complex. Phenotypes generally reflect the influence of multiple selective agents in addition to predators. Predation can influence phenotypic values both directly, through predator-mediated selection, and also indirectly through trait correlations and interactions with other selective agents (e.g. Gould and Lewontin 1979; Lande and Arnold 1983; Sih 1987; Koehl 1996; Pigliucci and Preston 2004). Most prey exhibit numerous types of antipredator traits, and correlations among these and other traits are ubiquitous for several reasons (e.g. physiological, architectonical, functional, or developmental causes). Thus, effects of predators on one trait can indirectly influence other traits which are not under selection via predation. In natural systems, predators might also often affect the selective regime of prey by altering their interactions with other selective agents, such as altering levels of competition or densities of other predators. Because of this potential for complex networks of direct and indirect effects of environmental factors (e.g. predation) on traits, researchers must employ a pluralistic approach assessing both direct and indirect effects of multiple environmental factors on multiple phenotypes to gain a thorough understanding of how predation influences prey evolution (DeWitt and Langerhans, 2003; Schaack and Chapman, 2003; Caumul and Polly, 2005; Hoverman et al. 2005).

Many traits exhibit tradeoffs, where a given trait value increases fitness in one respect, but decreases fitness in another. Such tradeoffs are common among antipredator traits and non-defensive traits alike. For example, reduced activity level might reduce foraging opportunities, morphological and chemical defenses can be energetically costly and reduce fecundity, capability for high-speed movement might reduce sustained locomotion capabilities, grouping behaviors might lead to increased competition, defenses against one predator might increase vulnerability to another, and sexually selected ornamental traits can increase susceptibility to predation (e.g. Sih 1987; Lima and Dill 1990; Andersson 1994; McCollum and Van Buskirk 1996; Rigby and Jokela 2000; Reimchen and Nosil 2002; Vamosi 2002; DeWitt and Langerhans 2003; Ghalambor et al. 2004; Caro 2005; Langerhans et al. 2005).

Predators can drive phenotypic differences between prey populations via several different mechanisms, and tradeoffs are not prerequisites for divergence. Three common ways that predators can drive divergence of prey are: fitness tradeoffs in prey traits across different predator regimes (Fig. 4A), competition for "enemy-free space" (Fig. 4B), and predators altering interactions of prey with other selective agents (Fig. 4C). The first case stems from tradeoffs, while the latter two mechanisms of divergence need not involve such tradeoffs. The common thread among all modes of divergence is that divergent selection—selection pulling trait means of different populations toward different adaptive peaks—is responsible for phenotypic differences in each case (Fig. 4D). Phenotypic differentiation can result from environmentally contingent (i.e. phenotypic plasticity) or environmentally independent phenotype production.

10.7.1 To induce or not to induce: tradeoffs can drive predator-induced plasticity

Many prey organisms have evolved adaptive predator-induced phenotypic plasticity, where particular phenotypes are only produced under the threat of predation, thus avoiding fitness costs in the absence of particular predators. Other organisms (and sometimes other populations of the same species) have instead evolved fixed phenotypes, and exhibit constitutive defenses against predators. Whether plasticity or fixed phenotype production evolves largely depends on the spatial and temporal variability of predation, and the ability to predict future predation levels. Plasticity will typically be favored in a fluctuating environment where the environmental state can be accurately predicted using environmental cues, while fixed phenotype production will typically be favored when the predator regime is constant and costs associated with plasticity and information acquisition are relatively high (e.g. Bradshaw 1965; Levins 1968; Scheiner 1993; Gotthard and Nylin 1995; DeWitt and Langerhans 2004).

The importance of predators in adaptive plasticity of prey has received considerable attention during the past two decades. Predator-induced defenses in animals and herbivore-induced defenses in plants are very common (reviewed in Havel 1987; Karban and Baldwin 1997; Chivers and Smith 1998; Agrawal et al. 1999; Tollrian and Harvell 1999). An obvious form of predator-induced plasticity is the widespread occurrence of predator-induced antipredator behaviors in animals (e.g. Sih 1987; Lima and Dill 1990; Werner and Anholt 1993). Induced morphologies, life histories, and chemical defenses also now appear common in many taxa (e.g. Dodson 1989; Karban and Baldwin 1997; Agrawal et al. 1999; Tollrian and Harvell 1999; Pigliucci 2001; DeWitt and Scheiner 2004). For example, some animals induce defensive morphologies such as unwieldy body shapes or spines in the presence of predators, some animals induce egg hatching in response to a nearby predatory attack, and many plants induce chemical defenses when herbivores initiate an attack. Such plastic responses to predation represent important and widespread evolutionary consequences of predation, however plasticity's influence on diversification rates is largely unknown. The role of plasticity as a catalyst for subsequent evolutionary divergence and speciation, and its role as an inhibitor of extinction, is now receiving considerable attention (West-Eberhard 1989; Schlichting and Pigliucci 1998; Pigliucci 2001; Robinson and Parsons 2002; Pigliucci and Murren 2003; Price et al. 2003; West-Eberhard 2003; Schlichting 2004).

10.7.2 Divergent selection between predator regimes: evolutionary divergence among prey

Performance tradeoffs can also drive evolutionary divergence among prey populations, rather than favor phenotypic plasticity. Because predation is heterogeneously distributed across space and time, and because many traits exhibit tradeoffs across predatory environments, divergent selection on prey traits across predator regimes may be very common in nature.

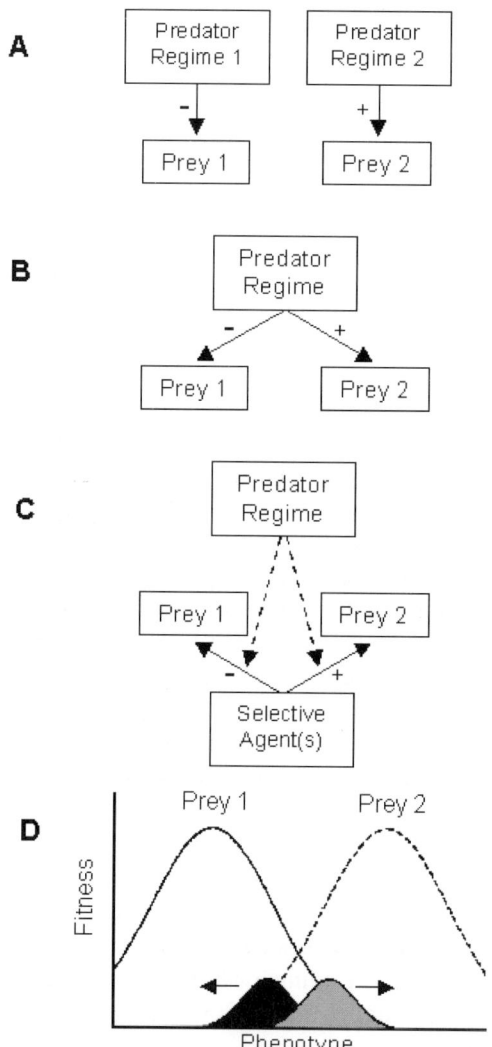

Fig. 4. Three common ways in which predators can drive phenotypic divergence among prey populations (or species). Solid arrows in A-C depict selection on a prey trait. The sign beside each arrow indicates the direction of the optimal phenotype in relation to the population mean. Divergent selection arising from (A) divergent predator regimes, (B) competition for enemy free space within a given predator regime, and (C) an interaction between predation and the prey's selective regime. (D) Hypothetical fitness functions resulting from each scenario depicted in A-C, with trait distributions for the two prey populations represented by the shaded areas. Arrows in D illustrate the direction selection is pulling trait means for each population

Empirical support for this proposition is growing (e.g. Reimchen 1994; McPeek 1995; McPeek et al. 1996; Reznick 1996; Reznick et al. 1997; Walker 1997; Conover and Munch 2002; Relyea 2002; Vamosi 2002; Vamosi and Schluter 2002; Langerhans et al. 2004; Reimchen and Nosil 2004; Vamosi and Schluter 2004; Langerhans et al. 2005).

One tradeoff that might be common in many organisms is the conflict between rapid bursts of movement (often important in capture deterrence) and endurance (often important for other essential activities, such as foraging) (e.g. Dohm et al. 1996; Reidy et al. 2000; Vanhooydonck et al. 2001; Wilson et al. 2002; Domenici 2003; Blake 2004). Aquatic organisms with coupled locomotor systems (i.e. using the same structures for more than one swimming mode), are predicted to generally experience a tradeoff between fast-start swimming performance and prolonged swimming performance, as these tasks require opposite suites of morphological and physiological traits. Specifically, a shallow anterior region and a deep caudal region are required to produce rapid fast-starts—a very important escape mechanism in most fishes and anurans (see Fig. 3)—however a deep anterior region and a shallow caudal region are required to optimize prolonged swimming performance—important for acquiring resources and mates (e.g. Blake 1983; Webb 1984; Webb 1986; Videler 1993; Vogel 1994; Walker 1997; Blake 2004). While the relationship between morphology and swimming performance can be complicated, this general tradeoff has been empirically confirmed when comparing across distantly related, morphologically disparate taxa (e.g. "accelerators" vs. "cruisers"), as well as within species (Reidy et al. 2000; Fig. 5). This tradeoff is predicted to generate divergent selection across low- and high-predation environments for prey populations. Recent empirical work supports this hypothesis, as a number of aquatic organisms have been found to exhibit the predicted morphological differences between populations (and species) inhabiting such divergent predator regimes (Walker 1997; Walker and Bell 2000; Langerhans and DeWitt 2004; Langerhans et al. 2004; Dayton et al. 2005).

In many cases, phenotypic divergence between predator regimes involves a suite of traits, rather than simply one form of capture deterrence. For example, many studies have investigated phenotypic differences between predatory environments (e.g. low vs. high levels of predation from predatory fish) in livebearing fish, particularly the Trinidadian guppy.

This work has revealed that divergent selection across predator regimes can drive the simultaneous divergence of a large number of traits, and livebearing fishes have become a model system to investigate predator-driven evolution.

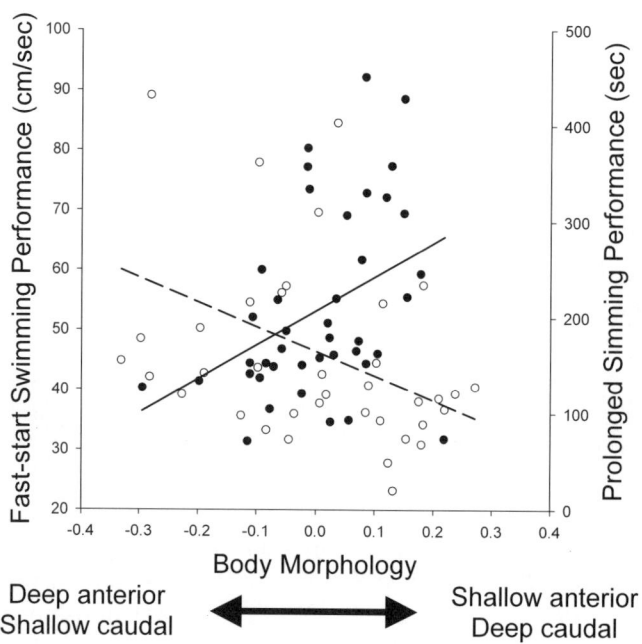

Fig. 5. Swimming performance tradeoff for body shape in male western mosquitofish (*Gambusia affinis*). Body shape represented by a canonical axis describing fish with a relatively deep anterior/head region and shallow caudal peduncle region at the negative end of the axis, and fish with an opposite morphology at the positive end. Solid line and filled symbols: fast-start swimming performance ($P = 0.006$); dashed line and open symbols: prolonged swimming performance ($P = 0.007$). Data from Langerhans et al. (2004) and RB Langerhans, MC Belk, and TJ DeWitt (unpublished data)

Guppies exhibit striking differences between predator regimes in body color, body shape, swimming performance, many types of behaviors, and many life history parameters (e.g. Seghers 1974; Farr 1975; Endler 1995; Magurran et al. 1995; Reznick 1996; Houde 1997; Reznick et al. 1997). Regarding only life history characters, guppies from high-predation populations are known to mature earlier and at a smaller size, produce more and smaller offspring per litter, reproduce more often, produce higher reproductive allotments, and exhibit longer reproductive lifespans than guppies from low-predation populations (Reznick and Endler 1982; Reznick et al. 1996; Reznick et al. 2006). Another livebearing fish, the Bahamas mosquitofish (*Gambusia hubbsi*), also exhibits striking phenotypic differences between predator regimes, including divergence in habitat use (Fig. 6A), body color (Fig. 6B), several life history parameters (Krumholz 1963; Sohn 1977; Downhower et al. 2000), male genital size (Langerhans et al. 2005), body shape (R.B. Langerhans unpublished data), and swimming performance (R.B. Langerhans unpublished data).

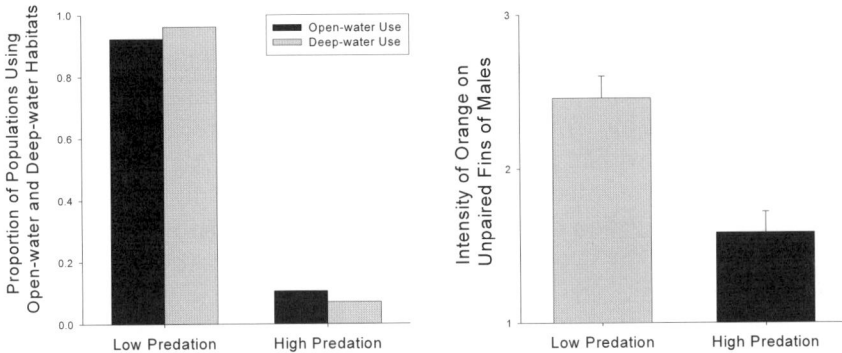

Fig. 6. Divergence in (A) habitat use and (B) body color among low- and high-predation environments in Bahamas mosquitofish (*Gambusia hubbsi*). Prey populations avoid dangerous habitat types (chi-square, $P < 0.0001$) and exhibit less conspicuous coloration (ANOVA, $P < 0.0001$) in high-predation populations (N = 55 populations; 25 low-predation, 30 high-predation). Habitat use was easily assessed using underwater visual survey, and orange intensity was assessed by ranking individuals from 1 to 3 (1: low, 2: medium, 3: high). Data from RB Langerhans (unpublished data)

Other well-studied examples of phenotypic divergence between predator regimes for closely related populations or species includes threespine sticklebacks (*Gasterosteus* spp.) and *Enallagma* damselflies. For instance, sticklebacks exhibit greater levels of defensive armor and schooling behaviors under relatively high predation intensities (e.g. Vamosi 2002; Doucette et al. 2004; Reimchen and Nosil 2004; Vamosi and Schluter 2004), and *Enallagma* damselflies exhibit larger caudal lamellae and greater arginine kinase activity (enhancing rapid retreat) with predatory dragonflies (e.g. McPeek et al. 1996; McPeek 1997; McPeek 1999). In many cases, the presumed tradeoffs have been tested and the adaptive significance of the trait shifts have been identified.

10.7.3 Divergent selection within predator regimes: the search for enemy-free space

Not only can divergent selection arise between alternative predator regimes, but predators can also drive divergent selection on prey traits within predator regimes (Fig. 4B). This is traditionally believed to stem from apparent competition—or competition for enemy-free space—which occurs when one species negatively impacts the density of another species, not through consumption of shared resources, but rather through a positive effect on the density of a shared predator. However, apparent competition—or even sympatry—is not required for predators to generate prey divergence within particular predator regimes (Abrams 2000). To date, this mode of predator-driven divergence has primarily received theoretical investigation (e.g. Fryer 1959; Holt 1977; Brown and Vincent 1992; Abrams 2000; Doebeli and Dieckmann 2000; Abrams and Chen 2002; Bowers et al. 2003), with empirical studies slow to venture into this field (e.g. Askew 1961; Clarke 1962; Owen 1963; Gilbert 1975; Ricklefs and O'Rourke 1975; Bond and Kamil 2002).

One way that trait divergence can be favored by predator-mediated selection within predator regimes is if predators are most efficient at detecting and consuming particular prey phenotypes, and deviations from these phenotypes in any direction would enhance prey fitness. For instance, if two prey populations differ in their average phenotype, with one slightly less and one slightly more than that most vulnerable to predation, then selection will drive the populations in opposite directions of phenotype space (Holt 1977). That is, there are multiple ways prey can reduce predation risk, but intermediate trait values exhibit low fitness. This type of divergence can further occur even when there is only one initial prey population, if predators are also allowed to evolve and prey additionally compete for limited resources (Brown and Vincent 1992). However, trait divergence is not the only possible outcome, as prey populations might also respond to predation in parallel manners (Abrams 2000). Although this scenario is typically discussed in a sympatric framework, trait divergence can actually occur via such a mechanism among either sympatric or allopatric prey populations. The particular adaptive peak a prey population might traverse can also be influenced by genetic drift or genetic constraints; in these cases, divergence can result from an interaction between predator-mediated natural selection and other evolutionary factors.

Schluter (2000) reviewed possible examples where divergent selection within predator regimes has resulted in prey divergence; examples include diversification of aspect diversity in moths (Ricklefs and O'Rourke 1975), leaf shape in passion flower vine species (*Passiflora* spp.) attacked by egg-laying *Heliconius* butterflies (Gilbert 1975), and gall shapes in oak gall wasps attacked by parasitoids (Askew 1961). A potentially common scenario in nature that might produce divergent selection within predator regimes is when multiple types of background environments exist, and selection favors different means of crypticity in prey to avoid detection (e.g. Bond and Kamil 2002; Nosil 2004; Bond and Kamil 2006). Importantly, such divergent selection within predator regimes (i.e. selection caused by the same predatory agent) need not result from apparent competition, as predator density can remain constant and selection coefficients can also be unaffected by predator density.

10.7.4 Divergent selection involving other selective agent(s): predation as a context shift

Rather than acting as a direct selective force, predation can alter the environmental context for prey interactions, thus changing their selective regime (Fig. 4C; Buckling and Rainey 2002; Doucette et al. 2004; Eklöv and Svanbäck 2006; Steets et al. 2006). This arises when predators cause changes in prey traits or densities, or attributes of other community members, that result in changes in selection (magnitude and/or nature) experienced by prey. For example, predators often cause behavioral shifts in prey (Kotler and Brown 1988; Lima and Dill 1990; Peacor and Werner 2001; Lingle 2002; Fig. 6A) which might alter a prey's selective environment, possibly strengthening divergent selection already in place. Interestingly, predators need not even consume any individuals to drive phenotypic divergence in prey under this scenario. Such non-lethal effects of predators represent an important, growing field of study in ecology (see chapter 17), however, this topic is only now beginning to receive considerable attention by evolutionists. So far, it is unknown whether predators might weaken divergent selection among prey populations as often as they strengthen it.

One type of prey interaction that predation might often influence is resource competition. Predation appears to play an important role in the divergence of benthic and limnetic species pairs of threespine stickleback in British Columbia (e.g. Vamosi 2002; Vamosi and Schluter 2004). In a pond experiment, divergent selection among benthic and limnetic stickleback was strengthened under high predation pressure from aquatic insects and cutthroat trout (*Oncorhynchus clarkii*), even though the level of resource competition actually decreased (Rundle et al. 2003). In this case, it seems that predators increased resource partitioning among divergent prey phenotypes, possibly by inducing changes in habitat use or foraging behavior, and caused competition to decline at a greater rate with phenotypic distance compared to an environment lacking predators. Interestingly, benthic and limnetic pairs of stickleback have apparently evolved only in lakes with cutthroat trout and no other fish species present. Stickleback have not evolved such phenotypic divergence within any of 16 nearby candidate lakes, which have an average of approximately three other fish species, and all but one have more potential predators and competitors than cutthroat trout (Vamosi 2003).

10.7.5 Predation as a driver of speciation: eating individuals, spitting out species

Until now, we have only discussed predation's influence on phenotype divergence within lineages, however predation can also affect the splitting of lineages (i.e. speciation), the elimination of lineages (i.e. extinction), and overall diversification rates (speciation minus extinction). Both theoretical and empirical work strongly confirm that divergent selection can lead to speciation (i.e. reproductive isolation between populations) as a byproduct of ecological adaptation (e.g. Mayr 1942; Dobzhansky 1951; Rice and Hostert 1993; Funk 1998; Rundle et al. 2000; Schluter 2001; Coyne and Orr 2004; Nosil et al. 2005; Rundle and Nosil 2005; Funk et al. 2006). For example, mate choice might involve traits under divergent selection, or selection against hybrids might favor assortative mating. Further, predator-mediated selection against migrants from alternative predator regimes can result in reproductive isolation. Thus, any of the mechanisms described above could result in reproductive isolation between divergent prey populations. But, has predation actually been an important driver of speciation? Given the apparent ubiquity of predation's influence on phenotype divergence, it seems reasonable to expect that it might often contribute to the process of speciation. Unfortunately, investigation into predation's role in the formation of species has received little attention to date (see Fig. 1). However, recent work is beginning to shed some light onto this question (e.g. McPeek and Wellborn 1998; Stoks et al. 2005; Vamosi 2005).

There is now accumulating evidence suggesting predation is a significant driver of speciation in *Heliconius* butterflies (e.g. McMillan et al. 1997; Mallet et al. 1998; Jiggins et al. 2001; Naisbit et al. 2003; Jiggins et al. 2004). Further, predation is strongly implicated in divergence and reproductive isolation between color pattern morphs of *Timema* walkingstick insects (e.g. Sandoval 1994; Crespi and Sandoval 2000; Nosil et al.

2002; Nosil et al. 2003; Nosil 2004; Nosil and Crespi 2006). In this case, predator-mediated divergent selection (by avian predators) across background environments (host plants) has apparently resulted in cryptic color pattern divergence and partial reproductive isolation between diverging prey populations. Ongoing work is also uncovering the importance of predation on reproductive isolation among populations of livebearing fish species inhabiting divergent predator regimes (RB Langerhans unpublished data).

Additionally, several general models of speciation might involve predation, although they may not typically focus on predation per se. For example, habitat selection (e.g. host-plant preference, oviposition site selection) can often be influenced by predator regime, and facilitate speciation by enhancing assortative mating. Further, many models of speciation involving frequency-dependent selection, or competition for resources, can be interpreted as consequences of predator-prey interactions (e.g. apparent competition).

10.8 Conclusions and future directions

Predation's general importance in the evolution of prey phenotypes is without question. In this chapter, I described a framework for understanding antipredator traits based on the manner in which traits influence prey fitness and the chronological sequence of predatory encounters. The framework illustrates the ubiquity of predation's influence on prey traits. However, many questions remain concerning the detailed nature of predation's role in generating the diversity of phenotypes and species we see today. This chapter highlights the need for further investigation into how reproductive strategies might reflect antipredator adaptations, and how predation might influence trait divergence and speciation. An important component of future research should involve a pluralistic approach to studies of phenotype evolution, where multiple selective agents and multiple phenotypes are examined simultaneously. While research into the significance of predation in producing evolutionary divergence of prey species experienced a slow start through most of the twentieth century, recent and ongoing research suggests the field is now very rapidly growing. Our understanding of the varied ways that predators might influence the evolution of prey phenotypes and reproductive isolation between prey populations should be greatly improved by the end of the current century.

10.9 Acknowledgements

I thank A. Elewa for the opportunity to contribute to this book. The chapter was improved by comments from David Reznick and an anonymous reviewer. The author was supported by Washington University and an Environmental Protection Agency Science to Achieve Results fellowship.

References

Aarnio K, Bonsdorff E (1997) Passing the gut of juvenile flounder, *Platichthys flesus*: differential survival of zoobenthic prey species. Marine Biology 129 (1): 11-14

Abrams PA (2000) Character shifts of prey species that share predators. American Naturalist 156 (4, Supplement): S45-S61

Abrams PA, Chen X (2002) The evolution of traits affecting resource acquisition and predator vulnerability: Character displacement under real and apparent competition. American Naturalist 160 (5): 692-704

Abrams PA, Rowe L (1996) The effects of predation on the age and size of maturity of prey. Evolution 50(3): 1052-1061

Acharya L, Fenton MB (1992) Echolocation behaviour of vespertillionid bats (*Lasiurus cinereus* and *Lasiurus borealis*) attacking airborne targets including arctiid moths. Canadian Journal of Zoology 70: 1292-1298

Addicott JF (1974) Predation and prey community structure: an experimental study of the effect of mosquito larvae on the protozoan communities of pitcher plants. Ecology 55: 475-492

Agrawal AA, Tuzun S, Bent E (1999) Induced plant defenses against pathogens and herbivores: biochemistry, ecology, and agriculture. American Phytopathological Society, St. Paul, MN

Almany GR, Webster MS (2004) Odd species out as predators reduce diversity of coral-reef fishes. Ecology 85 (11): 2933-2937

Andersson M (1994) Sexual selection. Princeton University Press, Princeton, NJ

Askew RR (1961) On the biology of the inhabitants of oak galls of Cynipidae (Hymenoptera) in Britain. Transactions of the Society for British Entomology 14: 237-268

Bates HW (1862) Contributions to an insect fauna of the Amazon valley. Lepidoptera: Heliconidae. Transactions of the Linnean Society of London 23:495-566

Beddard FE (1892) Animal coloration: an account of the principal facts and theories related to the colours and markings of animals. Swan Sonnenschein & Co., London

Bertram BC (1978) Predators and prey. In Krebs JR, Davies NB (eds) Behavioral ecology: an evolutionary approach. Blackwell, Oxford 64-96

Blackledge TA, Coddington JA, Gillespie RG (2003) Are three-dimensional spider webs defensive adaptations? Ecology Letters 6 (1): 13-18
Blake RW (1983) Fish locomotion. Cambridge University Press, Cambridge
Blake RW (2004) Fish functional design and swimming performance. Journal of Fish Biology 65 (5): 1193-1222
Bond AB, Kamil AC (2002) Visual predators select for crypticity and polymorphism in virtual prey. Nature 415 (6872): 609-613
Bond AB, Kamil AC (2006) Spatial heterogeneity, predator cognition, and the evolution of color polymorphism in virtual prey. Proceedings of the National Academy of Sciences of the United States of America 103 (9): 3214-3219
Bowers DM (1993) Aposematic caterpillars: life-styles of the warningly colored and unpalatable. In Stamp NE, Casey TM (eds) Caterpillars: ecological and evolutionary constraints on foraging. Chapman & Hall, New York 331–371
Bowers RG, White A, Boots M, Geritz SAH, Kisdi E (2003) Evolutionary branching/speciation: contrasting results from systems with explicit or emergent carrying capacities. Evolutionary Ecology Research 5 (6): 883-891
Boycott BB (1958) The cuttlefish - *Sepia*. New Biology 25: 98-118
Bradshaw AD (1965) Evolutionary significance of phenotypic plasticity in plants. Advances in Genetics 13: 115-155
Brakefield PM, Shreeve TM, Thomas JM (1992) Avoidance, concealment, and defence. In Dennis RLH (ed) The ecology of butterflies in Britain. Oxford University Press, Oxford 93-119
Brinkhof MWG, Cave AJ, Hage FJ, Verhulst S (1993) Timing of reproduction and fledging success in the coot *Fulica atra*: evidence for a causal relationship. Journal of Animal Ecology 62 (3): 577-587
Brodie ED, Jr. (1968) Investigations on the skin toxin of the adult roughskinned newt, *Taricha granulosa*. Copeia: 307-313
Brooke MD (1998) Ecological factors influencing the occurrence of 'flash marks' in wading birds. Functional Ecology 12 (3): 339-346
Brown JS (1999) Vigilance, patch use and habitat selection: foraging under predation risk. Evolutionary Ecology Research 1 (1): 49-71
Brown JS, Vincent TL (1992) Organization of predator-prey communities as an evolutionary game. Evolution 46 (5): 1269-1283
Buckling A, Rainey PB (2002) The role of parasites in sympatric and allopatric host diversification. Nature 420 (6915): 496-499
Cain AJ, Sheppard PM (1954) Natural selection in *Cepaea*. Genetics 39: 89-116
Caldwell JP (1982) Disruptive selection: a tail color polymorphism in *Acris* tadpoles in response to differential predation. Canadian Journal of Zoology 60: 2818-2827
Caro TM (2005) Antipredator defenses in birds and mammals. University of Chicago Press, Chicago
Caumul R, Polly PD (2005) Phylogenetic and environmental components of morphological variation: skull, mandible, and molar shape in marmots (*Marmota*, Rodentia). Evolution 59 (11): 2460-2472
Chapman LJ, Chapman CA, Chandler M (1996a) Wetland ecotones as refugia for endangered fishes. Biological Conservation 78 (3): 263-270

Chapman LJ, Chapman CA, Nordlie FG, Rosenberger AE (2002) Physiological refugia: swamps, hypoxia tolerance and maintenance of fish diversity in the Lake Victoria region. Comparative Biochemistry and Physiology A 133 (3): 421-437

Chapman LJ, Chapman CA, Ogutu-Ohwayo R, Chandler M, Kaufman L, Keiter AE (1996b) Refugia for endangered fishes from an introduced predator in Lake Nabugabo, Uganda. Conservation Biology 10 (2): 554-561

Chivers DP, Kiesecker JM, Marco A, DeVito J, Anderson MT, Blaustein AR (2001) Predator-induced life history changes in amphibians: egg predation induces hatching. Oikos 92 (1): 135-142

Chivers DP, Smith RJF (1998) Chemical alarm signalling in aquatic predator-prey systems: a review and prospectus. Ecoscience 5 (3): 338-352

Clarke B (1962) Balanced polymorphism and the diversity of sympatric species. Systematics Association Publications, London

Conover DO, Munch SB (2002) Sustaining fisheries yields over evolutionary time scales. Science 297 (5578): 94-96

Cott HB (1940) Adaptive coloration in animals. Methuen, London

Coyne JA, Orr HA (2004) Speciation. Sinauer Associates, Sunderland, MA

Crespi BJ, Sandoval CP (2000) Phylogenetic evidence for the evolution of ecological specialization in *Timema* walking-sticks. Journal of Evolutionary Biology 13 (2): 249-262

Crowl TA, Covich AP (1990) Predator-induced life-history shifts in a freshwater snail. Science 247 (4945): 949-950

Darling FF (1938) Bird flocks and breeding cycles. Cambridge University Press, Cambridge

Darwin C (1871) The descent of man and selection in relation to sex. John Murray, London

Dayton GH, Saenz D, Baum KA, Langerhans RB, DeWitt TJ (2005) Body shape, burst speed and escape behavior of larval anurans. Oikos 111 (3): 582-591

de Ruiter PC, Wolters V, Moore JC (2005) Dynamic food webs: multispecies assemblages, ecosystem development and environmental change. Academic Press, Burlington, MA

DeVito J, Chivers DP, Kiesecker JM, Marco A, Wildy EL, Blaustein AR (1998) The effects of snake predation on metamorphosis of western toads, *Bufo boreas* (Amphibia, Bufonidae). Ethology 104 (3): 185-193

DeWitt TJ (1996) Functional tradeoffs and phenotypic plasticity in the freshwater snail *Physa*. Binghamton University, Binghamton, New York

DeWitt TJ, Langerhans RB (2003) Multiple prey traits, multiple predators: keys to understanding complex community dynamics. Journal of Sea Research 49 (2): 143-155

DeWitt TJ, Langerhans RB (2004) Integrated solutions to environmental heterogeneity: theory of multimoment reaction norms. In DeWitt TJ, Scheiner SM (eds) Phenotypic plasticity. Functional and conceptual approaches. Oxford University Press, New York 98-111

DeWitt TJ, Robinson BW, Wilson DS (2000) Functional diversity among predators of a freshwater snail imposes an adaptive trade-off for shell morphology. Evolutionary Ecology Research 2 (2): 129-148

DeWitt TJ, Scheiner SM (2004) Phenotypic plasticity. Functional and conceptual approaches. Oxford University Press, New York

DeWitt TJ, Sih A, Hucko JA (1999) Trait compensation and cospecialization in a freshwater snail: size, shape and antipredator behavior. Animal Behaviour 58: 397-407

Dice LR (1947) Effectiveness of selection by owls of deer mice (*Peromyscus maniculatus*) which contrast in colour with their background. Contributions from the Laboratory of Vertebrate Biology, University of Michigan 34: 1-20

Dobzhansky T (1951) Genetics and the origin of species. Columbia University Press

Dodson S (1989) Predator-induced reaction norms. Bioscience 39 (7): 447-452

Doebeli M, Dieckmann U (2000) Evolutionary branching and sympatric speciation caused by different types of ecological interactions. American Naturalist 156(4, Supplement): S77-S101

Dohm MR, Hayes JP, Garland T (1996) Quantitative genetics of sprint running speed and swimming endurance in laboratory house mice (*Mus domesticus*). Evolution 50 (4): 1688-1701

Domenici P (2003) Habitat, body design and the swimming performance of fish. In Bels VL, Gasc J-P Casinos A (eds) Vertebrate biomechanics and evolution. BIOS Scientific Publishers Ltd, Oxford 137-160

Domenici P, Blake RW (1997) The kinematics and performance of fish fast-start swimming. Journal of Experimental Biology 200 (8): 1165-1178

Doucette LI, Skulason S, Snorrason SS (2004) Risk of predation as a promoting factor of species divergence in threespine sticklebacks (*Gasterosteus aculeatus* L.). Biological Journal of the Linnean Society 82 (2): 189-203

Downhower JF, Brown LP, Matsui ML (2000) Life history variation in female *Gambusia hubbsi*. Environmental Biology of Fishes 59 (4): 415-428

Driver PM, Humphries DA (1988) Protean behavior: the biology of unpredictability. Oxford University Press, Oxford

Edmunds M (1974) Defence in animals: a survey of anti-predator defences. Longman, New York

Ehrlich PR, Raven PH (1964) Butterflies and plants: a study in coevolution. Evolution 18: 586-608

Eisner T, Meinwald J (1966) Defensive secretions of arthropods. Science 153: 1341-1350

Eklöv P, Svanbäck R (2006) Predation risk influences adaptive morphological variation in fish populations. American Naturalist 167 (3): 440-452

Elton CS (1927) Animal ecology. Sidgwick and Jackson, London

Endler JA (1978) A predator's view of animal color patterns. Evolutionary Biology 11: 319-364

Endler JA (1991) Interactions between predators and prey. In Krebs JR, Davies NB (eds) Behavioural ecology: an evolutionary approach. Blackwell Scientific Publications, Oxford 169-196

Endler JA (1995) Multiple-trait coevolution and environmental gradients in guppies. Trends in Ecology & Evolution 10 (1): 22-29

Ewer RF (1966) Juvenile behaviour of the African ground squirrel, *Xerus erythropus* (E. Geoff.). Zeitschrift für Tierpsychologie 23: 190-216

Farr JA (1975) Role of predation in evolution of social behavior of natural populations of the guppy, *Poecilia reticulata* (Pisces: Poeciliidae). Evolution 29 (1): 151-158

Fryer G (1959) Some aspects of evolution in Lake Nyasa. Evolution 13: 440-451

Funk DJ (1998) Isolating a role for natural selection in speciation: host adaptation and sexual isolation in *Neochlamisus bebbianae* leaf beetles. Evolution 52 (6): 1744-1759

Funk DJ, Nosil P, Etges WJ (2006) Ecological divergence exhibits consistently positive associations with reproductive isolation across disparate taxa. Proceedings of the National Academy of Sciences of the United States of America 103 (9): 3209-3213

Galton F (1871) Gregariousness in cattle and men. MacMillan's Magazine 23: 353-357

Gauthier-Clerc M, Tamisier A, Cezilly F (1998) Sleep-vigilance trade-off in green-winged teals (*Anas crecca crecca*). Canadian Journal of Zoology 76 (12): 2214-2218

Ghalambor CK, Reznick DN, Walker JA (2004) Constraints on adaptive evolution: the functional trade-off between reproduction and fast-start swimming performance in the Trinidadian guppy (Poecilia reticulata). American Naturalist 164 (1): 38-50

Gilbert LE (1975) Ecological consequences of a coevolved mutualism between butterflies and plants. In Gilbert LE, Raven PH (eds) Coevolution of animals and plants. University of Texas Press, Austin 210–240

Gochfeld M (1982) Reproductive synchrony and predator satiation: an analogy between the Darling effect in birds and mast fruiting in plants. Auk 99: 586-587

Gotthard K, Nylin S (1995) Adaptive plasticity and plasticity as an adaptation: a selective review of plasticity in animal morphology and life-history. Oikos 74 (1): 3-17

Gould SJ, Lewontin RC (1979) The spandrels of San Marco and the panglossian paradigm: a critique of the adaptationist programme. Proceedings of the Royal Society of London Series B-Biological Sciences 205: 581-598

Greene HW (1988) Antipredator mechanisms in reptiles. In Gans C, Huey RB (eds) Biology of the reptilia. Alan R. Liss, Inc., New York 1-152

Hairston NG (1987) Diapause as a predator avoidance adaptation. In Kerfoot WC, Sih A (eds) Predation: direct and indirect impacts on aquatic communities. University Press of New England, Hanover, NH 281-290

Hale ME, Long JH, McHenry MJ, Westneat MW (2002) Evolution of behavior and neural control of the fast-start escape response. Evolution 56 (5): 993-1007

Hamilton WD (1971) Geometry for the selfish herd. Journal of Theoretical Biology 31 (2): 295-311

Hanifin CT, Yotsu-Yamashita M, Yasumoto T, Brodie ED (1999) Toxicity of dangerous prey: variation of tetrodotoxin levels within and among populations of the newt *Taricha granulosa*. Journal of Chemical Ecology 25 (9): 2161-2175

Härlin C, Härlin M (2003) Towards a historization of aposematism. Evolutionary Ecology 17 (2): 197-212

Harvey PH, Greenwood PJ (1978) Anti-predator defence strategies: some evolutionary problems. In Krebs JR, Davies NB (eds) Behavioural ecology: an evolutionary approach. Blackwell Scientific Publications, Oxford

Havel JE (1987) Predator-induced defenses: a review. In Kerfoot WC, Sih A (eds) Predation: direct and indirect impacts on aquatic communities. University Press of New England, Hanover, N.H. 263-278

Helfman G, Collette B, Facey D (1997) The Diversity of Fishes. Blackwell Science, Malden, MA

Hoff KV, Wassersug RJ (2000) Tadpole locomotion: axial movement and tail functions in a largely vertebraeless vertebrate. American Zoologist 40 (1): 62-76

Holling CS (1959) The components of predation as revealed by a study of small mammal predation of the European pine sawfly. Canadian Entomologist 91: 293-320

Holmes W (1940) The colour changes and colour paterns of *Sepia officinalis* L. Proceedings of the Zoological Society of London 110: 17-35

Holt RD (1977) Predation, apparent competition, and the structure of prey communities. Theoretical Population Biology 12: 197-229

Holt RD, Lawton JH (1994) The ecological consequences of shared natural enemies. Annual Review of Ecology and Systematics 25: 495-520

Honma A, Oku S, Nishida T (2006) Adaptive significance of death feigning posture as a specialized inducible defence against gape-limited predators. Proceedings of the Royal Society of London Series B-Biological Sciences 273 (1594): 1631-1636

Houde AE (1997) Sex, color, and mate choice in guppies. Princeton University Press, Princeton, N.J.

Hoverman JT, Auld JR, Relyea RA (2005) Putting prey back together again: integrating predator-induced behavior, morphology, and life history. Oecologia 144 (3): 481-491

Hoy R, Nolen T, Brodfuehrer P (1989) The neuroethology of acoustic startle and escape in flying insects. Journal of Experimental Biology 146: 287-306

Hubbs C (1996) Geographic variation in life history traits of *Gambusia* species. Proceedings of the Desert Fishes Council 27: 1-21

Huffaker CB (1958) Experimental studies on predation: dispersion factors and predator-prey oscillations. Hilgardia 27: 343-383

Humphries DA, Driver PM (1967) Erratic display as a device against predators. Science 156: 1767-1768

Humphries DA, Driver PM (1971) Protean defence by prey animals. Oecologia 5: 285–302

Ims RA (1990) On the adaptive value of reproductive synchrony as a predator-swamping strategy. American Naturalist 136 (4): 485-498

Isley FB (1938) Survival value of acridian protective coloration. Ecology 19 (3): 370-389

Istock CA (1967) The evolution of complex life cycle phenomena: and ecological perspective. Evolution 21: 592-605

Jackson DA, Peres-Neto PR, Olden JD (2001) What controls who is where in freshwater fish communities - the roles of biotic, abiotic, and spatial factors. Canadian Journal of Fisheries and Aquatic Sciences 58 (1): 157-170

Janzen DH (1981) Evolutionary physiology of personal defense. In Townsend CR, Calow P (eds) Physiological ecology: an evolutionary approach to resource use. Blackwell, Oxford 145-164

Janzen FJ, Tucker JK, Paukstis GL (2000) Experimental analysis of an early life-history stage: avian predation selects for larger body size of hatchling turtles. Journal of Evolutionary Biology 13 (6): 947-954

Jeffries MJ, Lawton JH (1984) Enemy free space and the structure of ecological communities. Biological Journal of the Linnean Society 23: 269-286

Jiggins CD, Estrada C, Rodrigues A (2004) Mimicry and the evolution of premating isolation in *Heliconius melpomene* Linnaeus. Journal of Evolutionary Biology 17 (3): 680-691

Jiggins CD, Naisbit RE, Coe RL, Mallet J (2001) Reproductive isolation caused by colour pattern mimicry. Nature 411 (6835): 302-305

Karban R, Baldwin IT (1997) Induced responses to herbivory. University of Chicago Press, Chicago

Kenward RE (1978) Hawks and doves: factors affecting success and selection in goshawk attacks on wood pigeons. Journal of Animal Ecology 47: 449-460

Kerfoot WC, Sih A (1987) Predation: direct and indirect impacts on aquatic communities. University Press of New England, Hanover, NH.

Kettlewell HBD (1956) Further selection experiments on industrial melanism in the Lepidoptera. Heredity 10: 287-301

Kiltie RA (1988) Countershading: universally deceptive or deceptively universal. Trends in Ecology & Evolution 3 (1): 21-23

Koehl MAR (1996) When does morphology matter? Annual Review of Ecology and Systematics 27: 501-542

Komárek S (1998) Mimicry, aposematism and related phenomena in animals and plants: bibliography 1800-1990. Vesmir, Prague

Kotler BP, Brown JS (1988) Environmental heterogeneity and the coexistence of desert rodents. Annual Review of Ecology and Systematics 19: 281-307

Kozlowski J, Uchmanski J (1987) Optimal individual growth and reproduction in perennial species with indeterminate growth. Evolutionary Ecology 1: 214-230

Krumholz L (1963) Relationships between fertility, sex ratio, and exposure to predation in populations of the mosquitofish, *Gambusia manni* at Bimini, Bahamas. Internationale Revue der gesamten Hydrobiologie 48: 201-256

Lack D (1954) The natural regulation of animal numbers. Clarendon, Oxford

Lande R, Arnold SJ (1983) The measurement of selection on correlated characters. Evolution 37 (6): 1210-1226

Langerhans RB, DeWitt TJ (2004) Shared and unique features of evolutionary diversification. American Naturalist 164 (3): 335-349

Langerhans RB, Layman CA, DeWitt TJ (2005) Male genital size reflects a tradeoff between attracting mates and avoiding predators in two live-bearing fish species. Proceedings of the National Academy of Sciences of the United States of America 102 (21): 7618-7623

Langerhans RB, Layman CA, Shokrollahi AM, DeWitt TJ (2004) Predator-driven phenotypic diversification in *Gambusia affinis*. Evolution 58 (10): 2305–2318

Levins R (1968) Evolution in changing environments. Princeton University Press, NJ.

Lev-Yadun S, Dafni A, Inbar M, Izhaki I, Ne'eman G (2002) Colour patterns in vegetative parts of plants deserve more research attention. Trends in Plant Science 7: 59–60

Lima SL, Dill LM (1990) Behavioral decisions made under the risk of predation: a review and prospectus. Canadian Journal of Zoology 68 (4): 619-640

Lindeman RL (1942) The trophic-dynamic aspect of ecology. Ecology 23: 377-418

Lingle S (2002) Coyote predation and habitat segregation of white-tailed deer and mule deer. Ecology 83 (7): 2037-2048

Lloyd DG (1987) Selection of offspring size at independence and other size-versus-number strategies. American Naturalist 129 (6): 800-817

Lloyd M, Dybas HS (1966) The periodical cicada problem. I. Population ecology. Evolution 20 (2): 133-149

MacArthur RH (1962) Some generalized theorems of natural selection. Proceedings of the National Academy of Sciences of the United States of America 48: 1893-1897

MacArthur RH, Wilson EO (1967) The theory of island biogeography. Princeton University Press, Princeton, New Jersey, USA

Magurran AE, Seghers, B. H., Shaw PW, Carvalho GR (1995) The behavioral diversity and evolution of guppy, *Poecilia reticulata*, populations in Trinidad. Advances in the study of behavior 24: 155-202

Mallet J, Joron M (1999) Evolution of diversity in warning color and mimicry: polymorphisms, shifting balance, and speciation. Annual Review of Ecology and Systematics 30: 201-233

Mallet J, McMillan WO, Jiggins CD (1998) Mimicry and warning color at the boundary between races and species. In Howard DJ, Berlocher SH (eds) Endless forms: species and speciation. Oxford University Press, New York 390-403

Mangel M, Clark CW (1988) Dynamic modeling in behavioral ecology. Princeton University Press, Princeton

Mayr E (1942) Systematics and the origin of species. Columbia University Press, New York

McCollum SA, Van Buskirk J (1996) Costs and benefits of a predator-induced polyphenism in the gray treefrog *Hyla chrysoscelis*. Evolution 50 (2): 583-593

McMillan WO, Jiggins CD, Mallet J (1997) What initiates speciation in passion-vine butterflies? Proceedings of the National Academy of Sciences of the United States of America 94 (16): 8628-8633

McPeek MA (1995) Morphological evolution mediated by behavior in the damselflies of two communities. Evolution 49 (4): 749-769

McPeek MA (1997) Measuring phenotypic selection on an adaptation: lamellae of damselflies experiencing dragonfly predation. Evolution 51 (2): 459-466

McPeek MA (1999) Biochemical evolution associated with antipredator adaptation in damselflies. Evolution 53 (6): 1835-1845

McPeek MA, Schrot AK, Brown JM (1996) Adaptation to predators in a new community: swimming performance and predator avoidance in damselflies. Ecology 77 (2): 617-629

McPeek MA, Wellborn GA (1998) Genetic variation and reproductive isolation among phenotypically divergent amphipod populations. Limnology and Oceanography 43 (6): 1162-1169

Mcphail JD (1969) Predation and the evolution of a stickleback (*Gasterosteus*). Journal of Fisheries Research Board of Canada 26: 3183-3208

Miller RC (1922) The significance of the gregarious habit. Ecology 3 (2): 122-126

Moran AL, Emlet RB (2001) Offspring size and performance in variable environments: field studies on a marine snail. Ecology 82 (6): 1597-1612

Morgan MJ, Godin J-GJ (1985) Antipredator benefits of schooling behaviour in a cyprinodontid fish, the banded killifish (*Fundulus diaphanus*). Zeitschrift für Tierpsychologie 70: 236-246

Mostler G (1935) Beobachtungen zur frage der wespenmimikry. Zeitschrift für Morphologie und Ökologie der Tiere 29: 381-454

Müller F (1879) *Ituna* and *Thyridia*; a remarkable case of mimicry in butterflies. Transactions of the Entomological Society of London: xx-xxix

Myers JH, Bazely D (1991) Thorns, spines, prickles, and hairs: are they stimulated by herbivory and do they deter herbivores? In Tallamy DW, Raupp MJ (eds) Phytochemical induction by herbivores. John Wiley and Sons, New York 325-345

Naisbit RE, Jiggins CD, Mallet J (2003) Mimicry: developmental genes that contribute to speciation. Evolution & Development 5 (3): 269-280

Newman RA (1988) Adaptive plasticity in development of *Scaphiopus couchii* tadpoles in desert ponds. Evolution 42 (4): 774-783

Norris KS, Lowe CH (1964) An analysis of background color-matching in amphibians and reptiles. Ecology 45 (3): 565-580

Nosil P (2004) Reproductive isolation caused by visual predation on migrants between divergent environments. Proceedings of the Royal Society of London Series B-Biological Sciences 271 (1547): 1521-1528

Nosil P, Crespi BJ (2006a) Ecological divergence promotes the evolution of cryptic reproductive isolation. Proceedings of the Royal Society of London Series B-Biological Sciences 273: 991-997

Nosil P, Crespi BJ (2006b) Experimental evidence that predation promotes divergence in adaptive radiation. Proceedings of the National Academy of Sciences of the United States of America 103 (24): 9090-9095

Nosil P, Crespi BJ, Sandoval CP (2002) Host-plant adaptation drives the parallel evolution of reproductive isolation. Nature 417 (6887): 440-443

Nosil P, Crespi BJ, Sandoval CP (2003) Reproductive isolation driven by the combined effects of ecological adaptation and reinforcement. Proceedings of the Royal Society of London Series B-Biological Sciences 270 (1527): 1911-1918

Nosil P, Vines TH, Funk DJ (2005) Perspective: reproductive isolation caused by natural selection against immigrants from divergent habitats. Evolution 59 (4): 705-719

Owen DF (1963) Polymorphism and population density in the African land snail *Limicolaria martensiana*. Science 140: 666-667

Paine RT (1966) Food web complexity and species diversity. American Naturalist 100 (910): 65-76

Palleroni A, Miller CT, Hauser M, Marler P (2005) Predation Prey plumage adaptation against falcon attack. 434 (7036): 973-974

Peacor SD, Werner EE (2001) The contribution of trait-mediated indirect effects to the net effects of a predator. Proceedings of the National Academy of Sciences of the United States of America 98 (7): 3904-3908

Pigliucci M (2001) Phenotypic plasticity. Beyond nature and nature. Johns Hopkins University Press, Baltimore, MD

Pigliucci M, Murren CJ (2003) Genetic assimilation and a possible evolutionary paradox: can macroevolution sometimes be so fast as to pass us by? Evolution 57 (7): 1455-1464

Pigliucci M, Preston K (2004) The evolutionary biology of complex phenotypes. Oxford University Press, New York

Pimm SL (1982) Food Webs. Chapman and Hall, London, England

Polis GA, Winemiller KO (1996) Food webs : integration of patterns & dynamics. Chapman & Hall, New York

Poulton EB (1890) The colours of animals: their meaning and use especially considered in the case of insects. Kegan Paul, Trench, Trübner and Co. Ltd., London

Poulton EB (1902) The meaning of the white undersides of animals. Nature 65: 596

Price TD, Qvarnstrom A, Irwin DE (2003) The role of phenotypic plasticity in driving genetic evolution. Proceedings of the Royal Society of London Series B-Biological Sciences 270 (1523): 1433-1440

Pulliam HR (1973) On the advantages of flocking. Journal of Theoretical Biology 38 (2): 419-422

Pulliam HR, Caraco T (1984) Living in groups: is there an optimal group size? In Krebs JR, Davies NB (eds) Behavioural ecology: an evolutionary approach. Sinauer, Sunderland, MA 122-147

Reidy S, Kerr S, Nelson J (2000) Aerobic and anaerobic swimming performance of individual Atlantic cod. Journal of Experimental Biology 203 (2): 347-357

Reimchen TE (1994) Predators and morphological evolution in threespine stickleback. In Bell MA, Foster SA (eds) The evolutionary biology of the threespine stickleback. Oxford University Press, Oxford 240-276

Reimchen TE, Nosil P (2002) Temporal variation in divergent selection on spine number in threespine stickleback. Evolution 56 (12): 2472-2483

Reimchen TE, Nosil P (2004) Variable predation regimes predict the evolution of sexual dimorphism in a population of threespine stickleback. Evolution 58 (6): 1274–1281

Relyea RA (2002) Local population differences in phenotypic plasticity: predator-induced changes in wood frog tadpoles. Ecological Monographs 72(1): 77-93

Reznick DN (1996) Life history evolution in guppies: a model system for the empirical study of adaptation. Netherlands Journal of Zoology 46 (3-4): 172-190

Reznick DN, Bryant M, Holmes D (2006a) The evolution of senescence and post-reproductive lifespan in guppies (*Poecilia reticulata*). PLoS Biology 4 (1): 136-143

Reznick DN, Bryga H, Endler JA (1990) Experimentally induced life-history evolution in a natural population. Nature 346 (6282): 357-359

Reznick DN, Butler MJ, Rodd FH, Ross P (1996) Life-history evolution in guppies (*Poecilia reticulata*). 6. Differential mortality as a mechanism for natural selection. Evolution 50 (4): 1651-1660

Reznick DN, Endler JA (1982) The impact of predation on life history evolution in Trinidadian guppies (*Poecilia reticulata*). Evolution 36: 160-177

Reznick DN, Schultz E, Morey S, Roff D (2006b) On the virtue of being the first born: the influence of date of birth on fitness in the mosquitofish, *Gambusia affinis*. Oikos 113 (1): 135-147

Reznick DN, Shaw FH, Rodd FH, Shaw RG (1997) Evaluation of the rate of evolution in natural populations of guppies (*Poecilia reticulata*). Science 275 (5308): 1934-1937

Rice WR, Hostert EE (1993) Laboratory experiments on speciation: what have we learned in 40 years? Evolution 47 (6): 1637-1653

Ricklefs RE, O'Rourke K (1975) Aspect diversity in moths: a temperate-tropical comparison. Evolution 29: 313-324

Rigby MC, Jokela J (2000) Predator avoidance and immune defense: costs and trade-offs in snails. Proceedings of the Royal Society of London Series B-Biological Sciences 267: 171-176

Riley AM, Loxdale HD (1988) Possible adaptive significance of "tail" structure in "false head" lycaenid butterflies. Entomologists Record and Journal of Variation 84: 349-365

Robinson BW, Parsons KJ (2002) Changing times, spaces, and faces: tests and implications of adaptive morphological plasticity in the fishes of northern postglacial lakes. Canadian Journal of Fisheries and Aquatic Sciences 59 (11): 1819-1833

Robinson MH (1969) Defenses against visually hunting predators. In Dobzhansky T, Hecht MK, Steere WC (eds) Evolutionary Biology. Meredith Corporation, New York

Roeder KD (1962) The behaviour of free flying moths in the presence of artificial ultrasonic pulses. Animal Behaviour 10: 300-304

Roeder KD (1965) Moths and ultrasound. Scientific American 212 (4): 94-102

Roff DA (1992) The evolution of life histories: theory and analysis. Chapman & Hall, New York

Roff DA (2002) Life history evolution. Sinauer Associates Inc., Sunderland, MA

Rundle HD, Nagel L, Boughman JW, Schluter D (2000) Natural selection and parallel speciation in sympatric sticklebacks. Science 287 (5451): 306-308

Rundle HD, Nosil P (2005) Ecological speciation. Ecology Letters 8 (3): 336-352

Rundle HD, Vamosi SM, Schluter D (2003) Experimental test of predation's effect on divergent selection during character displacement in sticklebacks. Proceedings of the National Academy of Sciences of the United States of America 100 (25): 14943-14948

Ruxton GD, Sherratt TN, Speed MP (2004) Avoiding attack: the evolutionary ecology of crypsis, warning signals and mimicry. Oxford University Press, Oxford

Rydell J, Jones G, Waters D (1995) Echolocating bats and hearing moths: who are the winners. Oikos 73 (3): 419-424

Saenz D, Johnson JB, Adams CK, Dayton GH (2003) Accelerated hatching of southern leopard frog (*Rana sphenocephala*) eggs in response to the presence of a crayfish (*Procambarus nigrocinctus*) predator. Copeia (3): 646-649

Sandoval CP (1994) Differential visual predation on morphs of *Timema cristinae* (Phasmatodeae: Timemidae) and its consequences for host range. Biological Journal of the Linnean Society 52 (4): 341-356

Schaack S, Chapman LJ (2003) Interdemic variation in the African cyprinid *Barbus neumayeri*: correlations among hypoxia, morphology, and feeding performance. Canadian Journal of Zoology 81 (3): 430-440

Scheiner SM (1993) Genetics and evolution of phenotypic plasticity. Annual Review of Ecology and Systematics 24 (35-68): 35-68

Schlichting CD (2004) The role of phenotypic plasticity in diversification. In DeWitt TJ, Scheiner SM (eds) Phenotypic plasticity. Functional and conceptual approaches. Oxford University Press, New York 191-200

Schlichting CD, Pigliucci M (1998) Phenotypic evolution: a reaction norm perspective. Sinauer, Sunderland, MA

Schluter D (2000) The ecology of adaptive radiation. Oxford University Press, Oxford

Schluter D (2001) Ecology and the origin of species. Trends in Ecology and Evolution 16 (7): 372-380

Seghers BH (1974) Schooling behavior in the guppy (*Poecilia reticulata*): an evolutionary response to predation. Evolution 28: 486-489

Shurin JB, Allen EG (2001) Effects of competition, predation, and dispersal on species richness at local and regional scales. American Naturalist 158 (6): 624-637

Sih A (1987) Predators and prey lifestyles: an evolutionary and ecological overview. In Kerfoot WC, Sih A (eds) Predation: Direct and Indirect Impacts on Aquatic Communities. University Press of New England, Hanover, NH 203-224

Sohn JJ (1977) Consequences of predation and competition upon demography of *Gambusia manni* (Pisces: Poeciliidae). Copeia: 224-227

Stanley SM (1979) Macroevolution, pattern and process. W.H. Freeman, San Francisco

Stearns SC, Crandall RE (1981) Quantitative predictions of delayed maturity. Evolution 35 (3): 455-463

Steets JA, Salla R, Ashman TL (2006) Herbivory and competition interact to affect reproductive traits and mating system expression in Impatiens capensis. American Naturalist 167 (4): 591-600

Stein RA (1979) Behavioral response of prey to fish predators. In Stroud RH, Clepper H (eds) Predator-prey systems in fisheries management. Sport Fishing Institute, Washington, D.C.

Stemberger RS, Gilbert JJ (1987) Defenses of planktonic rotifers against predators. In Kerfoot WC, Sih A (eds) Predation: direct and indirect impacts on aquatic communities. University Press of New England, Hanover, NH 227-239

Stoks R, Nystrom JL, May ML, McPeek MA (2005) Parallel evolution in ecological and reproductive traits to produce cryptic damselfly species across the holarctic. Evolution 59 (9): 1976-1988

Sumner FB (1934) Does protective coloration protect? Results of some experiments with fishes and birds. Proceedings of the National Academy of Sciences of the United States of America 20 (10): 559-564

Thayer AH (1896) The law which underlies protective coloration. Auk 13 (2): 124-129

Thomson AL (1950) Factors determining the breeding seasons of birds: an introductory review. Ibis 92: 173-184

Tollrian R, Harvell CD (1999) The ecology and evolution of inducible defenses. Princeton University Press, Princeton

Treat AE (1955) The response to sound in certain Lepidoptera. Annals of the Entomological Society of America 48: 272-284

Trehorne JE, Foster WA (1980) The effects of group size on predator avoidance in a marine insect. Animal Behaviour 28: 1119-1122

Vamosi SM (2002) Predation sharpens the adaptive peaks: survival trade-offs in sympatric sticklebacks. Annales Zoologici Fennici 39 (3): 237-248

Vamosi SM (2003) The presence of other fish species affects speciation in threespine sticklebacks. Evolutionary Ecology Research 5 (5): 717-730

Vamosi SM (2005) On the role of enemies in divergence and diversification of prey: a review and synthesis. Canadian Journal of Zoology 83 (7): 894-910

Vamosi SM, Schluter D (2002) Impacts of trout predation on fitness of sympatric sticklebacks and their hybrids. Proceedings of the Royal Society of London Series B-Biological Sciences 269 (1494): 923-930

Vamosi SM, Schluter D (2004) Character shifts in the defensive armor of sympatric sticklebacks. Evolution 58 (2): 376-385

Van Buskirk J, Anderwald P, Lupold S, Reinhardt L, Schuler H (2003) The lure effect, tadpole tail shape, and the target of dragonfly strikes. Journal of Herpetology 37 (2): 420-424

Vanhooydonck B, Van Damme R, Aerts P (2001) Speed and stamina trade-off in lacertid lizards. Evolution 55 (5): 1040-1048

Van Noordwijk AJ, Mccleery RH, Perrins CM (1995) Selection for the timing of great tit breeding in relation to caterpillar growth and temperature. Journal of Animal Ecology 64 (4): 451-458

Vermeij GJ (1982) Unsuccessful predation and evolution. American Naturalist 120 (6): 701-720

Vermeij GJ (1987) Evolution and Escalation. Princeton University Press, Princeton, NJ

Videler JJ (1993) Fish swimming. Chapman & Hall, London

Vine I (1971) Risk of visual detection and pursuit by a predator and the selective advantage of flocking behaviour. Journal of Theoretical Biology 30 (2): 405-422

Vinyard G (1979) An ostracod (*Cypriodopsis vidua*) can reduce predation from fish by resisting digestion. American Midland Naturalist 102 (1): 188-190

Viscido SV, Miller M, Wethey DS (2001) The response of a selfish herd to an attack from outside the group perimeter. Journal of Theoretical Biology 208 (3): 315-328

Vogel S (1994) Life in moving fluids. Princeton University Press, Princeton

Vonesh JR (2005) Egg predation and predator-induced hatching plasticity in the African reed frog, *Hyperolius spinigularis*. Oikos 110 (2): 241-252

Walker JA (1997) Ecological morphology of lacustrine threespine stickleback *Gasterosteus aculeatus* L. (Gasterosteidae) body shape. Biological Journal of the Linnean Society 61 (1): 3-50

Walker JA, Bell MA (2000) Net evolutionary trajectories of body shape evolution within a microgeographic radiation of threespine sticklebacks (*Gasterosteus aculeatus*). Journal of Zoology 252: 293-302

Walker JA, Ghalambor CK, Griset OL, McKenney D, Reznick DN (2005) Do faster starts increase the probability of evading predators? Functional Ecology 19 (5): 808-815

Wallace AR (1879) The protective colours of animals. In Brown R (ed) Science for all. Cassell, Petter, Galpin & Co., London 128-137

Warkentin KM (1995) Adaptive plasticity in hatching age: a response to predation risk trade-offs. Proceedings of the National Academy of Sciences of the United States of America 92 (8): 3507-3510

Waters DA (2003) Bats and moths: what is there left to learn? Physiological Entomology 28(4): 237-250

Webb PW (1984) Body form, locomotion, and foraging in aquatic vertebrates. American Zoologist 24: 107-120

Webb PW (1986) Locomotion and predator-prey relationships. In Lauder GV, Feder ME (eds) Predator-Prey Relationships. University of Chicago Press, Chicago 24-41

Wellborn GA, Skelly DK, Werner EE (1996) Mechanisms creating community structure across a freshwater habitat gradient. Annual Review of Ecology and Systematics 27: 337-363.

Welty JC (1934) Experiments in group behavior of fishes. Physiological Zoology 7: 85-128

Werner EE, Anholt BR (1993) Ecological consequences of the trade-off between growth and mortality rates mediated by foraging activity. American Naturalist 142 (2): 242-272

West-Eberhard MJ (1989) Phenotypic plasticity and the origins of diversity. Annual Review of Ecology and Systematics 20: 249-278

West-Eberhard MJ (2003) Developmental plasticity and evolution. Oxford University Press, Oxford

Wiklund C, Tullberg BS (2004) Seasonal polyphenism and leaf mimicry in the comma butterfly. Animal Behaviour 68 (3): 621-627

Wilbur HM (1980) Complex Life Cycles. Annual Review of Ecology and Systematics 11: 67-93

Will U (1986) Mauthner neurons survive metamorphosis in anurans: a comparative study HRP study on the cytoarchitecture of the Mauthner neuron in amphibians. Journal of Comparative Neurology 244: 111-120

Williams GC (1966) Adaptation and natural selection. Princeton University of Press, Princeton

Wilson RS, James RS, Van Damme R (2002) Trade-offs between speed and endurance in the frog *Xenopus laevis*: a multi-level approach. Journal of Experimental Biology 205 (8): 1145-1152

Worthington EB (1940) Geographical differentiation in freshwaters with special reference to fish. In Huxley J (ed The new systematics. Oxford University Press, Oxford 287-303

Yack JE, Fullard JH (2000) Ultrasonic hearing in nocturnal butterflies. Nature 403 (6767): 265-266

Zaret TM (1980) Predation and freshwater communities. Yale University Press, New Haven

11 Predation impacts and management strategies for wildlife protection

Michael J. Bodenchuk[1] and David J. Hayes[2]

[1]State Director, USDA-APHIS-Wildlife Services, P.O. Box 26976, Salt Lake City, UT 84126, [2]Environmental Coordinator, USDA-APHIS-Wildlife Services, P.O. Box 1938, Billings, MT 59103

11.1 Abstract

Wildlife management activities often serve to balance wildlife populations with perceived available habitat and agency management objectives or recovery goals. Predation management, while controversial in the public arena, is occasionally necessary to help balance populations of prey and predators. To conduct predation management in a responsible manner, the nature of predation impacts to prey populations must be examined and understood. The authors discuss: 1) various factors which affect predation impacts, 2) behavioral changes in prey populations which result in secondary predation impacts, and 3) strategies which may be implemented to facilitate prey populations to attain agency management objectives.

Keywords: Predation, secondary impacts, secondary predation, management, predator-prey relationships, predator avoidance behavior, carrying capacity.

11.2 Introduction

Predation is a naturally occurring event which influences both predator and prey populations, behaviors and densities. In a natural ecosystem, predation usually "balances" with prey in some manner[1]. However, ecosystems in the contiguous United States have been altered in ways that affect both predator and prey populations and behavior (Hecht and

---

[1] Because of the complexity of predator-prey relationships, the following is not intended to summarize the role of predation in wildlife population dynamics, but to summarize primary effects and introduce some secondary effects that predators have on prey species.

Nickerson 1999). Numerous factors can be responsible for wildlife population declines, including habitat loss or change, severe weather (e.g., drought, deep snow), starvation, change in age and sex structure, disease, predation, predation risk, competition with livestock and other wildlife species, hunting and interactions of these factors (Wallmo 1981; Hall 1984; Whittaker and Lindzey 1999). Connolly (1978), however, after reviewing 58 studies of predation on wild ungulate populations concluded that in 31 cases predation appeared to be a primary limiting factor.

While predation is a natural phenomenon that cannot and should not be eliminated, at times some prey species suffer from excessive predation rates or risk of predation (Morse 1980; Edwards 1983; Lima et al. 1985; Ferguson et al. 1988; Hoban 1990; Lima and Dill 1990; Schmitz et al. 1997; Kie 1999; Hecht and Nickerson 1999; Creel et al. 2005). Professional game managers may also have management objectives for wildlife populations other than population densities which would ordinarily exist in "balance." The reality is that without the intervention of a management strategy to reduce predation, the effects of severe weather, disease or human influence on natural ecosystems, we may be cheating ourselves into believing we are conserving natural communities (Hecht and Nickerson 1999). Early detection and active management to reduce predation or predation risk may be important for the maintenance or recovery of some species. For small or severely declining populations, the best hedge is to increase numbers[2] as quickly as possible and to reduce variability in productivity and survival rates (Hecht and Nickerson 1999).

Conflicts occur when predators adversely affect desired prey species' populations below management objective levels established by the responsible wildlife management agency. Predation may affect adult survival, neonatal survival and recruitment or nesting success. Completely understanding predation, ecological factors and predation effects is not an exact science, and in most cases negative effects are inferred from observations of prey populations. Predation management therefore addresses the observed conflicts in relation to agency management objectives or recovery goals.

Many wildlife management agencies have established policy which allows predation management to: 1) protect critical wildlife management areas (e.g., waterfowl management areas), 2) support reintroductions of native wildlife, and 3) protect seriously depressed wildlife populations.

---

[2] Habitat restoration and species restoration may be the final goal of management agencies but most habitat restoration projects take many years, if not decades to complete. Depredation management could insure that depressed prey species populations remain viable on desired habitats or in selected areas.

Management strategies designed to protect these resources may vary and may be related to observations of prey species' populations.

11.3 Predator-prey relationships

Habitat carrying capacity (K) is an important concept with many implications to evaluating predation, predator-prey relationships and need for predation management. For purposes of game and range management, K is usually defined as the maximum population of a given species that can be supported in a defined habitat without permanently impairing the productivity of that habitat. The K of an area is usually defined by limiting factors, such as water, vegetation, and cover. Biological studies of population change typically demonstrate that once the K of an ecosystem is exceeded, a crash or collapse of the population follows associated with environmental degradation. K, to this point, generally has not considered the impact or secondary effects that predators can have on prey and/or that K can be a "fluid" component of the environment depending on the influence that weather, vegetation and predators can have on a prey species in the area.

Ballard et al. (2001) described four current predator-prey relationship models with emphasis on mule deer (*Odocoileus hemionus*). In brief, these models are: 1) low density equilibria, 2) multiple stable states, 3) stable-limit cycles, and 4) recurrent fluctuations. In each of these models, the relationship between predator and prey is examined in relation to prey and vegetative *K*. In each model, predation is considered regulatory at low population levels[3] (low in relation to *K*) and non-regulatory as populations near K^4. Less obvious, however, is restricted use of habitat by prey species due to the risk of predation (Hoban 1990). For simplicity, the authors assume that without the influence of predators, wildlife populations occupy specific habitats because those are the habitats that provide their life history requirements. However, if wildlife managers just evaluate *K* and do not consider habitat availability, predation or predation risk influences on prey species, their evaluations could be seriously flawed.

In addition to the primary negative effects of predation (i.e., how many of the affected prey species are directly killed by predators), there is a growing body of evidence that points to significant secondary effects of

---

[3] When populations are below K, each mortality adds to total mortality and mortality factors are termed additive.
[4] At K, mortality factors are compensatory (*i.e.,* they replace each other so that total mortality remains constant).

predation (Wehausen 1996, Pitt 1999, Ripple and Larsen 2000, Ripple et al. 2001, Barber et al. 2004, Preisser et al. 2005). Secondary effects in this context are negative effects to prey populations because of species "displacement" or antipredator behavior in prey (i.e., predators cause adaptive shifts in prey through shifts in behavior or occupied habitats) caused by predators (Morse 1980, Edwards 1983, Risenhoover and Bailey 1985, Lima et al. 1985, Ferguson et al. 1988, Hoban 1990, Lima and Dill 1990, Schmitz et al. 1997, Pitt 1999, Kie 1999) or the risk from predators (Creel et al. 2005). Secondary predation can be thought of as a trade-off by prey to reduce predation risks, but possibly at the expense of utilizing more favorable foraging or cover habitat, shifting daily activities, reduced reproductive success or other life history requirements (Burk 1982, Lima and Dill 1990, Hecht and Nickerson 1999, Ballard et al. 2001, Preisser et al. 2005). A secondary effect of predation or predation risk could be the restriction of range utilization by prey species to areas adjacent to escape terrain/cover (Bergerud et al. 1983, Bergerud and Page 1987, Wehausen 1996, Bleich et al. 1997, Bleich et al. 1997, Kunkel and Pletsher 2000, Creel and Winnie 2005, Creel et al. 2005), interspecific competition with other prey species (Gill et al. 2001) and distribution of prey over their range (Messier and Barrette 1985, Molvar and Bowyer 1994). The behavioral response to predation or predation risk may result in reduced nutrient intake and lower offspring survival in prey species which can lead to a population decline or an animal in poor condition which may choose a foraging strategy more risky than an animal that is well fed (Skogland 1991a, Bliech et al. 1997).

Another consideration of secondary predation and the distribution of prey populations (i.e., herbivores) is the movement of prey species onto private lands (Gude and Garrott 2003) and the resultant grazing patterns (i.e., effects on plants) (Schmitz et al. 1997, Ripple and Larsen 2000, Ripple et al. 2001).

Morgantini and Hudson (1985) reported that undisturbed elk (*Cervus elaphus canadensis*) in Alberta, Canada preferred grazing to browsing but avoided open grasslands during hunting seasons (i.e., predation risk). Wolf restoration in Yellowstone National Park (YNP) has provided a unique opportunity to examine gray wolf (*Canis lupus*) predation risks and predation effects in a well-studied environment. For example, elk intensively occupied riparian and aspen habitats in YPN prior to wolf introductions however, following wolf reintroductions increased aspen regeneration in the Lamar Valley was noted, ostensibly because of reduced elk use of meadow habitat due to the risk of predation (Ripple and Larsen 2000, Ripple et al. 2001).

To further bolster this observation, Gude and Garrott (2003), Gude et al. (2005) and Creel and Winnie (2005) reported that after a successful wolf predation event or human hunting, elk moved from the area or few elk resided in open areas. Gude and Garrott (2003) reported elk were found in smaller groups sizes and their distribution was different (i.e., smaller elk herds generally sought out areas with more dense vegetation for cover). Creel and Winnie (2005) found that elk group size was also smaller and elk were closer to or in cover when wolves were detected and that hunting success outside of YPN was impacted because of behavior changes in elk from wolves. Creel et al. (2005) stated that when wolves were present or detected, elk moved into the protective cover of wooded areas and reduced their use of preferred grassland portion of their habitat. In addition, Mech (1977) suggested that temporal and spatial availability of prey also influences predator-prey relationships. In Minnesota, white-tailed deer (*Odocoileus virginianus*) use buffer zones between gray wolf packs, which wolves avoid to prevent intraspecific strife with neighboring packs.

Hamlin (2005) reported on several study areas in Montana where the combination of changed elk distribution and behavior caused "indirect" population level effects in elk because of predation risk. Morgantini and Hudson (1985) concluded that special winter hunting seasons cause a major shift in diet selection and that this shift resulted in lower diet digestibility and that it is reasonable to assume that during severe winters, when digestible energy is limiting, a decrease of dietary concentration would have a negative impact on the welfare of elk populations in west-central Alberta. In addition, along with a possible negative impact on the welfare of elk, competition between elk and mule deer for browse, particularly in winter, could have a negative impact on deer populations (Keegan and Wakeling 2003).

In most of these areas, concurrent predation rate studies do not implicate direct predation by wolves as a cause for decline in the elk or mule deer herds. However, if elk are using the habitat in a different manner (i.e., elk are using only a portion of the habitat with lower K) the secondary effect of predation may cause a decline in the herd (Hamlin 2005). Thus, it appears that predation and predation risk (i.e., secondary predation) affects elk distribution and possibly abundance, and possibly browsing species like mule deer.

Wehausen (1996) and Hayes et al. (2000) examined mortality patterns of bighorn sheep (*Ovis canadensis*). Their results indicate that even a small number of mountain lions (*Puma concolor*) may effect bighorn sheep survival, and population-level impacts may be exacerbated if adult female sheep are heavily preyed upon or displaced into less optimal habitat. Wehausen (1996) believed mountain lion predation was responsible for behavioral changes and winter range abandonment and a subsequent population crash of bighorn sheep in the Sierra Nevada. The bighorn population decline appeared to result from the indirect effects of mountain lion predation because of habitat selection by bighorn sheep due to predator avoidance behavior. Fecal nitrogen levels of bighorn sheep were higher before their winter range abandonment, suggesting that the sheep maintained higher nutritional levels and reproductive success when they seasonally migrated to their winter range. The reduced vigor of the bighorn sheep herd was attributed to direct predation by mountain lions plus the reduced survival and recruitment based on habitat selection (i.e., a secondary predation impact).

Hayes et al. (2000) suggested population-level effects were exacerbated when mountain lions killed reproductive-age females and their offspring. Sustained high levels of mountain lion predation apparently impeded recovery of the Peninsular Range sheep population in California. Bighorn sheep distribution and the numbers that their ranges support are dependent on the assortment of predators that confine them to those ranges (Wishart 2000).

Edwards (1983), Ferguson et al. (1988) and Kie (1999) suggest that some larger prey species modify their behavior and occupy habitats with poorer quality forage to avoid predators. Edwards (1983) further states that, *"Although a direct link between nutritional status and reproductive rate has not been demonstrated for moose, indirect evidence suggests that poor diet may increase mortality or lower reproductive success."* Robinette et al. (1955) and Julander et al. (1966) reported reproductive success in mule deer is directly related to the quality of summer forage. Ferguson et al. (1988) suggested that both male and female caribou (*Rangifer tarandus*) will sacrifice high quality forage and phytomass on a year-round basis to avoid a high risk environment even though, in some winters, the animals, especially the calves face starvation.

Secondary impacts may exist for any predator/prey relationship. West (2002) reported on predator exclusion fencing for waterfowl protection on Bear River Refuge in northern Utah. Prior to his investigations, waterfowl failed to initiate nesting on upland sites purchased and set aside for nesting. Within the mammalian exclusion fences, ducks initiated nesting at a rate up to one nest/2 acres, indicating that the risk of predation was affecting nesting habitat selection by the birds.

11.4 Habitat v. predators

Several research projects (Hornocker 1970, Bartmann et al. 1992, O'Brien et al 2005) examined predation impacts juxtaposed with habitat impacts. Other reviews (Ballard et al. 2001, Gill et al. 2001) reported that habitat plays more of a role than predators. In most cases, the assessment of predation impacts is limited to primary impacts. When the potential for secondary predation impacts is considered, it is difficult to assess whether predation or habitat are limiting, since one influences the other.

Habitat can be limiting and habitat management is necessary. The authors believe that habitat management is a process and not a goal for management agencies. Once habitat is manipulated it progresses towards a climax vegetative community. Wildlife biologists and landowners must commit to habitat management on a continual basis to meet the diverse needs of multiple wildlife species and humans.

Because habitat management is necessary, because predators can affect habitat selection and use and because predation management can benefit habitat projects, it is inappropriate to look at issues as "habitat v. predators." Predation management can play a role in assisting species within the confines of existing habitat and habitat management provides habitat for the future.

11.5 Predation and management effects

The authors recognize that predation may not be the limiting factor for all declining wildlife populations and other factors such as poor habitat quality, drought, disease, etc. may play a role in limiting populations. There are studies which indicate that predation is not the cause of prey population declines. However, there are numerous research studies which detail negative prey population impacts from predators. In fact, both cases may be true, although not at the same time in the same area for the same predator-prey relationship. The literature reviewed herein is meant only to provide examples of some documented predator-prey relationships and not as a comprehensive review of the literature on predation or other factors that may regulate prey populations. With an adaptive management approach to resource management, actual effects will be inferred and mitigated as they are observed.

Research data has shown that predation management has the potential to benefit populations of both game and non-game wildlife populations. Numerous, scientific reviews (Connolly 1978; Sinclair 1991; Skogland

1991b; Ballard et al. 2001) examined the role of predation as a regulator of prey populations and found a variety of effects. Connolly (1978) reviewed the effects of predation on ungulates, and stated that a selective review of the literature could reinforce almost any view on the role of predation. He concluded, however, that predators acting in concert with weather, disease, and habitat changes could have important effects on prey numbers. Since Connolly's (1978) review, scientists have continued to debate whether predation is a significant regulating factor on ungulate populations (Messier 1991; Sinclair 1991; Skogland 1991b; Boutin 1992; Van Ballenberghe and Ballard 1994).

Predation on game species, however, is well documented and can adversely affect survival and recruitment of individuals into a population. Under certain conditions, predators have been documented as having a significant adverse effect on deer, pronghorn antelope (*Antilocapra americana*), bighorn sheep, game bird populations and threatened and endangered (T/E) species, and this predation is not necessarily limited to sick or inferior animals (Pimlott 1970; Bartush 1978; USDI 1978, 1995; Hamlin et al. 1984; Neff et al. 1985; Wehausen 1996). Conversely, a lack of predator damage management could adversely affect certain species (Connolly 1978; Schmidt 1986). In addition, predation management undertaken to protect livestock can augment wildlife management objectives set by state wildlife commissions or management agencies and by the U.S. Fish and Wildlife Service (USFWS) regarding T/E species.

Ballard et al. (2001) summarized predation impacts to prey, reviewed studies primarily conducted since the mid-1970's and made recommendations where predation management may be beneficial to deer populations. They note that similarities existed in studies where predation management was effective, including predation management being implemented when: 1) deer populations were below K, 2) predation had been identified as a limiting factor, 3) predation management reduced predator populations enough to be effective, 4) predation management efforts were timed just prior to reproduction of predator or prey species, and 5) efforts were targeted at a focused scale. Conversely, predation management was not effective when deer populations were at or near K, when predation was not a limiting factor, where predator populations were not effectively reduced and where efforts were conducted on too large or too small a scale.

11.5.1 Deer

As noted above, the scientific literature can be divided on the effects of predators on deer populations. Some recent studies (Gill et al. 2001) identified nutritional or disease problems affecting fawn survival and recruitment. However, many other studies have identified predation as an important cause of mortality, especially to newborn fawns. Hamlin et al. (1984) in a study of mule deer fawn mortality in Montana observed that a minimum of 90% of the summer mortality of fawns was a result of coyote (*Canis latrans*) predation. Trainer et al. (1981) reported that heavy mortality of mule deer fawns during late fall and winter was limiting the ability of the deer population to maintain or increase itself (i.e., recruitment) in Oregon. Their study concluded that predation, primarily by coyotes, was the major cause for low fawn crops on Steens Mountain in Oregon. Garner (1976), Garner et al. (1976), and Bartush (1978) determined the mortality of radio-collared white-tailed deer fawns in the Wichita Mountains of Oklahoma to be 87.9 to 89.6% with predators being responsible for 88.4 to 96.6% of the mortality. Garner (1976) further stated that inter-specific behavioral observations indicated that coyotes may find fawns by searching near single does. Beasom (1974) stated that predators were responsible for 74% and 61% of the fawn mortality for two consecutive years on his study area. Teer et al. (1991) documented that coyote diets contain nearly 90% deer during May and June. They concluded from work conducted at the Welder Wildlife Refuge, Texas that, "*Unequivocally coyotes take a large portion of the fawns each year during the first few weeks of life.*" Cook et al. (1971) stated that, "*Apparently, the neonatal period is a critical one in the life*" of white-tailed deer. Remains of 4 to 8 week old fawns were also common in coyote scats (feces) in studies from Steele (1969), Cook et al. (1971), Holle (1977), Litvaitis (1978), Litvaitis and Shaw (1980). Other researchers have also observed that coyotes are responsible for the majority of fawn mortality during the first few weeks of life (Knowlton 1964; White 1967; Cook et al. 1971; Salwasser 1976; Trainer et al. 1981).

Mackie et al. (1976) documented high winter losses of mule deer due to coyote predation in north-central Montana and stated that coyotes were the cause of most overwinter deer mortalities. Mackie et al. (1976) suggested that predation by coyotes ranked high as a probable cause of loss of mule deer fawns in the fall, while direct evidence of coyote predation during winter suggested this to be the proximal cause in the loss of fawn and adult mule and white-tailed deer. During other studies, designed to examine the impact of coyote predation on deer recruitment or coyote food habits, similar observations were noted (Steele 1969; Cook et al. 1971; Holle

1977; Litvaitis 1978; Litvaitis and Shaw 1980). Bates and Welch (1999), in Utah, state that coyote and black bear (*Ursus americanus*) predation on fawns could be significant and slowed recovery of already depressed deer herds. They further state, that research showed mule deer to be the principal prey item of mountain lions and suggested mountain lion predation could contribute to slow recovery of depressed prey populations.

Guthery and Beasom (1977) demonstrated that after coyote predation management, deer fawn production was 70% greater after the first year and 43% greater after the second year on their study area. Stout (1982) increased deer production on three areas in Oklahoma by 262%, 92%, and 167% the first summer following coyote predation management and increased production 154% overall for the three areas. Mule deer fawn survival was significantly increased and more consistent inside a predator-free enclosure in Arizona (LeCount 1977; Smith and LeCount 1976; Arizona Department Game and Fish 2004). Garner (1976), Garner et al. (1976), LeCount (1977), and Teer et al. (1991) stated that predation management may increase annual deer recruitment and survivability, but that impacts from other causes (e.g., drought, disease, hunting, livestock grazing, etc.) play a major role in achieving management objectives. Knowlton and Stoddart (1992) reviewed deer productivity data from the Welder Wildlife Refuge following coyote predation management. Deer densities tripled compared with those outside the enclosure, but without harvest management, ultimately returned to original densities due primarily to malnutrition and parasitism.

11.5.2 Pronghorn antelope

More than five decades ago, Jones (1949) believed that coyote predation was the main limiting factor of pronghorn antelope in Texas. More recently, Neff and Woolsey (1979, 1980) determined that coyote predation on pronghorn antelope fawns was the primary factor causing fawn mortality and low pronghorn antelope densities on Anderson Mesa, Arizona. Neff et al. (1985) concluded from a 5-year radio telemetry study that most of the coyotes that killed pronghorn antelope fawns on Anderson Mesa were residents. This means that most of the depredating coyotes were present on the fawning grounds during fawning times. A 6-year radio telemetry study of pronghorn antelope in western Utah showed that 83% of all fawn mortality was attributed to predation (Beale and Smith 1973). Trainer et al. (1983) concluded that predation was the leading cause of pronghorn antelope fawn loss, accounting for 91% of the mortalities that occurred during a 1981-82 study in southeastern Oregon. They also stated that most pronghorn antelope fawns were killed by coyotes and that known probable coyote kills comprised 60% of fawn mortality. Coyote predation was a leading cause of antelope fawn mortality on the National Bison Range (NBR) at Moiese, Montana (Byers 1997). Major losses of pronghorn antelope fawns to predators have also been reported from other radio telemetry studies (Barrett 1978; Beale 1978; Bodie 1978; Von Gunten 1978; Tucker and Garner 1980).

Arrington and Edwards (1951) observed that following coyote predation management in Arizona, an increase in pronghorn antelope populations occurred to the point where antelope were again huntable, whereas on areas without coyote predation management this increase was not noted. Coyote predation management on Anderson Mesa, Arizona increased the herd from 115 animals to 350 in 3 years, and peaked at 481 animals in 1971 (Neff et al. 1985). After coyote predation management was discontinued, the pronghorn antelope fawn survival dropped to only 14 and 7 fawns/100 does in 1973 and 1979, respectively. Initiation of another coyote predation management program began with the removal of an estimated 22% of the coyote population in 1981, 28% in 1982, and 29% in 1983. As a result, fawn production increased from a low of 7 fawns/100 does in 1979 to 69 and 67 fawns/100 does in 1982 and 1983, respectively. Antelope population surveys on Anderson Mesa conducted in 1983 indicated a population of 1,008 antelope, exceeding 1,000 animals for the first time since 1960. In addition, a study in southeastern Oregon documented that in 1985, 1986 and 1987 an estimated reduction of 24%, 48%, and 58% of the spring coyote population on the study area resulted in an increase in antelope fawns from 4 fawns/100 does in 1984 to 34, 71, and 84 fawns/100 does in 1985, 1986, and 1987, respectively (Willis et al. 1993).

The USFWS and O'Gara (1994) conducted an aerial gunning operation on the NBR in 1985 that resulted in an increase in antelope fawn survival for several years and eventually dropped in subsequent years. Limited aerial gunning of coyotes was again conducted on the NBR in 1992 and in 1993 primarily on the bighorn sheep range for the protection of lambs and to a lesser degree on the adjacent pronghorn antelope habitat. However, these aerial gunning operations were conducted after coyotes had denned and very little follow-up coyote predation management was conducted during the crucial period of pronghorn antelope fawning and bighorn lambing. The autumn pronghorn antelope fawn survival was 8.2 fawns/100 does in 1992, dropping to 1.8 fawns/100 does in 1993 and 11.3 fawns/100 does in 1994. In 1995, Wildlife Services (WS) conducted a limited aerial gunning operation before most coyote denning activity and followed up with limited ground predation management to remove coyotes in bighorn sheep habitats during lambing and in antelope fawning areas during the fawning period. The autumn antelope fawn survival for 1995 was 87.5 fawns/100 does and the best survival of twins that had ever been documented on the NBR. Similar observations of improved pronghorn antelope fawn survival and population increases following coyote predation management have been reported by Riter (1941), Udy (1953), and Hailey (1979).

Coyote predation management for the protection of pronghorn antelope is also cost effective in pronghorn antelope management as shown by Smith et al. (1986).

11.5.3 Sage grouse

Sage grouse (*Centrocercus urophasianus*) populations have declined throughout much of the western United States during the last several decades due to a variety of environmental factors (Connelly and Braun 1997). Sage grouse populations occupying habitats that are highly fragmented or in poor ecological condition may exhibit relatively low nest success, low juvenile recruitment, and poor adult survival that may be related to increased predation (Gregg 1991). Populations of some of the most important prairie grouse predators have increased dramatically during the last 100 years, and even in areas of good habitat, predator populations can be so abundant that habitat alone may not suffice to allow grouse populations to increase (Bergerud 1988). Schroeder and Baydack (2001) suggested that as habitats become more fragmented and populations of prairie grouse become more threatened, it becomes more important to consider predation management as a potential management tool. Because deteriorated sagebrush (*Artemisia* spp.) habitats may take 15-30 years to recover, a predation management strategy that effectively increases nesting success and juvenile survival may be useful to offset some of the negative effects of poorer habitat. This approach might also allow a more rapid recovery of grouse populations following habitat recovery. In a survey of United States public attitudes regarding predators and their management to enhance avian recruitment, Messmer et al. (1999) found that given information suggesting predators are among the threats to a declining bird population, the public generally supports using predation management for the protection of bird populations.

Presnall and Wood (1953) documented an example of coyotes as predators of sage grouse. In tracking a coyote approximately 5 miles to its den in northern Colorado, they found evidence along the way that the coyote had killed three adult sage grouse and destroyed a sage grouse nest.

Examination of the stomach contents from an adult female coyote removed the next day revealed parts of an adult sage hen plus six whole newly-hatched sage grouse chicks. The area around the den was littered with sage grouse bones and feathers. No other prey remains were found around the den, and it appeared that the pups had been raised largely upon sage grouse.

In southeastern Idaho, Burkepile et al. (2001) radio-marked 31 chicks from 13 broods in 1999 and 44 chicks from 15 broods in 2000. Survival estimates for 1999 and 2000 were only 15% and 18%, respectively. Predators were responsible for 90% of the mortality in 1999 and 100% of the mortality in 2000. Red fox (*Vulpes vulpes*) were believed to be one of the primary chick predators, but predation was also confirmed by unidentified avian and other mammalian predators as well. In Utah, nesting success is not necessarily believed to be a limiting factor for sage grouse (D. Mitchell, UDWR 2002 pers. comm.), but low chick survival during the first 2-3 weeks after hatching has been identified as a potentially limiting factor. Bunnell and Flinders (1999) also documented significant predation by red fox on sage grouse in their study area in Utah, and recently revised sage grouse management guidelines (Connelly et al. 2000) suggest that red fox populations should be discouraged in sage grouse habitats. To the extent that red fox, coyotes, common ravens (*Corvus corax*) and other predators which prey on chicks are also preying on eggs, reducing the populations of these predators from sage grouse nesting and early brood-rearing areas has the potential to benefit both nesting success and chick survival.

Cote and Sutherland (1996) reviewed and analyzed the results from 20 published studies where predation management had been undertaken to assess its effects on bird populations. Their analysis suggested that predation management consistently had a large, positive effect on hatching success and significantly increased autumn densities of the target bird species. Their analysis also suggested that predation management did not consistently result in increased breeding populations in the year following management. They speculated that this might be due to the action of density-dependence on avian populations, but noted that this has yet to be documented and deserves further research. They further suggested the possibility that predation management does in fact increase breeding populations, but the increased breeding birds emigrate out of the area into nearby areas where population monitoring or predation management may not be occurring.

Keister and Willis (1986) suggested that the major factor in determining sage grouse population levels in their study area in southeastern Oregon was loss of nests and chicks during the first 3 weeks after hatching.

Coyotes and ravens were suspected as the primary nest predators. A coyote removal project was initiated on their study area, and sage grouse productivity increased dramatically from 0.13 chicks/hen to 2.45 chicks/hen in just 3 years. Willis et al. (1993) analyzed data on sage grouse and predator populations, weather, and habitat from an area of Oregon that had some of the best sage grouse habitat in the state. The only meaningful relationship they found was a significant negative correlation between coyote abundance and the number of sage grouse chicks produced per hen. They concluded that fluctuation in predator abundance was probably the single most important factor affecting annual productivity of sage grouse in their study area. Slater (2003) however, reported on the effects on sage grouse of coyote removal for livestock protection in Wyoming. Despite differences in predator abundance between study areas, no differences were observed in nest predation rates. However, in Utah, predation management specifically for sage grouse protection has greatly increased adult grouse survival from mammalian predators in an ungrazed rangeland site near Strawberry Reservoir (D. Mitchell, UDWR 2004 pers. comm.).

11.5.4 Ring-necked pheasants and turkeys

Dumke and Pils (1973) reported that ring-necked pheasant (*Phasianus colchicus*) hens were especially prone to predation during their nest incubation period. Trautman et al. (1974) examined the effects of predation management on pheasant populations in South Dakota by monitoring pheasant populations in similar 100 mi^2 areas with and without predation management. They examined two variations of predation management for 5 years, one targeting only red fox, and the other targeting badger (*Taxidea taxus*), raccoon (*Procyon lotor*), striped skunks (M*ephitis mephitis*) and red fox. They found pheasant densities were 19% and 132% higher in predation management areas than in non-management areas during fox removal and multiple predator species removal, respectively. Chesness et al. (1968) examined the effects of predation management on pheasant populations in paired treatment and non-treatment areas in Minnesota during 3 years by targeting primarily nest predators, including skunks, raccoons, and American crows (*Corvus brachyrhynchos*). They reported a 36% hatching success in predation management areas versus a 16% hatching success in non-management areas, as well as higher clutch sizes and chick production in management areas. Nohrenberg (1999) investigated the effects of limited predation management on pheasant populations on his study areas in southern Idaho and found consistently higher pheasant survival and productivity in management areas as compared to similar non-management areas. Frey et al. (2003) reported on the results of a 4-year study to protect ring-necked pheasants in Utah. Predation management of red fox, striped skunk and raccoons resulted in increased pheasant abundance on larger study sites (41.5 sq. km.) but not on smaller study sites (10.4 sq. km.).

Thomas (1989) and Speake (1985) reported that predators were responsible for more than 40% of nest failures of wild turkeys (*Meleagris gallopavo*) in New Hampshire and Alabama, respectively. Everret et al. (1980) reported that predators destroyed seven of eight nests on his study area in northern Alabama. Lewis (1973) and Speake et al. (1985) reported that predation was also the leading cause of mortality in turkey poults, and Kurzejeski et al. (1987) reported in a radio-telemetry study that predation was the leading cause of mortality in hens. Wakeling (1991) reported that the leading natural cause of mortality among older turkeys was coyote predation, with the highest mortality rate for adult females occurring in winter. Other researchers report that hen predation is also high in spring when hens are nesting and caring for poults (Speake et al. 1985, Kurzejeski et al. 1987, Wakeling 1991). Williams et al. (1980) reported a 59% hatching success for turkeys prior to a predator poisoning campaign, versus a 72% hatching success following a predator poisoning campaign.

11.5.5 Waterfowl

In a study of waterfowl nesting success in Canada, researchers found that eggs in most nests were lost to predators such as red foxes, coyotes, striped skunks, raccoons, Franklin's ground squirrels (*Spermophilus franklinii*), badgers, black-billed magpies (*Pica pica*) and American crows (Johnson et al. 1988). Cowardin et al. (1985) determined that predation was by far the most important cause of nest failure in mallards (*Anas platyrhynchos*) on their study area. Various studies have shown skunks and raccoons to be a major waterfowl nest predator resulting in poor nesting success (Keith 1961; Urban 1970; Bandy 1965). On the Bear River Refuge in Utah, striped skunks, red fox, raccoons and ravens were documented as common predators of nesting ducks (West 2002).

In documenting an extensive study of the effects of red fox predation on waterfowl in North Dakota, Sargeant et al. (1984) concluded that reducing high levels of predation was necessary to increase waterfowl production. Balser et al. (1968) determined that predation management resulted in 60% greater production in waterfowl in areas with predation management as compared to areas without management. He also recommended that when conducting predation management, the entire complex of potential predators should be targeted or compensatory predation may occur by a species not under management, a phenomena also observed by Greenwood (1986). Rohwer et al. (1997) documented a 52% nesting success for upland nesting ducks in an area receiving predation management, versus only a 6% nesting success in a similar non-treatment area. Garrettson and Rohwer (1994) likewise documented dramatically higher duck nesting success in areas where predation management occurred during the nesting season as compared to areas where no management occurred, and noted that the annual nature of predation management allowed for greater management flexibility than most habitat management efforts.

11.6 Factors affecting predation rates

Within predator-prey relationships, a number of factors exist which can influence predation rates and effects. In some cases prey population sizes, trends and age structure are required to understand the effects of predators on the prey community. Predators removing the same number of prey from a small population as from a large population will not have the same effect, just as the removal of juveniles or males may not have the same population-level effects as removing the same number of adult, reproductively viable females. However, the relative number of predators or prey found in an area may not be the most important factor. Additionally, these factors can work together to affect impacts to prey and predator populations.

11.6.1 Habitat factors

Habitat loss/degradation, disease and other factors have resulted in declines in many species throughout their ranges. To compound this threat to, especially, T/E species, some predators have experienced unnatural population increases as a result of human development, elimination of natural predators, ecosystem imbalances, garbage, supplemental feeding, and other factors.

Some components of habitat may make prey species more or less vulnerable to predation. Some habitat is essentially linear in shape, cover is reduced and edge is increased which makes the prey more vulnerable to predators (Wilcove et al. 1986; Paton 1994). In linear habitats, predators can more effectively search the habitat. Linear habitat includes dikes constructed in waterfowl management areas, which serve as nesting habitat, and riparian corridors which are important for fawning and for sage grouse brood rearing.

In desert environments, limited access to water (due to limited water sources or drought) can concentrate predator and prey species in a limited amount of habitat. Artificial water sources, some specifically built for wildlife, may increase the suitability of habitat for prey but also predators. Such may be the case with desert ranges where bighorn sheep existed at low densities prior to construction of water catchment devices. Following the creation of water sources, mountain lions may be sustainable in these same areas and, without predation management, may impact sheep populations (Wehusen 1996, Hayes et al. 2004). Water sources developed for upland game have been visited by coyotes (R. Pope, Uinta National Forest 2004 pers. comm.) and availability of water in desert environments may sustain coyote populations through summer droughts thus increasing predator abundance.

Fires or other open habitats, which can enhance habitat for some species that depend on lower successional stages, can increase predation impacts in the short-term (Pierce et al. 2004). Possible impacts from wildfires include decreases hiding cover for prey, concentration of both predator and prey species in smaller pockets of unburned habitat, and the removal of rodent populations which serve as alternate prey.

There may be a correlation between drought and lower fawn survival rates for mule deer and pronghorn antelope. The exact cause of this correlation, however, has not been determined. Predation related impacts may include reduced fawn weights (extending the period of vulnerability of fawns), reduced cover for fawns, increased duration of feeding bouts (increasing the risk of predation) and/or decreased alternative prey. Regardless, drought is one recurring environmental factor which affects habitat and can compound the impacts of predation.

While cover is essential for predation avoidance in some species (e.g., pheasants, ducks, deer fawns) excessive cover can be detrimental for other species which rely on vision for predation detection. Pronghorn antelope rely on sight to detect coyotes and other predators, and increases in vegetation height can increase vulnerability to predation (Goldsmith 1990). It has been reported that prairie dogs (*Cynomys* spp.), including Utah prairie dogs *(C. parvidens)*, also rely on sight for predator detection and tall grasses, forbs and shrubs can increase predation (King 1955; Koford 1958; Slobodchikoff and Coast 1980).

Perches, including artificial perches such as power transmission lines, can also increase predation by raptors or ravens (Coates, in press). Both direct predation and nest predation by ravens may increase because of the presence of powerlines. Powerlines also provide nesting structures for ravens and fragment habitat (Rowland 2004, Coates, in press).

Finally, availability of escape cover can affect predation effects. It is generally accepted that rugged terrain provides escape habitat for bighorn sheep to avoid being pursued by predators (Wishart 1978). Thick cover, such as cattails (*Typha* spp.) and *Phragmites australis* provide escape habitat for pheasants. A lack of escape cover can either increase predation or cause otherwise suitable habitat to go unused (Risenhoover and Bailey 1985; Frey et al. 2003).

11.6.2 Prey factors

While predation helped to form prey species evolution, certain aspects of prey populations make them more or less vulnerable to excessive predation impacts.

Among other factors, depressed prey populations are vulnerable to negative predation effects. It is generally accepted that predation effects are compensatory mortality at or above K (Ballard et al. 2001) and additive at some point below K. Populations may be depressed due to human factors, environmental conditions or a combination. Depressed populations in this context also include recently reintroduced populations, such as black-footed ferrets (*Mustela nigripes*) or bighorn sheep.

Breeding synchrony, and thus synchronized births, influence predation effects. "Flooding" predators with fawns or calves is an evolutationary strategy to ensure that predators cannot kill all of the prey before they outgrow their vulnerability to predation (Geist 1982). At high population levels, prey populations can generally flood predator populations. However, at low population levels predators can kill a larger percentage of the fawns or calves, or if low male:female ratios exist, breeding can be extended and births spread over a longer time period, increasing the timeframe when prey species are vulnerable to predation impacts.

Group size and composition affect an individual's vulnerability to predation and may affect overall predation impacts if those individuals are critical to the viability of the population. Hornocker (1970) noted that mule deer bucks were more vulnerable to predation by mountain lions and theorized it was because of smaller groups and more rugged terrain inhabited by older bucks and mountain lions. Mooring et al. (2004) noted similar risks for bighorn rams. Conversely, large groups can increase the likelihood of predator detection and decrease the individual predation risk for some species (Geist 1982).

The availability of alternative prey can affect predation impacts in two opposing directions. Alternative prey may support increased predator populations and prevent decreases in predator numbers in a downward cycle of the prey species of concern. Cunningham et al. (1995) and Shaw (1989) noted that mountain lion populations supported by other prey did not fluctuate with declines in mule deer populations. Cattle (*Bos taurus*) (both domestic and feral) supported mountain lions in studies in Arizona (Cunningham et al. 1995) and the availability of cattle increased mountain lion impacts on deer (Shaw 1977). Wagner and Stoddart (1972) and Clark (1972) noted that coyote populations increased with jack rabbit (*Lepus californicus*) abundance in study areas in northern Utah and southeastern Idaho and livestock predation increased during downward cycles in jack rabbit populations. Presumably predation on mule deer or pronghorn fawns would similarly increase, especially in the absence of livestock or other prey as buffer. Conversely, when available, increased numbers of alternative prey can diminish predation impacts on any one species, mitigating some predation impacts (Connolly 1978). Predation on deer or pronghorn antelope fawns may be lessened during upward cycles in rodent and rabbit populations.

11.6.3 Predator factors

Just as habitat and prey factors affect predation impacts, prey may be affected by characteristics in predator populations. In addition to simple numbers of predators, other characteristics, such as a social structure of the predator population, likely have a correlation with predation rates.

Breeding pairs of coyotes which have been on established territories for several years have extensive knowledge of the territories and may be able to impact fawns to a greater extent than a younger, newly established pair. Wagner and Conover (1999) noted an apparent "residual effect" of livestock protection in the year following intensive aerial coyote removal, which may be explained by fewer experienced "alpha" coyotes and tenancy on the territory.

In livestock predation, the vast majority of domestic lamb (*Ovis aries*) losses to coyotes are attributed to breeding (alpha) pairs (which represent <50% of coyote populations) (Connolly et al. 1976; Gese and Grothe 1995; Bromley and Gese 2001). In wildlife predation, the authors suspect a similar relationship may exist. Mule deer and pronghorn antelope fawns and all ground nesting birds are vulnerable (and can be impacted) during pup rearing periods for coyotes as a result of the increased food requirements of raising young (Till 1983; Till and Knowlton 1992).

Predator population demographics are critical to understanding predation impacts. For example, in most systems coyotes do not breed their first spring and form monogamous, territorial pairs during their second year, with females first whelping on their second birthday (Bekoff 1978). In addition, coyotes are monestrous with only the dominant breeding pair typically producing a single litter per territory each spring (Kennelly and Johns 1976); beta females may also produce offspring but this rarely occurs (Gese et al. 1996). Some researchers believe food abundance regulates coyote numbers by influencing reproduction, survival, dispersal, space-use patterns, and territorial density (Gier 1968; Knowlton 1972; Todd et al. 1981; Todd and Keith 1983; Mills and Knowlton 1991; Gese et al. 1996). In contrast, Crabtree and Sheldon (1999) suggested that litter size at birth (among coyotes) appears relatively invariant with respect to changes in prey abundance, and that litter size at birth appears largely unaffected by levels of human exploitation. Red fox breed their first year and whelp on their first birthday (Creed 1960; Storm et al. 1976). Essentially, in most systems there are significant numbers of non-breeding coyotes in the population during the late winter and spring, while there are practically no non-breeding red fox (Strom et al. 1976).

Some predator species are novel in the ecosystems they currently occupy, and prey in these systems may not have evolved with adequate defense strategies. For example, red fox and raccoons are considered invasive in Utah. Raccoons can exist at high densities due, in part, to their lack of territoriality, and artificial food sources (Twichell and Dill 1949; Yeager and Rennels 1943; Urban 1970; Sonenshine and Winslow 1972; Hoffman and Gottschang 1977; Rivest and Bergerson 1981).

Territoriality affects the density of predators and can affect predation rates through population regulation. Raccoons and striped skunks do not appear to be territorial and are regulated by food availability (Twichell and Dill 1949; Yeager and Rennels 1943, Urban 1970, Sonenshine and Winslow 1972, Hoffman and Gottschang 1977; Rivest and Bergeron 1981; Rosatte 1987; Storm and Tzilkowski 1982). Red fox are territorial during breeding, but can be compressed into very small territories if food is readily available (Harris 1977; MacDonald and Newdick 1982; Harris and Rayner 1986). Breeding coyotes are territorial with non-breeders existing in spaces between territories or in unoccupied breeding territories (Bekoff 1978). Mountain lions are perhaps the most territorial and early studies indicated that territoriality regulated mountain lion populations. More recent studies suggest this is not the case and current studies have shown a large degree of territory overlap (J. Hart, USGS 2005 pers. comm.). Territories may exist as a breeding strategy or to limit access to food resources, and the degree to which territories overlap can also influence overall predator abundance and predation rates.

Predator populations are often supported by prey other than the species of concern, which allows them to exert increased impacts on recovering populations. Mountain lion populations may be supported by elk populations and cattle numbers (Shaw 1989; Cunningham et al. 1995). Raven populations have increased range-wide since the late 1960's (Sauer et al. 2005) and are likely supported by anthropogenic food sources (Coates in press). Additionally, only a portion of these increased raven populations are breeding birds, while much of the population exists as non-breeding "murders" which are not bound to a single territory during nesting and brood rearing seasons (Goodwin 1986).

Individual behavior cannot be discounted as an operative mechanism for predation impacts. Some mountain lions seem to specialize on bighorn sheep and horse foals (*Equus caballus*) as principal prey and appear to hunt them at disproportionally high rates (Turner et al. 1992; Wehausen 1996; Kamler et al. 2002; Mooring et al. 2004). Mooring et al. (2004) also noted that some mountain lions also concentrate on elk, especially in late winter and through calving season. Likewise, black bears appear to concentrate on cervid calving areas and doubtlessly some bears specialize on preying on cervid calves and fawns during a short period of the year (Wilton 1984; Wilton et al. 1984).

Finally, the combination of predator species may have an increased impact on a single resource. Predation or predation risk from bears, wolves or mountain lions alone may not impact elk herds, but the complex of predators may impact survival in some systems, particularly if different predators impact prey at different stages. In Utah, coyote, mountain lion and, in some areas, bears may collectively impact mule deer fawn survival. Red fox predation management for sage grouse protection aided adult grouse survival in Strawberry Valley, but chick:hen ratios remained chronically low until raven removals were incorporated into the strategy (J. Flinders, BYU 2005 pers. comm.).

As an example of the complexity of effects in multiple predator systems, consider mountain lion predation on mule deer. Hornocker (1970) and Logan and Sweanor (2001) indicated that mountain lions kill male deer in excess of their abundance in the population while they kill female and fawn deer in relative proportion to their abundance to each other. When fawns are abundant, predation by mountain lions is distributed between does and fawns and relatively few breeding age does are killed (compensatory mortality). However, if coyotes or other predators reduce fawn abundance, mountain lion predation increasingly becomes concentrated on adult does (additive mortality). This shift in mortality is not predicated on the deer's relationship to K and may happen at any level.

11.7 Specific strategies

11.7.1 Non-lethal strategies

Predation management does not necessarily mean only lethal predator removal. Non-lethal strategies may include making habitat less attractive to specific predator species, predator exclusion devices (i.e., fencing), predator aversion techniques, live trapping and translocation and other techniques. Each strategy, including lethal predator removal has strengths and weaknesses that vary not only with the technique, but with the specific application. It is important to note that non-lethal strategies are those which are not directly lethal to the predator. However, some non-lethal methods, such as making habitat less suitable or translocation may result in mortality for predators that are forced to disperse or compete with resident individuals for limited resources.

While numerous non-lethal strategies exist for livestock protection, few have been identified for wildlife protection. Certainly, habitat enhancement which increases K may be beneficial if it allows prey species populations to increase by reducing density dependent predation. Habitat improvements must be designed to preclude creating linear habitat, creating perching habitat or other features that make prey more vulnerable to predators. Further, habitat enhancement may take many years or decades and significant financial commitments before it becomes attractive and sufficient to support larger prey species populations.

Habitat enhancement to increase K is generally expensive and may not address predation impacts. Predation management should be designed to increase wildlife populations to the level of the current habitat and not to artificially elevate prey populations above current habitat limits. Habitat enhancement designed to address water or escape cover availability or to mitigate for fire impacts may be beneficial by limiting negative predation impacts. Habitat enhancement which reverses encroachment of junipers (*Juniperus* spp.) or sage brush may not mitigate predation impacts. In fact, if the design of the habitat project leaves isolated pockets of habitat, it may exacerbate predation by making the habitat easier for the predators to hunt.

Of the features identified as prey factors above, only breeding synchrony may be addressed as non-lethal predation management technique. Increased buck:doe ratios may increase breeding and fawning synchrony and may limit predation impacts.

Factors identified as predator factors are most often addressed through selective predator removal. In the case of endangered predators (i.e., Canada lynx (*Lynx canadensis*) or gray wolves), capture and translocation may be a more appropriate management strategy, due to limited numbers of the predator. However, for more common, non-listed predators, selective lethal management may be the most cost-effective and biologically sound strategy.

11.7.2 Mule deer protection strategies

Predation management for mule deer protection may be necessary when mule deer populations are substantially below established management objectives. Predation management may be implemented when populations are depressed in relation to the objectives and the trend is stable to declining. Fawn:doe ratios of less than 50:100 indicate recruitment is a limiting factor in stabilizing the population. Because a complex of predators may be impacting mule deer populations, mountain lion and black bear harvest through sport hunting may also be implemented in an area with a depressed deer population. Mountain lions appear to kill mule deer does and fawns in relation to their relative abundance (Hornocker 1970; Logan and Sweanor 2001), so mountain lion predation may be indicated by declining populations overall, but not in low fawn:doe ratios. Black bears appear to concentrate on fawning areas and some bears appear to specialize on preying on calves and fawns (Wilton 1984; Wilton et al. 1984). If predation is limiting recruitment, low fawn:doe ratios generally indicate neonatal coyote predation (Connolly 1978).

As noted above, the authors believe territorial, breeding coyotes play a disproportionate role in neonatal fawn predation due, in part, to their need to provision pups during fawning season (Till 1982, Till and Knowlton 1992). Therefore, the current strategy is to remove breeding coyotes where their territories overlap critical fawning grounds, preferably before whelping. Aerial gunning during the coyote breeding season is more selective for breeders where ground crews elicit howling and direct the aircraft to responding coyotes; non-breeding coyotes generally do not howl during this season (Gese and Ruff 1998).

Biologically, coyote populations may lend themselves to certain management strategies. Gantz (1990) noted that breeding coyotes remained on their mountainous territories during January-March even at elevations above 7500 feet. Pair bonds are established in December and

January, with breeding in late January through February. One coyote predation management strategy is to remove coyotes after pair bonds are established, but before whelping in late March and April. Tracking and shooting coyotes from helicopters on high elevation fawning range between January and March targets territorial coyotes whose territories overlap fawning areas.

In desert ranges, fawning habitat may be described more by access to water than by an elevation or geographical boundaries. In these areas, removal of the territorial coyote pairs preying on deer, including fawns using the same water sources is more critical than coyote removal in the general area when deer populations are low and identified in need of protection.

Ballard et al. (2001) recommended removal of 70% of the coyotes in an area to be effective. The authors reject this as a requirement for addressing primary predation impacts and point to effective livestock protection with a much lower removal rate. If coyote removal was non-selective, removal of 70% may be necessary to ensure removal of the breeding pairs. However, with selectively targeting breeding coyotes (and dens in some areas), removal of as little as 30% of the overall coyote population may be adequate to protect fawns during their first 1½ months of vulnerability.

The exception to this may be where secondary effects of predation keep deer populations chronically below management objectives. If the objective is realistically set based on K, increased predation management may be necessary to allow more deer to survive through winter. In an adaptive management strategy, if predation management on fawning range does not appear to increase recruitment, additional or a different management strategy may be necessary to address possible secondary effects.

Coyote predation management will also need to deal with seasonal work loads of field personnel. Ideally, coyote predation management should begin in late December and continue through June. Management after April 1 will likely be ground work as a follow-up to aerial gunning because of the lack of snow; female coyotes are more sedentary with pups and possible work load increase for field personnel. Because of the quantity of management actions to be done, management may need to begin in November, with follow-up flights or ground work to remove breeding coyotes that re-occupy vacant territories.

11.7.3 Pronghorn antelope protection strategies

Like deer, pronghorn antelope are more vulnerable during the first 1½ months of life and may be in need of protection if management objectives are not being met. An exception would be predation on adult pronghorn during drought (i.e., where coyotes and pronghorn may be concentrated near limited water sources) or following a transplant effort, where adults may be stressed or wander into areas where predator detection is limited. Successful pronghorn protection projects have addressed only primary predation impacts.

Exceptions aside, an effective strategy for pronghorn protection is to remove breeding coyote pairs before whelping in April (pronghorn fawns are born in May) and removal of 30% or less of the overall coyote population is documented to be adequate to protect pronghorn fawns during their first 1½ months of vulnerability (Neff et al. 1985). Waiting until female coyotes whelp can render management strategies less effective. Removing only breeding males will likely result in limited pronghorn protection as females will continue to provision pups. Field observations indicate that coyotes concentrate near domestic cattle calving pastures during calving, presumably because of the availability of cattle placentas after birthing and vulnerable calves. Where calving pastures overlap pronghorn antelope fawning areas, integrated predation management will protect both resources, as cattle are generally born before pronghorns and before coyotes whelping. Because pronghorn exist in open habitat, fixed-wing aerial gunning is usually adequate, although some ground efforts may be necessary in foothill areas or deep, brush filled ravines.

11.7.4 Bighorn sheep protection strategies

In limited circumstances (i.e., transplanted populations) predation management may be necessary to protect bighorn sheep populations, especially where winter snow limits access to escape cover. In most bighorn sheep populations, coyote predation does not limit bighorn sheep. Mountain lions however, can exert significant predation pressure on bighorn sheep, both primary predation and secondary effects in habitat selection (Wehausen 1996). For example, some big horn sheep management plans call for minimum viable population size of 125 sheep (UDWR Statewide Management Plan for Bighorn Sheep 1999). Predation management may be necessary if predation impacts a population below that number or where little alternative prey exists.

In some dessert sheep units, little alternative prey exist to support a mountain lion (M. Bodenchuk, WS 2006 pers. comm.). In these areas, any mountain lion existing in the area likely depends on bighorn sheep. When it is determined that bighorn sheep in these areas are in need of protection, any mountain lion found near the sheep should be considered for removal since they could prey on and impact the local bighorn sheep population. Mountain lion predation management may be implemented using trailing hounds, foot or neck snares and, in limited areas, with helicopter aerial gunning on fresh snow.

11.7.5 Sage grouse protection strategies

Sage grouse protection is an example of adaptive predation management. In some areas that is no information to indicate that predation is negatively affecting sage grouse populations. However, in many of these areas, Wildlife Services conducts predation management for livestock protection and these efforts must be considered part of the existing baseline. In other areas, predation (either primary or secondary effects) is limiting sage grouse populations (Bodenchuk 2006).

As an example, in the Strawberry Valley, Utah, sage grouse experienced significant declines, despite the elimination of grazing in core habitat areas. Red fox predation was identified as a significant cause of mortality, and red fox predation management has reduced fox predation (Bunnell and Flinders 1999). However, chick:hen ratios were not yet at objective and raven predation management, using DRC-1339 treated eggs, significantly reduced raven predation in the area (i.e., 90% reduction based on transects). Chick:hen ratios then increased towards management objective of 2:1 (R. Baxter, BYU 2005 pers. comm.).

One avenue remaining to investigate is the effect predation has on expanding populations including primary and secondary effects in dispersal habitat. Sage grouse populations typically inhabit large, unbroken expanses of sagebrush (Rowland 2004). Predation or the risk of predation in areas without active predation management may limit pioneering efforts by birds, thus limiting use of habitat with otherwise adequate life history requirements of the birds. An experimental predation management effort may be justified in areas determined to have suitable habitat to see if the birds colonize and re-establish populations in these areas.

Predation management may also be justified in some areas due to the vulnerability of the populations. Gunnison's sage grouse (*Centrocercus minimus*) in San Juan County, Utah exist in only four known leks, and any mortality may jeopardize these populations (San Juan County 2000).

Further, predation management supports investments in habitat improvement in these areas, as part of an adaptive management strategy (Bodenchuk 2006). The costs of predation management (i.e., both ecological and financial) may be less than the cost of detailed research projects to scientifically evaluate if management is justified.

Where justified, predation management should address areas of suitable habitat, especially areas within the nesting radius (i.e., 5 miles) of known leks (Rowland 2004). Removal of red fox, adult coyote or badger would reduce adult and nest predation by mammals and weekly treatment of ravens using DRC-1339 treated eggs would reduce raven egg and chick predation (Coates and Delehanty 2004). This predation management strategy could effectively reduce both primary and secondary predation effects and should be conducted from initiation of sage grouse breeding through the end of June (Coates and Delehanty 2004).

11.7.6 Ring-necked pheasant protection strategies

Pheasants appear vulnerable in winter when concentrated in limited cover, and through the nesting season. In large enough areas of habitat in Utah, predation management before the nesting season doubled ring-necked pheasant abundance (Frey et al. 2003). These efforts included removal of striped skunks, raccoons and red fox. Residual effects of management were noted on the 16.5 sq. mi. experimental areas for raccoon and striped skunks (i.e., lower abundance), but red fox reoccupied these areas within a year. However, pheasant populations remained higher in treatment areas than in non-treatment areas one year after management actions were terminated (T. Messmer, Utah State University 2003 pers. comm.).

Mammalian predation management was accomplished mostly by using snares and traps, and with limited spotlight shooting from December through June. While predation management was effective in increasing pheasant abundance, and subsequently pheasant harvest, many areas that pheasants inhabit are on private ownership and public access may be limited.

11.7.7 Waterfowl and shore bird protection strategies

Significant populations of ducks, geese and shore birds use declining wetland areas as nesting areas. For example, earlier reports indicated that the Bear River Delta, Utah represents a significant source population for American avocets (*Recurvirostra americana*) and cinnamon teal (*Anas cyanoptera*), and prior to 1983, an estimated 80,000 ducklings were produced at Bear River Refuge (West 2002). Since the flood in the early 1980's and subsequent habitat and water management modifications (i.e., creation of dikes, invasion of raccoons and red fox), few ducklings have been produced (West 2002). To reverse this decline in productivity, management strategies have focused on reducing red fox, raccoons and striped skunks predation along nesting habitat from late March through June.

Significant nest predation exists from ravens which can easily hunt linear habitat and remove eggs (A. Trout, USFWS 2004 pers. comm.). Nesting ravens at this time of year occupy territories and the authors believe damage from these birds is limited. However, non-nesting ravens make up a substantial portion of the raven population and these birds can move between areas and effectively hunt nesting habitat (Goodwin 1986). The Environmental Protection Agency registration label for DRC-1339 egg baits allows for protection of federally listed Threatened or Endangered Species *or other wildlife in need of special protection* (emphasis added). Raven predation management for the protection of nesting shorebirds and waterfowl maybe effective using DRC-1339.

Implementing predation management prior to and through nesting season addresses both primary and secondary effects. West (2002) noted sharp increases in nesting as a result of predation management (i.e., protection from mammalian predators) indicating nest site selection may be influenced by predation risk. Secondary predation effects (i.e., reduced nest site availability) may be more limiting than primary effects, but a strategy which addresses both primary and secondary predation would be more effective for reaching waterfowl and shore bird recruitment objectives.

11.7.8 Endangered and threatened species protection strategies

Numerous examples exist for species that benefit from predation management. Examples in the western United States include: Utah Prairie dogs, desert tortoise (*Gopherus agassizii*) and black-footed ferrets. Other examples are the five species of sea turtles that inhabit the Atlantic and Gulf Coasts of the United States. These species include the: loggerhead (*Caretta caretta*), green (*Chelonia mydas*), leatherback (*Dermochelys coriacea*), hawksbill (*Eretmochelys imbricata*), and Kemp's ridley (*Lepidochelys kempii*). Additional there are seven subspecies of beach mice that inhabit the Atlantic and Gulf Coasts of the United States that could benefit from predation management.

Utah prairie dogs, like other prairie dog species, are vulnerable during translocation attempts and badger management is warranted through their first summer (Coffen and Pederson 1993, Truett et al. 2001, K. McDonald, UDWR 1997 pers. comm., B. Bonebreak, BLM 2005 pers. comm.). Coyotes have also been noted in these areas and selective predation management of coyotes may also be warranted (T. Bonzo, UDWR 2005 pers comm.). Coyote predation management may be accomplished through aerial gunning but badger predation management would require the use of conibear and foot-hold traps, neck snares (where the colony is fenced) and spotlight shooting. Traps must be equipped with pan-tension devices when set near prairie dog colonies to exclude prairie dogs and other lighter weight non-target species (Turkowski et al. 1984; Phillips and Gruver 1996).

Young desert tortoise are vulnerable to raven and coyote predation (USFWS 1990, F. Rowley, BLM 1987 pers comm.). Predation by ravens is associated with nesting ravens, but proactive predation management in Clark County, Nevada addresses all raven populations which are supported by anthropogenic food sources (R. Beach, WS 2003 pers. comm.). Coyote predation is primarily limited to juvenile tortoises and may be a factor in April when tortoises are above ground and coyotes are provisioning pups (G. Larson, WS unpubl. rep. 1988). Prior to the removal of livestock, Wildlife Services conducted significant coyote predation management projects in tortoise habitat.

Captive-bred, reintroduced populations of black-footed ferrets appear to be vulnerable to predation (Colorado-Utah Black-footed Ferret Recovery Working Team 2001), and in South Dakota the 30-day post release survival rate without predation management was only about 30% while it approached 80% with predation management in place (Badlands National Park, unpubl. rep.).

In the cases above, predation management also provided biological samples for disease monitoring which is critical for both prairie dog and ferret reintroduction and recovery.

All turtle species listed are protected under the U. S. Endangered Species Act, international agreements, and state laws. Heavy predations from a variety of predators and other activities have significantly decreased the breeding success of sea turtles. It has been determined that the most significant predators of sea turtle nests are raccoons, red foxes, coyotes, feral/free-ranging dogs (*Canis familiaris*), feral hogs (*Sus scrofa*), and ghost crabs (*Ocypode* spp.). Recently, in some areas of the southwestern Florida, coyotes have learned to excavate and feed on sea turtle eggs. The nine-banded armadillo (*Dasypus novemcinctus*), has also been observed to excavate and consume sea turtle eggs along some beaches; apparently, this is a new development in armadillo learned behavior. It has become critical for the continued existence of these T/E sea turtles that nest predation is actively monitored and predation management initiated.

Six federally listed T/E species of mice, two species of endangered rats, and one species of endangered rabbit are found along the Florida's coastal regions and include the following: Perdido Key beach mouse (*Peromyscus polionotus trissyllepsis*), Saint Andrews beach mouse (*Peromyscus polionotus peninsularis*), Anastasia Island beach mouse (*Peromyscus polionotus phasma*), Choctawhatchee beach mouse (*Peromyscus polionotus allophrys*), Key Largo cotton mouse (*Peromyscus gossypinus allapaticola*), Key Largo Woodrat (*Neotoma floridana smalli*), Southeastern beach mouse (*Peromyscus polionotus niveiventris*), silver rice rat (*Oryzomys palustris natator*), and the Lower Keys marsh rabbit (*Sylvilagus palustris hefneri*). The suspected and potential predators of these endangered mammals include feral/free-ranging house cats, bobcats (*Felis rufus*), foxes, coyotes, feral/free-ranging dogs, black rats (*Rattus rattus*), raccoons, skunks (*Mephitis mephitis and Spilogale putorius*), armadillos, owls (*Tytonidae and Strigidae*), hawks (*Accipitridae*), great blue herons (*Ardea herodias*) and snakes (*Masticophis flagellum, Coluber constrictor, and Elaphe* spp.) (USFWS 1999).

In general big game protection objectives are to increase overall recruitment in a specific area. However, T/E species protection is conducted to establish populations, increase populations in specific areas or protect newly recolonizing populations. It is considered inadequate to remove a predator as it leaves a critical habitat or reintroduction area after killing a reintroduced T/E species. Preventive predation management is designed to be intensive in scope, but usually limited to a relatively small area (i.e., reintroduction site) and usually for a short period of time (i.e., usually just prior to the release of a T/E species).

11.8 Conclusion

Biologists, researchers and academia will continue to debate whether predation is a regulatory or a limiting factor (Sinclair 1991; Skoland 1991b; Boutin 1992; Van Ballenberghe and Ballard 1994), but to wildlife managers, responsible for managing and maintaining wildlife populations, the distinction between regulatory or limiting may not be obvious or of great importance. Conditions that allow predation to become a factor limiting recruitment are dynamic in natural ecosystems with a full complement of predators which complicates management. Managers need to determine when most "mortalities" occur and whether primary predation, secondary predation impacts, or other factors are an important cause.

Further studies may be required to determine if secondary predation effects are adversely affecting wildlife populations (Lima and Dill 1990) as the secondary effects of predation are often difficult to observe and understand. These studies may result in redefining "carrying capacity" as currently used or better describe the role predators play in the utilization of resources by prey species. While the models described by Ballard et al. (2001) can incorporate secondary effects of predation, the secondary effects would have the effect of lowering K based on the prey populations' use of the habitat. Habitat is currently thought of as food, water, cover and space. Interspecies relationships between predators and prey certainly affect the availability of food, water, cover and space and may make habitat availability more difficult for prey species to use and for managers to interpret. If predators drastically influence prey behavior and displace those prey into less optimal habitats or increase prey species energy requirements, managers have a more difficult job and must consider the secondary effects of predation to effectively manage wildlife populations under their responsibility.

Managers are best served to approach predation management with an open mind, remembering that the goal is active management and conservation of wildlife. Reducing predation is sometimes a necessary component of strategies to accomplish management or recovery objectives. The best overall predation management strategy is an adaptive management approach that monitors many factors, considers a full range of predation management techniques, including nonlethal and lethal, continually evaluates method potential and actual effectiveness, and makes appropriate adjustments (USDA 1997).

Predation management is a critical component of wildlife management, and when done in as targeted a manner as possible, can be accomplished without negative environmental impacts. Predation management should be designed as a component of an adaptive management strategy, considering both non-lethal and lethal techniques, where populations are not meeting objectives or recovery goals. The predation management strategy must address the critical components (i.e., adult survival, recruitment, nest success) identified by management as lacking in performance so that management objectives can be reached or maintained as effectively as possible.

11.9 Acknowledgement

We are indebted to W. Pitt and D. Bergman for providing comments and editorial suggestions. Our sincere gratitude is extended to M. Howell and L. Paulik, National Wildlife Research Center, Fort Collins, Colorado for their diligent work and providing support with published literature that was used to develop this review. This review was supported by USDA-APHIS-Wildlife Services.

References

Arizona Department of Game and Fish (2004) Press Release on Heart Bar exclosures and predator impacts during drought. Phoenix, Arizona

Arrington ON, Edwards AE (1951) Predator control as a factor in antelope management. Transactions North American Wildlife Conference 16: 179-193

Ballard WB, Lutz D, Keegan TW, Carpenter LH, deVos JC Jr (2001) Deer-predator relationships: a review of recent North American studies with emphasis on mule and black-tailed deer. Wildlife Society Bulletin 29: 99-115

Balser DS, Dill DH, Nelson HK (1968) Effect of predator reduction on waterfowl nesting success. Journal of Wildlife Management 32: 669-682

Bandy LW (1965) The colonization of artificial nesting structures by wild mallards and black ducks. MS Thesis, Ohio State University, Columbus 67 pp

Barber SM, PJ White, Mech LD (2004) Multi-trophic level ecology of wolves, elk, and vegetation in Yellowstone National Park: elk calf mortality study. NRPP Project #71604, Annual Accomplishment Report, PO Box 168 Yellowstone Center for Resources, Yellowstone National Park, Mammoth, Wyoming

Barrett MW (1978) Pronghorn fawn mortality in Alberta. Proceedings Pronghorn Antelope Workshop 8: 429-444

Bartmann RM, White GC, Carpenter LH (1992) Compensatory mortality in a Colorado mule deer population. Wildlife Monographs 121 pp

Bartush WS (1978) Mortality of white-tailed deer fawns in the Wichita Mountains, Comanche County, Oklahoma, Part II. MS Thesis Oklahoma State University, Stillwater, Oklahoma 161 pp

Bates B, Welch M (1999) Managing mule deer recovery through the use of predator management plans. Proceedings Deer/Elk Workshop. Salt Lake City, Utah in press

Beale DM (1978) Birth rate and fawn mortality among pronghorn antelope in western Utah. Proceedings Pronghorn Antelope Workshop 8: 445-448

Beale DM, Smith AD (1973) Mortality of pronghorn antelope fawns in western Utah. Journal of Wildlife Management 37: 343-352

Beasom SL (1974) Relationships between predator removal and white-tailed deer net productivity. Journal of Wildlife Management 38: 854-859

Bekoff M (editor) (1978) Coyotes: Biology, Behavior and Management, Academic Press, New York
Bergerud AT (1988) Increasing the numbers of grouse. In Bergerud AT, Gratson MW (eds) Adaptive Strategies and Population Ecology of Northern Grouse. University of Minnesota Press, Minneapolis, Minnesota 686-731
Bergerud AT, Page RE (1987) Displacement and dispersion of parturient caribou calving as antipredator tactics. Canadian Journal of Zoology 65: 1597-1606
Bergerud AT, Butler HE, Miller DR (1983) Antipredator tactics of calving caribou: dispersion in mountains. Canadian Journal of Zoology 62: 1566-1575
Bliech VC, Bowyer RT, Wehausen JD (1997) Sexual segregation in mountain sheep: resources or predation. Wildlife Monographs134: 1-50
Bodenchuk MJ (2006) Sage grouse and predation management. The Communicator: A Quarterly publication of Utah's Community Based Conservation Program, Vol 2. Issue 1. Utah State University, Logan, Utah
Bodie WL (1978) Pronghorn fawn mortality in the upper Pahsimeroi River drainage of central Idaho. Proceedings Pronghorn Antelope Workshop 8: 417-428
Boutin S (1992) Predation and moose population dynamics: a critique. Journal of Wildlife Management 56: 116-127
Bromley C, Gese EM (2001) Effects of sterilization on territory fidelity and maintenance, pair bond, and survival of free-ranging coyotes. Canadian Journal of Zoology 79: 386-392
Bunnell KD, Flinders JT (1999) Restoration of sage grouse in Strawberry Valley, Utah 1998-99 report. Unpublished report to: Utah Reclamation Mitigation and Conservation Commission. Brigham Young University, Provo, Utah
Burk T (1982) Evolutionary significance of predation on sexually signaling males. Florida Entomologist 65: 90-104
Burkepile NA, Reese KP, Connelly JW (2001) Mortality patterns of sage grouse chicks in southeast Idaho. Abstract of presentation made at 2001 Annual Meeting of the Idaho Chapter of The Wildlife Society, Boise, Idaho
Byers JA (1997) American pronghorn: social adaptations and the ghost of predators past. University Chicago Press, Chicago, Illinois
Chessness AA, Nelson MM, Longley WH (1968) The effect of predator removal on pheasant reproductive success. Journal of Wildlife Management 32: 683-697
Clark FW (1972) Influence of jackrabbit density on coyote population change. Journal of Wildlife Management 36: 343-356
Coates PS (2006) Efficacy of chicken egg baits treated with DRC-1339 to remove common ravens. Proceedings Vertebrate Pest Conference in press
Coates PS, Delehanty DJ (2004) The effects of raven removal on sage grouse nest success. Proceedings Vertebrate Pest Conference 21: 17-20
Coffen M, Pedersen J (1993) Techniques for the transplant of Utah prairie dogs. In Management of Prairie Dog Complexes for the Black-Footed Ferret. Oldemeyer J, Biggins D, Miller B (eds) Biological Rep 13, USFWS, Washington, DC 60-66

Colorado-Utah Black-footed Ferret Recovery Working Team (2001) Colorado-Utah Black-footed Ferret Allocation Proposal. Colorado Division of Wildlife, 6060 Broadway, Denver, Colorado

Connelly JW, Braun CE (1997) Long-term changes in sage grouse *Centrocercus urophasianus* populations in western North America. Wildlife Biology 3; 229-234

Connelly, JW, Schroeder MA, Sands AR, Braun CE (2000) Guidelines to manage sage grouse populations and their habitats. Wildlife Society Bulletin 28: 967-985

Connolly GE (1978) Predators and Predator Control. In Schmidt JL, Gilbert DL (eds) Big Game of North America: Ecology and Management. Wildlife Management Institute 369-394

Connolly GE, Timm RM, Howard WE, Longhurst WM (1976) Sheep killing behavior of captive coyotes. Journal of Wildlife Management 40: 400-407

Cook, RS, White M, Trainer DO, Glazener WC (1971) Mortality of young white-tailed deer fawns in south Texas. Journal of Wildlife Management 35: 47-56

Cote IM, Sutherland WJ (1996) The effectiveness of removing predators to protect bird populations. Conserv Biol 11: 395-405

Cowardin LM, Gilmer DS, Shaiffer CW (1985) Mallard recruitment in the agricultural environment of North Dakota. Wildlife Monographs 92: 1-37

Crabtree RL, JW Sheldon (1999) Coyotes and canid coexistence in Yellowstone. In Clark TW, Curlee AP, Minta SC, Karieva PM (eds) Carnivores in Ecosystems, The Yellowstone Experience. Yale University Press, New Haven, Connecticut 127-163

Creed RFS (1960) Gonad changes in the wild red fox (*Vulpes vulpes crucigera*). Journal of Physiology (London) 151: 19-20

Creel S, Winnie J, Jr (2005) Response of elk herd size to fine-scale spatial and temporal variation in the risk of predation by wolves. Animal Behavior 69 pp

Creel S, Winnie J, Jr, Maxwell B, Hamlin K, Creel M (2005) Elk alter habitat selection as an antipredator response to wolves. Ecology in press

Cunningham SC, Haynes LA, Gustavson C, Haywood DD (1995) Evaluation of the interaction between mountain lions and cattle in the Aravipa-Klodyke area of Southeast Arizona Final Report. Arizona Game and Fish Department, Research Branch. Technical Report #17: 64 pp

Dumke RT, Pils CM (1973) Mortality of radio-tagged pheasants on the Waterloo wildlife Area. Wisconsin Department of Natural Resources Technical Bulletin 72: 52 pp

Everett DD, Speake DW, Maddox WK (1980) Natality and neonatality of a north Alabama wild turkey population. Proceedings National Wild Turkey Symposium 4: 117-126

Edwards J (1983) Diet shifts in moose due to predator avoidance. Oecologia 60: 185-189

Ferguson SH, Bergerud AT, Ferguason R (1988) Predation risk and habitat selection in the persistence of a remnant caribou population. Oecologia 76: 236-245

Frey SN, Majors S, Conover MR, Messmer TA, Mitchell DL (2003) Effect of predator control on ring-necked pheasant populations. Wildlife Society Bulletin 31: 727-735

Gantz G (1990) Seasonal movement pattern of coyotes in the Bear River Mountains of Utah and Idaho. MS Thesis, Utah State University, Logan, Utah. 67 pp

GAO (United States General Accounting Office) (2001) Wildlife Services Program: Information on activities to manage wildlife damage. Report to Congressional Committees. GOA-02-138 71

Garner GW (1976) Mortality of white-tailed deer fawns in the Wichita Mountains, Comanche County, Oklahoma. PhD Thesis, Oklahoma State University, Stillwater, Oklahoma 113 pp

Garner GW, Morrison JA, Lewis JC (1976) Mortality of white-tailed deer fawns in the Wichita Mountains, Oklahoma. Proceedings Annual Conference Southeast Association Fish and Wildlife Agencies 13: 493-506

Garrettson PR, Rohwer FC (1994) Effects of mammalian predator removal on nest success of upland ducks on conservation reserve program land in North Dakota. Delta Waterfowl Wetlands Research Station, 1994 Project Report 7

Geist V (1982) Adaptive behavioral strategies. In Thomas JW, DE Toweill (eds) Elk of North America, Ecology and Management, A Wildlife Management Institute Book, Stackpole, Books, Harrisburg, Pennsylvania 218-277

Gese EM, Grothe S (1995) Analysis of coyote predation on deer and elk during winter in Yellowstone National Park, Wyoming. American Midland Naturalist 133: 36-43

Gese EM, Ruff RL (1998) Howling by coyotes (*Canis latrans*): variation among social classes, seasons, and pack sizes. Canadian Journal of Zoology 76: 1037-1043

Gese EM, Ruff RL, Crabtree R L (1996) Social and nutritional factors influencing the dispersal of resident coyotes. Animal Behaviour 52: 1025-1043

Gier JT (1968) Coyotes of Kansas. Kansas Agricultural Experiment Station Bulletin 393, Kansas State University, Manhattan, Kansas

Gill RB, Beck TDI, Bishop CJ, Freddy DJ, Hobbs NT, Kahn RH, Miller MW, Pojar TM, White GC (2001) Declining mule deer populations in Colorado: reasons and responses a report to the Colorado Legislature, November 1999). Colorado Division of Wildlife Special Report 77

Goldsmith AE (1990) Vigilance behaviour of pronghorns in different habitats. Journal of Mammalogy 71: 460-462

Goodwin D (1986) Crows of the World. Raven. British Museum of Natural History. Cornell University Press, Ithaca, New York 138-145

Greenwood RJ (1986) Influence of striped skunk removal on upland duck nest success in North Dakota. Wildlife Society Bulletin 14: 6-11

Gregg MA (1991) Use and selection of nesting habitat by sage grouse in Oregon. Thesis, Oregon State University, Corvallis, Oregon

Gude J, Garrott B (2003) Lower Madison Valley wolf-ungulate research project 2002-2003 annual report. Montana State University, Bozeman, Montana

Gude JA, Garrott RA, Borkowski JJ, King FJ (Sumitted 2005) Prey risk allocation in a grazing ecosystem. Ecol Appl

Guthery FS, Beasom SL (1977) Responses of game and nongame wildlife to predator control in south Texas. Journal of Range Management 30: 404-409

Hailey TL (1979) A handbook for pronghorn management in Texas. Fed Aid Wildlife Restor Rept Ser No 20. Texas Parks and Wildl. Dep., Austin, Texas 59 pp

Hall LK (ed) (1984) White-tailed deer ecology and management. Stackpole, Harrisburg, Pennsylvania

Hamlin K (2005) Monitoring and assessment of wolf-ungulate interactions and population trends within the Greater Yellowstone area, southwestern Montana and Montana statewide. Montana Department of Fish, Wildlife and Parks, Helena, Montana

Hamlin KL, Riley SJ, Pyrah D, Dood AR, Mackie RJ (1984) Relationships among mule deer fawn mortality, coyotes, and alternate prey species during summer. Journal of Wildlife Management 48: 489-499

Harris S (1977) Distribution, habitat utilization and age structure of a suburban fox (*Vulpes vulpes*) population. Mammal Review 7: 25-39

Harris S, JMV Rayner (1986) Urban fox (*Vulpes vulpes*) population estimates and habitat requirements in several British cities. Journal of Animal Ecology 55: 575-591

Hayes CL, Rubin ES, Jorgensen MC, Boyce WM (2000) Mountain lion predation of bighorn sheep in the Peninsular Ranges, California. Journal of Wildlife Management 64: 954-959

Hecht A, Nickerson PR (1999) The need for predator management in conservation of some vulnerable species. Endangered Species Update 16: 114-118

Hoban PA (1990) A review of desert sheep in the San Andres Mountains, New Mexico. Desert Bighorn Sheep Council Transactions 34: 14-22

Hoffmann CO, Gottschang JL (1977) Numbers, distribution, and movements of a raccoon population in a suburban residential community. Journal of Mammalogy 58: 623-636

Holle DG (1977) Diet and general availability of prey of the coyote (*Canis latrans*) at the Wichita Mountains National Wildlife Refuge, Oklahoma. MS Thesis. Oklahoma State University, Stillwater, Oklahoma 59 pp

Hornocker MG (1970) An analysis of mountain lion predation upon mule deer and elk in the Idaho Primitive Area. Wildlife Monographs 21: 1-39

Johnson DH, Sargeant AB, Greenwood RJ (1988) Importance of individual species of predators on nesting success of ducks in the Canadian prairie pothole region. Canadian Journal of Zoology 67: 291-297

Jones PV, Jr (1949) Antelope management. Coyote predation on antelope fawns: main factor in limiting increase of pronghorns in the upper and lower plains areas in Texas. Texas Game and Fish 7: 4-5, 18-20

Julander O, Robinette WL, Jones DA (1961) Relation of summer range conditions to mule deer herd productivity. Journal of Wildlife Management 25: 54-60

Kamler JF, Lee RM, deVos JC, Jr, Ballard WB, Whitlaw HA (2002) Survival and cougar predation of translocated bighorn sheep in Arizona. Journal of Wildlife Management 66: 1267-1272

Keegan TW, Wakeling BF (2003) Elk and deer competition. In deVos JC, Jr, Conover MR, Headrick NE (eds) Mule Deer Conservation: Issues and Management Strategies. Berryman Institute Press, Utah State University, Logan, Utah 139-150

Keister, GP, Willis MJ (1986) Habitat selection and success of sage grouse hens while nesting and brooding. Progress report. Pitman Robinson Project W-87-R-2. Oregon Department of Fish and Wildlife, Portland, Oregon

Keith LB (1961) A study of waterfowl ecology on small impoundments in southeastern Alberta. Wildlife Monographs 6 pp

Kennelly JJ, Johns BE (1976) The estrous cycle of coyotes. Journal of Wildlife Management 40: 272-277

Kie JG (1999) Optimal foraging and risk of predation: effects on behavior and social structure in ungulates. Journal of Mammalogy 80: 1114-1129

King JA (1955) Social behavior, social organization and population dynamics in a black-tailed prairie dog town in the Black Hills of South Dakota. Contribution of the Laboratory of Vertebrate Biology 67. University Michigan, Ann Arbor, Michigan 123 pp

Knowlton FF (1964) Aspects of coyote predation in south Texas with special reference to white-tailed deer. PhD Thesis, Purdue University Lafayette, Indiana 147 pp

Knowlton FF (1972) Preliminary interpretation of coyote population mechanics with some management implications. Journal of Wildlife Management 36: 369-382

Knowlton FF, Stoddart LC (1992) Some observations from two coyote-prey studies. In Boer AH (ed) Ecology and Management of the Eastern Coyote. University of New Brunswick, Fredericton, New Brunswick, Canada 101-121

Koford CB (1958) Prairie dogs, whitefaces, and blue grama. Wildlife Monographs 3: 1-78

Kunkel KE, DH Pletsher (2000) Habitat factors affecting vulnerability of moose to predation by wolves in southeastern British Columbia. Canadian Journal of Zoology 78: 150-157

Kurzejeski EW, Vangilder LD, Lewis JB (1987) Survival of wild turkey hens in north Missouri. Journal of Wildlife Management 51, 188-193

LeCount A (1977) Causes of fawn mortality. Final Report, Federal Aid Wildlife Restoration Project W-78-R, WP-2, J-11. Arizona Game and Fish Department, Phoenix, Arizona 19 pp

Lewis JC (1973) The world of the wild turkey. JB Lippincott Co, New York, New York 158 pp

Lima SL, Dill LM (1990) Behavioral decisions made under the risk of predation: a review and prospectus. Canadian Journal of Zoology 68: 619-640

Lima SL, Valone TJ, Caraco T (1985) Foraging-efficiency predation-risk trade-offs in the grey squirrel. Animal Behavior 33: 155-165

Litvaitis JA (1978) Movements and habitat use of coyotes on the Wichita Mountains National Wildlife Refuge. MS Thesis, Oklahoma State University, Stillwater, Oklahoma 70 pp

Litvaitis JA, Shaw JH (1980) Coyote movements, habitat use, and food habits in southwestern Oklahoma. Journal of Wildlife Management 44: 62-68

Logan KA, LL Sweanor (2001) Desert Puma, Evolutionary Ecology and Conservation of an Enduring Carnivore. Island Press, Washington, DC

MacDonald DW, Newdick MT (1982) The Distribution and Ecology of Foxes. *Vulpes vulpes* (L.) in Urban Areas. In Bornkamm R, Lee JA, Seaward MRD (eds) Urban Ecology. Blackwell Scientific Publication, Oxford, United Kingdom 123-135

Mackie CJ, Hamlin KL, Knowles CJ, Mundinger JG (1976) Observations of Coyote Predation on Mule and White-tailed deer in the Missouri River Breaks, 1975-76. Montana Deer Studies, Montana Department of Fish and Game, Federal Aid Project 120-R-7 117-138

Mech LD (1977) Wolf-pack buffer zones as prey reservoirs. Science 198: 320-321

Messier F (1991) The significance of limiting and regulating factors on the demography of moose and white-tailed deer. Journal of Animal Ecology 60, 377-393

Messier F, Barrette C (1985) The efficacy of yarding behaviour by white-tailed deer as an antipredator strategy. Canadian Journal of Zoology 63: 785-789

Messmer TA, Brunson MW, Reiter D, Hewitt DG (1999) United States public attitudes regarding predators and their management to enhance avian recruitment. Wildlife Society Bulletin 27: 75-85

Mills LS, Knowlton FF (1991) Coyote space use in relation to prey abundance. Canadian Journal of Zoology 69: 1516-1521

Molvar EM, Bowyer RT (1994) Costs and benefits of group living in a recently social ungulate: the Alaskan moose. Journal of Mammalogy 75: 621-630

Mooring MS, Fitzpatrick TA, Nishihira TT, Reisig DD (2004) Vigilance, predation risk, and the allee effect in desert bighorn sheep. Journal of Wildlife Management 68: 519-532

Morgantini LE, RJ Hudson (1985) Changes in diets of wapiti during a hunting season. Journal of Range Management 38: 77-79

Morse DH (1980) Behavioral mechanisms in ecology. Harvard University Press, Cambrigde Massachusetts 383 pp

Neff DJ, Woolsey NG (1979) Effect of predation by coyotes on antelope fawn survival on Anderson Mesa. Arizona Game and Fish Department, Special Report Number 8. Phoenix, Arizona 36 pp

Neff DJ, Woolsey NG (1980) Coyote predation on neonatal fawns on Anderson Mesa, Arizona. Proceedings Biennial Pronghorn Antelope Workshop 9: 80-97

Neff DJ, Smith RH, Woolsey NG (1985) Pronghorn antelope mortality study. Arizona Game and Fish Department Research Branch Final Report Federal Aid Wildlife Restoration Project W-78-R 22

Nohrenberg GA (1999) The effects of limited predator removal on ring-necked pheasant populations in southern Idaho. Thesis, University of Idaho, Moscow, USA

O'Brien CS Boyd, HM, Krausman PR, Ballard WB, Cunningham SC, deVos JC, Jr (2005) Influence of wildfire and coyote presence on habitat use by collard peccaries. Wildlife Society Bulletin 33: 365-375

O'Gara BW (1994) Statements about aerial gunning of the National Bison Range National Wildlife Refuge and the resulting increase in pronghorn antelope fawn recruitment. Research Biologist, USFWS (retired) and Wildlife Professor, University Montana, Missoula, Montana

Paton PWC (1994) The effect of edge on avian nest success: how strong is the evidence? Conservation Biology 8: 17-26

Phillips RL, Gruver KS (1996) Selectivity and effectiveness of the Paw-I-Trip pan tension device on 3 types of traps. Wildlife Society Bulletin 24: 119-122

Pierce BM, Bowyer RT, Bleich VC (2004) Habitat selection by mule deer: forage benefits or risk of predation? Journal of Wildlife Management 68: 533-541

Pimlott DH (1970) Predation and productivity of game populations in North America. Transactions of the International Congress of Game Biologists 9: 63-73

Pitt WC (1999) Effects of multiple vertebrate predators on grasshopper habitat selection: trade-offs due to predation risk, foraging, and thermoregulation. Evolutionary Ecology 13: 499-515

Preisser EL, Bolnick DI, Benard MF (2005) Scared to death? The effects of intimidation and consumption in predator-prey interactions. Ecology 86: 501-509

Presnall CC, Wood A (1953) Coyote predation on sage grouse. Journal of Mammalogy 34, 127

Ripple WJ, Larsen EJ (2000) Historic aspen recruitment, elk, and wolves in northern Yellowstone National Park, USA. Biological Conservation 95: 361-370

Ripple WJ, Larsen EJ, Renkin RA, Smith DW (2001) Trophic cascades among wolves, elk and aspen on Yellowstone National Park's northern range. Biological Conservation 102: 227-234

Risenhoover KL, Bailey JA (1985) Foraging ecology of mountain sheep: implications for habitat management. Journal of Wildlife Management 49: 797-804

Riter WE (1941) Predator control and wildlife management. Transactions of the North American Wildlife Conference 6: 294-299

Rivest P, Bergeron JM (1981) Density, food habits, and economic importance of raccoons (*Procyon lotor*) in Quebec agrosystems. Canadian Journal of Zoology 59: 1755-1762

Robinette WL, Gashweiler JS, Jones DA, Crane HS (1955) Fertility of mule deer in Utah. Journal of Wildlife Management 19: 115-135

Rohwer FC, Garrettson PR, Mense BJ (1997) Can predator trapping improve waterfowl recruitment in the Prairie Pothole region? Proceedings Eastern Wildlife Damage Management Conference 7: 12-22

Rosatte RC (1987) Striped, spotted, hooded and hog-nosed skunks. In Novak M, Baker JA, Obbard ME, Malloch B (eds) Wild Furbearer Management and

Conservation in North America. Ministry of Natural Resources, Ontario, Canada 599-613

Rowland M (2004) Effects of management practices on grassland birds: Greater sage-grouse. Northern Prairie Wildlife Research Center, Jamestown, ND 47

Salwasser H (1976) Man, deer and time on the Devil's Garden. Proceedings Western Association of State Game Fish Commissioners 56: 295-318

San Juan County Gunnison Sage-Grouse Working Group (2000) Gunnison Sage Grouse (*Centrocercus minimus*) Conservation Plan. Utah State University, Logan, Utah

Sargeant AB, Allen SH, Eberhardt RT (1984) Red fox predation on breeding ducks in midcontinent North America. Wildlife Monographs 89 pp

Sauer JR, Hines JE, Fallon J (2005) The North American Breeding Bird Survey, Results and Analysis 1966 - 2004. Version 2005.2

Schmitz OJ, Beckerman AP, O'Brien KM (1997) Behaviorally mediated trophic cascades: effects of predation risk on food web interactions. Ecology 78: 1388-1399

Schmidt RH (1986) Community-Level Effects of Coyote Population Reduction. Special Technical Publication 920, American Society for Testing and Materials. Philadelphia, Pennsylvania

Schroeder MA, Baydack RK (2001) Predation and the management of prairie grouse. Wildlife Society Bulletin 29: 24-32

Shaw HG (1977) Impact of mountain lion on mule deer and cattle in northwestern Arizona. In Phillips RL, Jonkel C (eds) Proceedings 1975 Predator Sym., For Cons Exp Sta, University of Montana, Missoula, Montana 17-32

Shaw HG (1989) Soul Among Lions - The Cougar as Peaceful Adversary. Johnson Books, Boulder, Colorado 140 pp

Slobodchikoff CN, Coast R (1980) Dialects in the alarm calls of prairie dogs. Behavioral Ecology and Sociobiology 7, 49-53

Sinclair ARE (1991) Science and practice of wildlife management. Journal of Wildlife Management 55: 767-773

Skogland T (1991a) Ungulate foraging strategies: optimization for avoiding predation or competition for limiting resources? Trans International Union of Game Biologists Congress 18: 161-167

Skogland T (1991b) What are the effects of predators on large ungulate populations? Oikos 61: 401-411

Slater SJ (2003) Sage-grouse (*Centrocerus urophasianus*) use of different-aged burns and the effects of coyote control in southwestern Wyoming. MS Thesis, University Wyoming, Laramie, Wyoming 177 pp

Smith RH, LeCount A (1976) Factors affecting survival of mule deer fawns. Final Report, Federal Aid Project Wildlife Restoration W-78-R, WP-2. J-4. Arizona Game and Fish Dept. Phoenix, Arizona

Smith RH, Neff DJ, Woolsey NG (1986) Pronghorn response to coyote control - A benefit:cost analysis. Wildlife Society Bulletin 14: 226-231

Sonenshine DE, Winslow EL (1972) Contrasts in distribution of raccoons in two Virginia localities. Journal of Wildlife Management 36: 838-847

Speake DW (1985) Wild turkey population ecology on the Appalachian Plateau region of northeastern Alabama. Fed Aid Proj No. W-44-6, Final Report Alabama Game and Fish Division, Montgomery, Alabama

Speake DW, Metzler R, McGlincy J (1985) Mortality of wild turkey poults in northern Alabama. Journal of Wildlife Management 49, 472-474

Steele JL Jr (1969) An investigation of the Comanche County deer herd. Oklahoma Dept Wildlife Conservation Federal Aid Fish and Wildlife Restoration Project W-87-R 20 pp

Storm GL, Tzilkowski MW (1982) Furbearer population dynamics: a local and regional management perspective. In Anderson GC (ed) Midwest Furbearer Management. Proceedings Sym 43rd Midwest Fish and Wildlife Conf, Wichita, Kansas 69-90

Storm GL, Andrews RD, Phillips RL, Bishop RA, Siniff DB, Tester JR (1976) Morphology, reproduction, dispersal, and mortality of midwestern red fox populations. Wildlife Monographs 49: 1-82

Stout GG (1982) Effects of coyote reduction on white-tailed deer productivity on Fort Sill, Oklahoma. Wildlife Society Bulletin 10: 329-332

Teer JG, Drawe DL, Blankenship TL, Andelt WF, Cook RS, Kie J, Knowlton FF, White M. (1991) Deer and coyotes: The Welder Experiments. Transactions of the North American Wildlife Natural Resources Conference 56: 550-560

Thomas GE (1989) Nesting ecology and survival of hen and poult eastern wild turkeys in southern New Hampshire. Thesis, University of New Hampshire, Durham, New Hampshire

Till JA (1992) Behavioral effects of removal of coyote pups from dens. Proceedings of the Vertebrate Pest Conference 15: 396-399

Till JA, Knowlton FF (1983) Efficacy of denning in alleviating coyote depredations upon domestic sheep. Journal of Wildlife Management 47: 1018-1025

Todd AW, Keith LB (1983) Coyote demography during a snowshoe hare decline in Alverta. Journal of Wildlife Management 47, 394-404

Todd AW, Keith LB, Fisher CA (1981) Population ecology of coyotes during a fluctuation of snowshoe hares. Journal of Wildlife Management 45: 629-640

Trainer CE, Willis MJ, Keister GP Jr, Sheehy DP (1983) Fawn mortality and habitat use among pronghorn during spring and summer in southeastern Oregon, 1981-82. Oregon Department of Fish and Wildlife, Wildlife Research Report Number 12: 117 pp

Trainer CE, Lemos JC, Kister TP, Lightfoot WC, Toweill DE (1981) Mortality of mule deer fawns in southeastern Oregon, 1968-1979. Oregon Department Fish Wildlife Research Development Section Wildlife Research Report 10: 113 pp

Trautman CG, Fredrickson L, Carter AV (1974) Relationships of red foxes and other predators to populations of ring- necked pheasants and other prey, South Dakota. Transactions of North American Wildlife Natural Resources Conference 39: 241-252

Truett JC, Dullum JLD, Matchett MR, Owens E, Seery D (2001) Translocating prairie dogs: a review. Wildlife Society Bulletin 29: 863-872

Tucker RD, Garner GW (1980) Mortality of pronghorn antelope fawns in Brewster County, Texas. Proceedings West Conference Game Fish Commission 60: 620-631
Turkowski FJ, Armistead AR, Linhart SB (1984) Selectivity and effectiveness of pan tension devices for coyote foothold traps. Journal of Wildlife Management 48: 700-708
Turner JW Jr, Wolfe ML, Kirkpatrick JF (1992) Seasonal mountain lion predation on a feral horse population. Canadian Journal of Zoology 70: 929-934
Twichell AR, HH Dill (1949) One hundred raccoons from one hundred and two acres. Journal of Mammalogy 30: 130-133
UDWR Statewide Management Plan for Bighorn Sheep (1999) UDWR, 1594 W North Temple, Salt Lake City, Utah 84116
Udy JR (1953) Effects of predator control on antelope populations. Utah Dept Fish and Game. Salt Lake City, Utah Publication Number 5: 4 pp
Urban D (1970) Raccoon populations, movement patterns, and predation on a managed waterfowl marsh. Journal of Wildlife Management 34: 372-382
USDA (U.S. Department of Agriculture) (1997 revised) Animal Damage Control Program Final Environmental Impact Statement. USDA, APHIS, ADC Operational Support USDA Staff, 4700 River Road, Unit 87, Riverdale, Maryland 20737
USDI (U.S. Department of the Interior) (1978) Predator damage in the West: a study of coyote management alternatives. US Fish and Wildlife Services (FWS), Washington, DC 168 pp
USDI (1995) Report of effects of aircraft overflights on the National Park System. USDI-National Park Service D-1062, July, 1995
USFWS (1990) Endangered and threatened wildlife and plants; determination of threatened status for the Mojave population of desert tortoise. 55 Code of Federal Regulations 12178, Vol. 55 No. 63
USFWS (1999) South Florida multi-species recovery plan. Atlanta, Georgia. 2172 pp
Van Ballenberghe V, Ballard WB (1994) Limitation and regulation of moose populations: the role of predation. Canadian Journal of Zoology 72: 2071-2077
Von Gunten BL (1978) Pronghorn fawns mortality on the National Bison Range. Proceedings Pronghorn Antelope Workshop 8, 394-416
Wagner FH, Stoddart LC (1972) Influence of coyote predation on black-tailed jackrabbit populations in Utah. Journal of Wildlife Management 36: 329-342
Wagner KK, Conover MR (1999) Effect of preventive coyote hunting on sheep losses to coyote predation. Journal of Wildlife Management 63: 606-612
Wakeling BF (1991) Population and nesting characteristics of Merriam's turkey along the Mongolon Rim, Arizona. Arizona Game and Fish Department, Tech Rep No 7, Phoenix, Arizona
Wallmo OC (ed) (1981) Mule and black-tailed deer of North America. University Nebraska, Lincoln, Nebraska

Wehausen JD (1996) Effects of mountain lion predation on big horn sheep in the Sierra Nevada and Granite mountains of California. Wildlife Society Bulletin 24: 471-479

West BC (2002) The influence of predator enclosures and livestock grazing on duck production at Bear River Migratory Bird Refuge, Utah. PhD Dissertation Utah State University, Logan Utah

White M (1967) Population ecology of some white-tailed deer in south Texas. Dissertation, Purdue University, Lafayette, Indiana 215 pp

Whittaker DG, Lindzey FG (1999) Effect of coyote predation on early fawn survival in sympatric deer species. Wildlife Society Bulletin 27: 256-262

Wilcove DS, McClellan CH, Dobson AP (1986) Habitat fragmentation in the temporate zone. In Conservation Biology, Soule ME (ed) Sinauer, Sanderlands, Massachusetts 237-256

Williams LE, Austin DH, Peoples TE (1980) Turkey nesting success in a Florida study area. Proceedings of the National Wild Turkey Symposium 4: 102-107

Willis MJ, Keister GP, Immell DA, Jones DM, Powell RM, Durbin KR (1993) Sage grouse in Oregon. Wildlife Research Report Number 15. Oregon Department of Fish and Wildlife, Portland, Oregon

Wilton ML (1984) Black bear predation on young cervids. Alces 19: 136-147

Wilton ML, Carlson DM, McCall CI (1984) Occurrence of neonatal cervids in the spring diet of black bears in south central Ontario. Alces 20: 95-105

Wishart W (1978) Bighorn sheep. In Schmidt JL, Gilbert DL (eds) Big Game of North America Ecology and Management. Wildlife Management Institute 161-171

Wishart W (2000) A working hypothesis for Rocky Mountain bighorn sheep management. In Thomas AE, Thomas HL (eds) Transaction North American Wild Sheep Conference 2: 47-52

Yeager LE, Rennels RG (1943) Fur yield and autumn foods of the raccoon in Illinois river bottom lands. Journal of Wildlife Management 7: 45-60

12 Invasive Predators: a synthesis of the past, present, and future

"…if all the animals and plants of Great Britain were set free in New Zealand, a multitude of British forms would over the course of time become thoroughly naturalized there, and would exterminate many of the natives." Darwin 1872

William C. Pitt[1] and Gary W. Witmer[2]

[1]USDA/APHIS/WS, National Wildlife Research Center, Hawaii Field Station, P.O. Box 10880, Hilo, Hawaii 96721; [2]USDA/APHIS/WS, National Wildlife Research Center, 4101 LaPorte Avenue, Fort Collins, Colorado 80521

12.1 Abstract

Invasive predators have had devastating effects on species around the world and their effects are increasing. Successful invasive predators typically have a high reproductive rate, short generation times, a generalized diet, and are small or secretive. However, the probability of a successful invasion is also dependent on the qualities of the ecosystem invaded. Ecosystems with a limited assemblage of native species are the most susceptible to invasion provided that habitat and climate are favorable. In addition, the number of invasion opportunities for a species increases the likelihood that the species will successfully establish. The list of routes of entry or pathways into many ecosystems continues to grow as transportation of goods into even the remotest areas become common. Species may enter new areas accidentally (e.g., hitchhikers on products) or as intentional introductions (e.g., sport fish). Pet releases, either accidental or intentional, are a growing area of concern as exotic pets become common and the desire for new or different species grows. Several invasive predators have had major effects on prey populations around the world (e.g., black rats, feral cats, mongoose) or have had devastating effects in isolated areas (e.g., brown treesnakes, Nile perch). Although management of established species has been a priority, eradication has been extremely difficult once a species has become widely distributed. However, little resources are directed toward interdiction efforts, removing incipient populations, or preventing new introductions. The regulation of animal movement in most countries and the inspection of products being moved were not developed to protect native ecosystems. Thus, species may be moved with relative ease between regions and countries. The most cost effective approach to invasive species management is to prevent new species from becoming established by providing funding for interdiction efforts, research prior to a species becoming widespread, and restricting the movement of species.

Keywords: Amphibians, birds, invasive species, fish, mammals, management, predation, regulation, reptiles.

12.2 Introduction

Invasive species are species nonnative to a specific ecosystem that cause or may cause ecological harm, negative economic effects, or harm to human health and safety (National Invasive Species Council 2001). Although some nonnative species may be viewed as beneficial (e.g., crops), many have had dramatic effects on the ecosystems invaded. In particular, invasive predators have had catastrophic effects on numerous species during the past several hundred years (Savidge 1988; Witte et al. 1992; Vitousek et al. 1996). These effects likely will increase as more predators are moved, existing habitats are reduced, and the pressure placed on ecosystems is increased. Each new predator introduced increases the chances that additional species will be lost to extinction (Blackburn et al. 2004). This chapter is an attempt to synthesize the effects of invasive predators on terrestrial ecosystems and to present the current status and emerging trends. We have limited the chapter's coverage to invasive vertebrate predators because they are often overlooked, management may be controversial due to competing interests, and their effects are increasing worldwide (Simberloff 1996; Lockwood 1999).

In the last 200 years, many species have been decimated or reduced to extinction by invasive predators, but in the last 30 years as transportation to even the most remote location has become commonplace, the number of invasive predators has increased and their effects are increasing (Simberloff 1996; Mooney and Hobbs 2000; Long 2003). Successful invasive predators generally share several common characteristics, beyond being abundant, widespread, and tolerant of a wide range of abiotic conditions (Lockwood 1999). They typically have a high reproductive rate and short generation times so the populations can grow quickly and rebound from stochastic events (Lockwood 1999). They have a generalized diet to take advantage of locally abundant resources and may switch from preferred prey once prey becomes rare (Murdoch 1969). Prey switching can ultimately lead to extinction of the preferred prey because the predator population is no longer tied to the abundance of the preferred prey (Murdoch 1969). Thus, predator numbers do not decrease as the preferred prey numbers decrease because alternative prey populations support the predator population. This has been observed several times with invasive predators, such as brown treesnakes systematically eliminating the avifauna of Guam (Savidge 1987). In addition, their effects go undetected at first and they are easily transported because they are small or secretive (e.g., snakes), they are ignored by local authorities as innocuous (e.g., coqui frogs), they are purposefully moved or released (e.g., pets), commensal with humans (e.g., rats) or there is resistance to control measures (e.g., feral cats). This lack of understanding and detection allows incipient populations to become established and makes eradication difficult or impossible. Species that have all of these attributes tend to be the most successful at colonizing new habitats (Lockwood 1999).

The probability of a successful invasion is also dependent on the qualities of the ecosystem invaded (Simberloff and Von Holle 1999). Beyond a suitable climate and habitat, ecosystems with a limited assemblage of resident species are the most susceptible to invasion. The lack of resident species decreases the number of potential competitors and predators. Last but not least, the number of invasion opportunities for a species increases the likelihood that the species will successfully establish. Island ecosystems are more susceptible than mainland areas because they have few predators or competitors, they have a lot of air and sea traffic, and they typically provide a favorable climate for many species (Elton 1958, Simberloff 1995). The increased susceptibility of insular populations to extinction compared to mainland areas has been clearly delineated. Since 1600, 93 percent of the land and freshwater birds that have gone extinct worldwide were insular forms (King 1985). In addition, predation by invasive species is considered second only to habitat loss as the leading cause of avian extinctions and declines on islands, with rats (Rattus spp., 56%) and domestic cats (Felis catus, 26%) implicated in most avian extinctions caused by invasive predators (King 1985; Griffin et al. 1989). As remaining habitat patches mirror islands, invasive predators may have similar effects.

The number of pathways invasive species may arrive is varied and likely increasing. Generally, species are either accidentally or intentionally transported. Accidental movements include hitchhikers on agricultural products (e.g., brown treesnakes, coqui frogs) and pet escapes (e.g., pythons and Nile monitors). Pet escapes or releases are especially disconcerting because managers typically are not looking for species that have such a low probability of detection and released populations may remain tied to a particular location or semi-captive until the population is well established. Much of the importation of exotic wildlife is due to the enormous pet industry (Ruesink et al. 1995; Witmer and Lewis 2001). Intentional releases include those that were intended to provide food for people (e.g., feral pigs, bullfrogs), to combat other species (e.g., mongoose, feral cats, cattle egrets, cane toads), or for aesthetic or recreational reasons (e.g., sport fish, feral pigs). Although many of the intentional releases had altruistic intentions, some are for insidious or financial reasons. Species smuggled and released for the pet trade are an - increasing threat and difficult to prevent because heightened security measures and the realignment of customs inspections are not focused on invasive species.

12.3 Species profiles

Several species have become widely publicized for their overall effect as invasive species or as successful invaders in multiple areas. Most of the highlighted species were listed as the worst invasive predators by Lowe et al. (2004) but three potentially predatory species on the list were not included because they are not that widespread or their primary effects are not from predation. Brushtail possums (Trichosurus vulpecula) do prey on invertebrates and birds but their primary effects are as a disease vector and herbivore (Clout and Ericksen 2000; Cowan 2001). The effects of common mynah are as a nuisance and agricultural pest, although they may prey and compete with native birds (Long 1981; Pell and Tideman 1997). The red-eared slider (Trachemys scripta) has been introduced around the world through the pet trade. These omnivorous turtles may compete with native turtles, prey on invertebrates, forage on vegetation, and occasionally take birds (Luiselli et al. 1997; Chen and Lue 1998). We added a few species to highlight emerging issues; these include Burmese pythons, cattle egrets, barn owls, and Nile monitors. Most invasive birds are not predators but cause a myriad of agricultural and human health threats, however, these two species (barn owls and cattle egrets) were included to highlight their increasing range expansion and predation effects. Nile monitors and Burmese pythons highlight the ever increasing problem of the pet trade in establishing invasive species. The source of many of the new invasive predators are from the pet trade where people release unwanted pets or attempt to naturalize them so they may breed in the wild and supply demand (Ruesink et al. 1995; Cassey et al. 2004; Enge et al. 2004). In an attempt to understand the effects of invasive predators and potential problems with control efforts, we provide a brief summary of several noteworthy species and attempts at control.

12.3.1 Mammals

Black rats

One of the most widespread and destructive predators is the black, ship or roof rat (Rattus rattus), introduced around the world from the late 1600s to 1800s) (Long 2003). Black rats have become so ubiquitous and widespread that little attention was paid to this species, whereas new invasions receive more attention and eventually funding for research and control. Black rats are arboreal and in addition to causing significant damage to plants, black rats are efficient predators of many species, especially birds. A large majority of the recorded vertebrate extinctions since 1600 have been on islands and introduced mammals are responsible for the vast majority of these extinctions (Groombridge 1992). Further, black rats have been implicated in many of the documented extinction events, such as honeycreepers in Hawaii, United States (Atkinson 1977), small mammals in the Caribbean (Seidel and Franz 1994) land birds and a bat on Big South Cape Island, New Zealand (Atkinson 2001), and several vertebrates and invertebrates on Lord Howe Island (King 1985; Case and Bolger 1991). Rats have been the most destructive invasive species accounting for losses of numerous species around the world.

Numerous techniques have been developed to control rat populations from introducing other predators, to trapping, to fencing, to a variety of poisons. The introduction of other predators, such as mongoose, owls, or cats have had little success and usually just increased the predation pressure on native fauna. Trapping has had limited success in small areas but rats are highly mobile and may become trap shy. Fencing options for rats over large areas has not been used effectively until recently (Clapperton and Day 2001). However, fences must be combined with other techniques to initially remove rats. The most effective way to control or eradicate rats has been with the use of toxicants, primarily anticoagulants. During the last 15 years, efforts to eradicate rodents from islands have increased and many successful eradication projects have been completed using commercially available rodenticides (Myers et al. 2000; Atkinson 2001; Veitch and Clout 2002).

Feral cats

Wild populations of domesticated cats are distributed throughout the world, wherever humans are present (Long 2003). However, in areas with reduced predator populations, feral cats often become the dominant predator and often exist at much higher densities than native predators (Van't Woudt 1990). In the United States, the feral cat population has been estimated at over 30 million and that these feral animals kill about 465 million birds per year (Pimental et al. 2000). Pimental et al. (2000) estimated the value of those birds at $17 million. In the United Kingdom, the feral cat population may exceed 5 million and kill as many as 70 million wild animals per year (Churcher and Lawton 1987). The diet of feral and free-ranging cats varies depending on availability, abundance, and geographic location. Foods may be naturally occurring, but also include those made available by people, whether intentional or unintentional (Long 2003). In a survey of New Zealand scientific literature, Fitzgerald (1990) concluded that prey selection of feral and free-ranging cats is dependent on availability. The author found that cats on mainland situations fed most heavily on mammals; whereas, cats on islands fed almost exclusively on birds (particularly seabirds). Feral and free-ranging cats are known to prey on birds as large as mallard ducks (Figley and VanDruff 1982) and young brown pelicans (Anderson et al. 1989) and mammals as large as hares and rabbits. Many of these cat populations rely heavily on humans, either for handouts or waste food stuffs, especially when prey populations are low.

Effects of predation on native species by feral cat populations are widespread and significant (Whittaker 1998). Cats have been one of the most important biological factors (excluding humans) causing the depletion or extinction of both island and mainland bird species (Nogales et al. 2004). In isolated environments such as islands, feral cats are directly responsible for a number of extinctions and extirpations worldwide and across multiple taxa (Towns et al. 1990; Veitch 2001; Long 2003). Jackson (1978) reports cats as the most significant factor, next to habitat destruction, contributing to the extinction of bird species. He reports that at least 33 species have become extinct as a result of cat predation; most of these are on islands.

Another significant problem created by cats is that they are reservoirs and transmitters of various diseases and parasites to both domestic and wild animal species, as well as to humans. Cats serve as reservoirs or hosts for dermatomycoses, fleas, scabies, gram-positive bacterial infections, cat scratch fever, distemper, histoplasmosis, leptospirosis, mumps, plague, rabies, ringworm, salmonellosis, toxoplasmosis, tularemia, and various endo- and ecto-parasites (Warner 1984; Fitzwater 1994).

If feral cats are so destructive to wildlife, especially on islands, why is there not a greater effort to control feral cat populations? The control of feral cats is a very controversial area as many members of the public and some advocacy groups are strong supporters of cats and are against the killing of feral cats. These persons and groups often prefer the trap-neuter-release approach to feral cat management (Castillo and Clarke 2003). Some groups actually maintain feeding stations for feral cat colonies. These more socially acceptable methods of cat control have had limited success at reducing predation by feral cats, so most wildlife professionals and governmental agencies advocate the strict control or elimination of feral cat populations (Pech 2000; Parkes and Murphy 2002). The most commonly used methods to control or eliminate feral cats were trapping and shooting, although some countries also use toxic baits (Eason et al. 1992; Veitch 2001; Short et al. 2002; Wood et al. 2002; Hess et al. 2004). Nogales et al. (2004) identified 48 successful eradication efforts on islands. Most of these eradication efforts were on small unpopulated islands where the cat population is closed and the number of nontarget animals was low. In addition, seabirds can form extremely dense nesting colonies and the removal of predators can have dramatic effects.

Mongoose

Small Indian mongooses (Herpestes javanicus, synonymous with H. auropunctatus) were native to India, Pakistan, southern China, Java, Iran, and Iraq (Corbet and Hill 1992). Mongooses were introduced to combat rats in sugarcane fields during the late 1800s to early 1900s and snakes in Asia (Gorman 1975; Sugimura et al. 2005). As sugarcane production spread from the Caribbean and South America (Jamaica, Puerto Rico, and Cuba, etc.), to the Pacific (Hawaiian and Fiji islands), and then to other parts of the world, mongoose introductions followed (Nellis and Everard 1983; Long 2003). While they may kill some rodents, mongooses are mainly diurnal whereas rats are mainly nocturnal. Hence, mongooses are basically useless as a means of rodent damage control. Mongooses use many habitats from forests to open grasslands and the edges of villages and feed on a wide variety of vertebrate, invertebrate, and plant foods (Nowak 1991). Mongoose proved to be ineffective at controlling rats but were serious predators of native ground nesting birds, as well as other vertebrate species (Gorman 1975; Tomich 1986). Mongooses have been implicated in the demise of ground nesting birds and ground nesting bird reproduction has ceased in cases where mongooses are present (Baker and

Russell 1979; Stone et al. 1994; Long 2003). In addition to the extinction or local extirpation of ground nesting birds worldwide, they have been implicated in the demise of frogs in Fiji, ground lizards and snakes on St. Croix, turtles on St. John, and small mammals in Japan and Puerto Rico (Seaman and Randall 1962; Gorman 1975; Nellis and Small 1983; Coblentz and Coblentz 1985; Vilella 1998; Sugimura et al. 2004). The successful reintroduction of endangered species where mongooses were the primary predator has been dependent on eradication of mongooses on select islands or in small areas (USFWS 1999). Beyond native wildlife, mongooses may have a great effect on poultry production and are a reservoir of rabies, leptospirosis and other diseases (Everard and Everard 1988; Pimental et al. 2000; Long 2003). Pimental et al. (2000) estimated that the mongoose causes about $50 million in damages each year in Hawaii and Puerto Rico alone.

Trapping and toxicant baits have been used in attempts to eradicate mongoose or reduce high populations of mongooses near and around native bird nesting habitats (Smith et al. 2000; Roy et al. 2002). Although mongooses are easily trapped and are susceptible to several rodenticides, mongoose eradication has proven extremely difficult with few successes (Roy et al. 2002; Long 2003; Sugimura et al. 2004). If mongooses can be eradicated locally, fences may be an option to prevent reinvasion (Clapperton and Day 2001). Mongooses are long lived and have high reproductive capacity with a gestation period of 42 days and 1-4 offspring in each litter (Nowak 1991). Further, where mongooses have been introduced, they have few predators or competitors to restrict populations.

Stoat or short-tailed weasel

The stoat (Mustela ermine) was native to northern parts of Eurasia and North America and was recently introduced into New Zealand and has spread to several offshore islands to control rabbits (King 1989). Although invasive predators have already reduced many of New Zealand native species, the stoat has had significant effects on kiwi and forest birds (O'Donnell et al. 1996; Basse et al. 1999; McDonald and Murphy 2000). The species differs from mongoose in that stoats are more arboreal than the former and thus they may affect cavity nesting birds, as well as other vertebrates (Basse et al. 1999). Techniques for stoat control remain similar to mongoose control (Alterio et al. 1999; McDonald and Larivière 2001)

Red fox

The red fox (Vulpes vulpes) is native to a large part of the northern hemisphere, but has been introduced to other parts of the world, notably Australia and many islands such as the Aleutian Islands of the United States (Long 2003). Their rapid range expansion throughout Australia was probably facilitated by the large prey base provided by previously introduced European rabbits. Foxes have been introduced for the fur industry and for sport hunting. They were introduced to islands off of Massachusetts (east coast of the United States) to control herring gull colonies and was so successful that foxes died from lack of food (Kadlec 1971). Foxes are adaptable and can use a wide range of habitats. Foxes are efficient predators, but will also consume fruit and vegetables. They prey on a wide array of small mammals and birds, but also eggs, young livestock and poultry, invertebrates, and carrion (Doncaster et al. 1990). They also feed on crustaceans and fish (Witmer and Lewis 2001). They have had substantial impacts on grounding nests bird populations, both in seabird colonies on islands and game bird populations on mainland situations (Witmer and Lewis 2001; Long 2003). In Australia, they have been implicated in the decline of several species of native marsupials (Kinnear et al.2002). Foxes also play a significant role in rabies epizootics (Anderson et al.1981).

Red foxes are managed with a variety of methods, including trapping, shooting, and poisonous baits. All of these methods were employed to eradicate red foxes from most of the Aleutian Islands (Ebbert 2000). Interestingly, a biological control method was successfully used on two small islands that had introduced arctic fox populations. Sterilized red foxes were put on those islands and the larger red foxes eliminated the arctic foxes and then eventually died out (Ebbert 2000).

Feral pig

Pigs (Sus scrofa) originated in Eurasia, were domesticated as livestock, and then moved around the world as an important food source (Long 2003). The lengthy list of introductions to continents and islands provided by Long (2003) clearly suggest that pigs are one of the most widely introduced mammalian species in the world. They were introduced to Florida in 1539, but had been brought much earlier to the islands of Hawaii and the West Indies (Long 2003). They have more recently been introduced to areas for sport hunting (Witmer et al. 2003). Captive pigs may escape captivity and successfully establish or supplement wild populations (Witmer et al. 2003). In the United States, feral swine occur in over 23 states and their numbers are estimated to exceed 4 million (Seward et al. 2004). They are the most abundant introduced ungulate in North America and their populations continue to expand (Sweeney et al. 2003). In addition to predation problems, feral pigs also cause substantial environmental damage (Seward et al. 2004; Sweeney et al. 2003) and pose significant disease hazards to livestock, humans, and wildlife (Witmer et al. 2003).

Feral pigs are omnivorous and will feed on a very wide variety of foods, both plant and animal (Henry and Conley 1972; Challies 1975; Seward et al. 2004). Plant materials include grasses, forbs, leaves, roots, seeds, shoots, fruits, and fungi. They also feed on a wide variety of cultivated crops and can cause substantial crop losses. Animal materials include fish, lizards, frogs, salamander, snakes, turtles, bird eggs and chicks, small rodents and rabbits, fawns, and small livestock. They also feed on a wide variety of invertebrates, including crabs, earthworms, leeches, snails, slugs, grasshoppers, centipedes, beetles, and many other insects. This broad range of foraging results in competition for food with wildlife (e.g., wild turkeys) and livestock, especially through the voracious consumption of mast (e.g., acorns). Nest destruction of the nests and eggs of ground nesting birds and sea turtles by feral pigs is significant in some areas (Seward et al. 2004). Feral pigs cause substantial losses to lamb production in Australia and in parts of the United States (California, Texas; Seward et al. 2004). Feral pigs are responsible for reducing many plant and animal populations resulting in these species being listed as endangered (Seward et al. 2004). On islands to which they have been introduced, they threaten ground-nesting seabirds, penguins, iguanas, and tortoises (Challies 1975; Wiewandt 1977; Long 2003; Seward et al. 2004). In Florida, they have destroyed up to 80% of sea turtle nests (Seward et al. 2004)

There were a variety of methods used to manage or eliminate feral pig populations, although eradication is difficult (Seward et al. 2004; Sweeney et al. 2003). Methods include trapping, shooting, pursuit with dogs, aerial shooting, night shooting over bait piles, exclusion fencing, and the use of toxicants. The use of toxicants is very limited in the United States because of non-target hazards, but they have been used extensively in Australia where there are many invasive mammals and nontarget hazards are minimal. Research is needed in management techniques such as population monitoring and oral delivery systems for disease vaccines, fertility control agents, and toxicants (Sweeney et al. 2003).

12.3.2 Birds

Cattle egret

The cattle egret (Bulbulcus ibis) was originally native to Africa, southern Europe and eastward through southeastern Asia and northern Australia. Prior to 1900, the species began an enormous range expansion and arrived in South America in 1877 and in the United States in about 1941 (Telfair 1994). The species currently occurs throughout the continental United States, South America, and somewhat into Canada. Cattle egrets were introduced into Hawaii in 1959 to help control flies around homes and cattle pastures; they were introduced to the Seychelles, Frigate, and Praslin islands for the same reason (Long 1981). The species range continues to expand, potentially throughout the Pacific basin. The birds are well adapted to forage in grasslands occupied by large grazers. Human conversion of large areas to livestock pasture has probably facilitated the range expansion of cattle egrets. Cattle egrets also use urban-suburban parks and aquatic habitats, although they are not dependent upon the latter.

Cattle egrets are voracious active foragers (Telfair 1994). They usually feed in loose aggregations of 10 to 100 birds. They are opportunistic feeders, feeding mainly on invertebrates including grasshoppers, crickets, spiders, beetles, ticks, flies, moths, katydids, roaches, earthworms, millipedes, centipedes, crayfish and may feed on prawns at aquaculture facilities (Grubb 1976; Hancock and Elliott 1978; Telfair 1994). They will also eat small vertebrates, including frogs, toads, lizards, snakes, mice, the eggs and chicks of nesting birds, and even exhausted small migrant birds along shorelines. In Hawaii, they prey upon native waterbird and seabird chicks, including the native black-necked stilt (Stone and Anderson 1988). When feeding their chicks, an adult egret can consume over 50% of its body weight each day. These birds often forage near grazing livestock, wild ungulates or by farm machinery. They often forage in newly plowed or burned fields. They are often seen using the backs of large ungulates for perches. These "hosts" make foraging by egrets much more efficient. However, the cattle egret foraging strategy varies depending on the size of prey they are focusing on and they are not reliant on the these "hosts" to effectively forage (Grubb 1976). They have been known to scavenge food in tern colonies and even force tern chicks to regurgitate for them. Because of their voracious and diversified feeding habits, and because they forage in sizable groups, cattle egrets could have impacts on the populations of various native or endemic species but these effects have been poorly documented. Additional problems caused by egrets include bird strike hazards at airports because they forage in large groups in grasslands common to airports (Fellow and Paton 1988). Due to the continued range expansion and movement of egrets, they may be ideal carriers of disease organisms and large rookeries may be sanitation hazard near developed areas.

A variety of methods can be used to move cattle egrets from areas they are not wanted. These include shooting, harassment/scare devices, trapping, netting and shooting (Fellow and Paton 1988; Telfair 1994). Because cattle egrets are a migratory non-game bird, they receive protection under state and federal laws at most locations and so control options are limited.

Future research needs include a better understanding of interspecies interactions and why certain areas are selected for foraging, and continued study of parasites and potential disease transmission (Telfair 1994). A quantification of their impacts on rare, endemic faunal species is needed (Stone and Anderson 1988).

Barn owl

The barn owl (Tyto alba) is the most widespread of all owl species being found on all continents except Antarctica (Marti 1992). It has been introduced to various islands (Hawaii, Seychelles, St. Helena) and has colonized other islands on its own (Long 1981). They were introduced to Hawaii in 1958-1963 with the hope that they would control rats in sugarcane plantations (Long 1981). Barn owls use a wide array of habitats, especially grasslands and agricultural areas with nesting cavities nearby. They will readily nest in many human structures.

Barn owls primarily feed on small mammals, bats, and some birds (Speakman 1991; Marti 1992). Lizards and invertebrates are found only in trace amounts in the diet. It is probably safe to assume that the diet is variable, depending on prey species availability. For example, significant predation on bats was noted in Bolivia and the British Isles (Speakman 1991; Vargas et al. 2002). They consume about 10% of their body weight per day. Barn owls are known to prey on seabirds and probably compete with Hawaii's native short-eared owl and Hawaiian owl (Stone and Anderson 1988). In the Seychelles, they preyed on numerous native birds, especially fairy and bridled terns (Long 1981; Bowler et al.2002). A successful barn owl control program has greatly reduced barn owl predation since 1996 (Bowler et al.2002).

A variety of methods can be used to move barn owls from areas they are not wanted. These include shooting, harassment/scare devices, trapping, netting and shooting. Because barn owls are a non-game species and a migratory bird species, they receive protection under state and federal laws at most locations. Future research needs include a better understanding of interspecies interactions and a quantification of barn owl impacts on rare, endemic faunal species is needed (Stone and Anderson 1988).

12.3.3 Reptiles

Brown treesnakes

Brown treesnakes (Boiga irregularis) were accidentally introduced into Guam shortly after World War II from their native range in Australia and Papua New Guinea and Australia. The snakes are slender and arboreal with a typically adult length of about 2 m. They have reached extremely high population levels (> 40 per hectare) on Guam because of the abundance of food and lack of abundant predators. The large snake population levels have resulted in the extirpation of most of Guam's native forest birds (9 of 11), extirpation of native lizard populations (9 of 12), and extirpation of two of the three native bats (Savidge 1987; Savidge 1988; Rodda and Fritts 1992; Rodda et al. 1997). Beyond the severe ecological effects, brown treesnakes have been a threat to human health and safety, agriculture, and cause frequent power outages. The snakes are poisonous rear-fanged snakes, thus they are unlikely to cause harm to adults. However, they may affect small children. Data from a single hospital in Guam suggests that there may be more than 26 bites per year (OTA 1983). Pets and poultry also are frequent prey items of the snakes. The largest economic impact from the snakes is the disruption of power systems. The arboreal snake frequently climbs utility poles, power lines, and other structures as travel corridors. Thus, snakes ground out these systems when they cross from grounded to live structures causing an estimated 1.4 million in damages from power outages (Vice and Pitzler 2002).

A variety of methods are employed to control snakes and restrict their access to aircraft and cargo leaving the islands including fence searches, trapping with mice, and searching with detector dogs (Vice et al. 2005). Other potential methods to control snakes include the use of toxicants, repellents, reproductive inhibition, and barriers but these have yet to be deployed over large areas for eradication.

Burmese pythons

Burmese pythons (Python molurus bivittatus) became established in Everglades National Park during the 1990s as the result of unwanted or accidentally released pets (S. Snow, National Park Service, pers. comm.).

Burmese pythons are large snakes (>7 m) with high reproductive rates. Originally from Southeast Asia, pythons are common pets in the United States (Pough et al. 1998). Pythons may compete with native snake species, prey on many native mammals and birds, and transmit disease to native reptiles. The number of snakes removed has increased during the past few years and this could represent a rapidly increasing population (S. Snow, unpubl. data.). Biological information on pythons is limited but potential habitat includes much of the Southeastern United States. Sources of mortality for the snakes in the Everglades National Park include motor vehicles, mowing equipment, fire, and possibly alligators (S. Snow, unpubl. data). Currently, management actions center on mechanical control and education efforts to prevent further introductions. Mechanical control techniques include trapping, hand capture, and early detection using dogs.

Nile monitor lizard

The Nile monitor lizard (Varanus niloticus) is native to Africa where it is the longest (2.1 m) lizard (Enge et al. 2004). They are imported for the pet industry, but their size and aggressive temperament probably limits their value as pets. They were first observed in the wild in southern Florida in 1990; since that time there have been 146 sightings or captures with all size classes present, suggesting a reproducing population (Enge et al. 2004). The lizard has a high reproductive capability, laying up to 60 eggs in a clutch (de Buffrenil and Rimblot-Baly 1999).

Nile monitor lizards are voracious predators. In their native range, they feed on wide array of freshwater, marine, and terrestrial prey, including shellfish and other invertebrates, fish, amphibians, reptiles, mammals, birds, and bird eggs (Enge et al. 2004). Cooperative hunting and nest robbing have been observed in the species. They readily inhabit human settlements and even forage around garbage dumps (Enge et al. 2004). In Africa, they compete with dwarf crocodiles, with crabs being the main prey of both species (Luiselli et al. 1999). This suggests that they could compete with the American alligator and the American crocodile (an endangered species) in Florida. Furthermore, the extensive canal systems of southern Florida provide ideal dispersal corridors for the lizards. Other native species that could be threatened by the monitor lizards should their population and range increase in Florida include sea turtles and diamondback terrapins because of egg predation, brown pelicans (a threatened species), burrowing owls, and gopher tortoises (Enge et al. 2004).

An eradication strategy for the Nile monitor lizard in Florida has been proposed (Campbell 2005). There would require an extensive and intensive trapping effort over a minimum of two years (Campbell 2005). At present, detection, monitoring, and trapping strategies are rudimentary, limiting efforts to control this species.

12.3.4 Amphibians

Bullfrogs

Bullfrogs (Rana catesbeiana) from the eastern United States were widely introduced from 1900-1940 into many western states including Hawaii as food resource. Bullfrogs have had significant ecological effects and have been difficult to control because they are highly mobile, have generalized eating habits, and have high reproductive capacity (Moyle 1973). Bullfrogs may cause the extirpation of other species due to intense predation and competition (Kats and Ferrer 2003). Management of bullfrog populations is difficult, due to commingling with native species in aquatic habitats. Adult frogs are removed by trapping or hand captures and tadpoles are destroyed by draining ponds or chemical treatment where feasible. Fencing may also be used to limit frog movements away from infested habitats.

Cane toads

Giant neotropical (Bufo marinus) or cane toads were widely introduced from Central America into sugar cane producing regions worldwide to control beetles causing damage to crops (McKeown 1978). However, the effort had very limited success because the beetles could climb into the vegetation away from the toads. Cane toads may compete with native species for food, compete with native amphibians for breeding sites, and prey on a variety of invertebrate and vertebrate species (McCoid 1995; Williamson 1999). Cane toads can be very active nest predators of birds and have a significant effect on native fauna (Boland 2004) Further, native species preying on cane toads may be poisoned by the toad's parotoid gland secretions (McCoid 1995).

The frogs also may be a nuisance when large numbers congregate for breeding in ponds or water features. Australia has been aggressively pursuing control options but has had little success in developing new methods (Luntz 1998). Currently, the only effective strategies are pond drying, hand capture, and trapping.

Coqui frogs

The coqui frog (Eleutherodactylus coqui) was introduced into Hawaii during the late 1980s likely from infested plant shipments from Puerto Rico (Kraus et al. 1999). Sizeable populations are now found on the islands of Hawaii, Maui, Oahu, and Kauai and the frog threatens Hawaii's multi-million dollar floriculture, nursery, real estate, and tourist industries, as well as its unique ecological systems (Beard and Pitt 2005). Most of the coqui affects stem primarily from a piercing call (80-90 dBA at 0.5 m) and from extremely high population densities that have exceeded 50,000 individuals ha^{-1} in Hawaii (Beard and Pitt 2005). Beyond being a noise nuisance, the loud nighttime choruses of frogs has affected real estate values because people desire a coqui free property (Kaiser et al. 2006). The floriculture industry may also be affected by refused shipments, reduced sales, and costs associated with control and quarantine efforts. Moreover, the high densities of frogs may effect native insect populations, forest nutrients, compete with native birds and bats, and alter ecosystem processes (Beard and Pitt 2005). The frogs may also benefit other invasive predators, but there is little evidence that rats, mongoose, or cane toads benefit from frogs as prey (Beard and Pitt 2006). Brown treesnakes typically require small prey as juveniles and the presence of another abundant food source in the Hawaii could increase the chance of brown treesnakes establishing a population if they arrive on the islands. However, there is already abundant food resources in the Hawaiian Islands, including geckos, birds, and small mammals (Shine 1991).

Due to the high densities of frogs and their present range, few options exist for control of wild populations. Mechanical controls include hand capturing, habitat alteration, and trapping. These mechanical methods only work on a small scale with a few populations. However, some success has been documented using hot water treatments for plant shipments. A hot (>45 °C) water treatment for at least 3 minutes will kill adult frogs and eggs (UH 2006). Biological control or the release of organisms to combat the frog likely will have little success and could have many unintended consequences. Unfortunately, disease organism have a low potential for controlling coqui frogs in Hawaii, primarily because viruses and diseases are most effective when applied to small populations of species with low reproductive capacity (Brauer and Castillo-Chavez 2001, Daszak et al. 2003). In large populations, diseases may initially induce temporary population declines, but subsequently surviving resistant individuals may lead to population levels similar to those prior to treatment. In addition, frogs could carry a virus or disease to other parts of the world where frog conservation is the priority (Angulo and Cooke 2002). Another important consideration is that most of the major frog diseases infect tadpole stages (Daszak et al. 2003). Because coqui frogs do not have a tadpole stage, they are less likely to be effected. Although many frogs are quite susceptible to a variety of chemicals, the terrestrial coqui frog has been unaffected by a wide range of potential pesticides. Currently, only citric acid and hydrated lime have proven to be effective and registered to use to combat the frogs (Pitt and Sin 2004a). Although these chemicals are effective if sprayed directly on the frogs, there are several limitations with these products including varying effectiveness due to weather conditions, potential phytotoxicity to plants, the cost of repeated spraying large areas, access to remote or private land, and other factors (Pitt and Sin 2004b).

12.3.5 Fish

Humans have moved fish around the world at least back to the time of the Romans (Moyle 1986). Moyle (1986) reviewed fish introductions in North America and noted that at least 150 species have been involved. Fish are introduced for various reasons, including as a source of food, for recreational fishing, as ornamentals, and to help with aquatic insect and plant control. Unfortunately, some of the species are voracious predators and can inflict great harm on native aquatic fauna. The salmonids and perches are perhaps the most significant predators in this group. Recently, some states and countries have only been stocking sterile fish to prevent breeding with native stocks and to restrict population growth. The list of the world's 100 worst invasive alien species includes brown trout (Salmo trutta) and rainbow trout (Oncorhynchus mykiss) from the first group and large-mouth bass (Micropterus salmoides) and Nile perch (Lates niloticus) from the second group (Lowe et al. 2004). A more recent threat is transporting and releasing fish through the pet trade (McNeely and Schutyser 2003). Aquarium fish represent a huge reservoir of potential invasive species with more than 5000 fish species traded globally and little is known of their potential effects (McDowall 2004). In the United States, up to 65% of the established nonnative fish populations species likely originated from the aquarium fish trade (Courtenay and Stauffer 1990). In Australia, 77% of nonnative fish originated from the aquarium fish trade (Koehn and MacKenzie 2004). The walking catfish (Clarias batrachus) is a voracious predator that has been transported to the United States and other countries via the pet trade and for aquaculture. Once introduced, they may spread throughout adjacent waterways and may significantly reduce native fish populations; many other species currently in the pet trade have similar potential (Simberloff et al. 1997).

Predacious fish have broad food habits and will consume invertebrates, amphibians, reptiles, and small fish. Drastic changes in a fish fauna can occur when the native fishes are not adapted to the style of predation of the introduced fish and extinctions and severe declines in the native species usually results (Moyle 1986; Moyle and Cech 1996). The Nile perch after arriving in Lake Victoria in the early 1960s and within 30 years more than 200 hundred fish species had disappeared and the perch became the main fishery species in the lake in the 1990s (Witte et al. 1990; Kitchell et al. 1997). Presumably, competition from introduced fish also causes declines in native fishes, but is more difficult to demonstrate (Moyle and Cech 1996). In Japan, large-mouth bass introduced to ponds reduced fish, shrimp, crayfish, and insects number (Maezono et al. 2005). Negative impacts of introduced predacious fish on native amphibian populations have been documented in Russia (Reshetnikov 2003), Australia (Gillespie 2001), Europe (Martinez-Solano et al. 2003), and North America (Bull and Marx 2002). With removal of the introduced fish, some amphibian populations recover relatively quickly (Hoffman et al. 2004).

Introduced fish species are often difficult to control or eliminate once established. Gill nets are used in some situations (Hoffman et al. 2004). In extreme pond or lake situations, a chemical toxicant such as rotenone is used to kill all fish; then restocking with native species can occur. More effective and species-specific methods are needed for managing or eliminating introduced predacious fish.

12.4 Regulation of invasive species

The regulation of wildlife, in general, and introduced species in particular varies by country and even within regions, territories, provinces or states of a specific country (Witmer and Lewis 2001). In general, the regulatory authority to manage wildlife is held at a fairly local level (e.g., state or province or territory). The central governments of many countries often retain regulatory authority in some situations, such as migratory species, endangered species, and species that might cause significant economic harm. In the United States, many federal laws exist that have some involvement with invasive species, but the federal government very limited legal authority to manage the transportation of vertebrate invasive species across state boundaries or the resources to implement regulations restricting invasive species movement (National Invasive Species Council 2001).

Unfortunately, most species of exotic animals are considered "innocent until proven guilty" in many countries. There has been debate over the use of "white lists" and "black lists" in the regulation of animal imports. After conducting risk assessments, one can list which species are allowed entry into the country (white lists), or one can list only those species that are categorically excluded from entry (black list, Ruesink et al.1995, National Invasive Species Council 2001). Currently, in the United States, the latter approach is used, and only a few vertebrates are categorically excluded as "injurious wildlife." These include hedgehogs, brush-tailed possums, and brown tree snakes. Many federal and state agencies and international and national non-governmental organizations have put forth guidelines and policy statements on invasive wildlife (including the need for white-black-gray lists), but these have only been implemented in a few countries (see discussion in Witmer and Lewis 2001). Currently, there are procedures in place for the listing species that are known to be invasive; such listings may be petitioned and involve stakeholders and the public in the course of the rule-making process (National Invasive Species Council 2001). Other countries, such as New Zealand, have white lists, which are ultimately more effective at stemming the tide of invasive species. However, there are problems with this approach as well. Many of the species listed on white lists are actually genera in New Zealand, thus one genera listed could contain more than 800 species with many species having unknown effects (McDowall 2004). Further, government agents must be able to accurately identify the species, hybrid, or subspecies in all stages development to effectively restrict or allow importation. Unfortunately, until better regulations are in place and adequate funds are made available for inspections and management, we can expect many more invasive species situations to arise.

12.5 Priorities of invasive species

The priorities of invasive species management may be cleanly divided by the point that a species is established. Prior to establishment of a population, research and funding should go to prevention and early detection to decrease the potential for species becoming a problem (Park 2004). To increase the effectiveness of limited funding, a risk analysis should be performed to determine the threat from nonnative species and promote awareness of species that could cause significant effects. Further, coordination and cooperation among state and local agencies decreases the potential for duplicated efforts and increases the response efforts for incipient species. After a species has become established, research and funding is shifted to documenting effects of the species on ecological services, agriculture, and local economies. Development of control strategies and public awareness are priorities after establishment to control the effects of the new species.

Unfortunately, the line that separates the priorities before and after establishment may be referred to as the money line. Prior to a species becoming firmly established, the cost to control a species is low and the probability of success is high (Simberloff 2003; Park 2004). However, the amount of funding available and the public interest in dealing with the potential problem is extremely low at this time. Funding for research and interdiction efforts prior to species establishment is low and only secured with public support pressuring public officials. After the species is established, funding typically becomes more available and public interest in dealing with the issue is higher. Conversely, the costs of control sky rocket and the probability of success drops precipitously. This same scenario has been repeated in many areas with many new species. A recent example is the above mentioned case of the coqui frog in Hawaii. Although the species became established by the late 1980s in a few locations, no funding was available even though the potential to eradicate was still fairly high. The primary public opinion was that this was not a major problem and there were likely to be few negative consequences of this introduction and control efforts could be harmful. This attitude existed even after repeated warnings by scientists (see Kraus and Campbell 2002 for a full discussion). Fifteen years later, the public opinion is extremely supportive of dealing with the issue and several studies have documented the effects of the frogs on ecological communities, real estate, agriculture, and human health (Beard and Pitt 2005; Kaiser et al. 2006). However, the likelihood of complete eradication now is low and would require extensive resources.

In conclusion, invasive predators are an increasing problem throughout the world and these effects are becoming magnified as available habitat is lost. These predators cause a diverse array of problems, cannot be easily predicted, and may cause more significant problems on island ecosystems than mainland areas. The number of new introductions is likely to escalate if the many pathways of invasion are not controlled. Currently, there are few options to control established invasive species and the cost for control efforts is high once a species becomes widespread and causes significant effects. The most cost effective approach to invasive species management is to secure funding for research and interdiction efforts prior to a species becoming widespread.

References

Alterio N, Moller H, Brown K (1999) Trappability and densities of stoats (Mustela erminea) and ship rats (Rattus rattus) in a South Island Nothofagus forest, New Zealand. New Zealand. Journal of Ecology 23: 95–100

Anderson DW, Kieth JO, Trapp GR, Gress F, Moreno LA (1989). Introduced small ground predators in California brown pelican colonies. Colonial Waterbirds 12: 98-103

Anderson RM, Jackson HC, May RM, Smith AM (1981) Population dynamics of fox rabies in Europe. Nature 289: 765–771.

Angulo E, Cooke B (2002) First synthesize new viruses then regulate their release? The case of the wild rabbit. Molecular Ecology 11: 2703-2709

Atkinson IAE (1977) A reassessment of factors, particularly Rattus rattus L., that influenced the decline of endemic forest birds in the Hawaiian Islands. Pacific Science 31: 109–133

Atkinson IAE (2001) Introduced mammals and models for restoration. Biological Conservation 99: 81-96

Baker RH, Russel CA (1979) Mongoose predation on nesting nene. 'Elapaio 40: 51-52

Basse B, McLennan JA, Wake GC (1999) Analysis of the impact of stoats, Mustela erminea, on northern brown kiwi, Apteryx mantelli, in New Zealand. Wildlife Research 26: 227–237

Beard KH, Pitt WC (2005) Ecological consequences of the coqui frog invasion in Hawaii. Diversity and Distributions 11: 427-433

Beard KH, Pitt WC (2006) Potential predators of an invasive frog (Eleutherodactylus coqui) in Hawaiian forests. Journal of Tropical Ecology 22: 1-3

Blackburn TM, Cassey P, Duncan RP, Evans KL, Gaston KJ (2004) Avian extinction and mammalian introductions on oceanic islands. Science 305: 1955-1958

Boland, CRJ (2004) Introduced cane toads Bufo marinus are active nest predators and competitors of rainbow bee-eaters Merops ornatus: observational and experimental evidence. Biological Conservation 120: 53-62

Bowler J, Betts M, Bullock I, Ramos JA (2002) Trends in Seabird Numbers on Aride Island Nature Reserve, Seychelles 1988-2000. Waterbirds 25: 26-38.

Brauer F, Castillo-Chavez C (2001) Mathematical models in population biology and epidemiology. Texts in applied Mathematics. Springer Verlag, New York, United States

Bull EL, Marx DB (2002) Influence of fish and habitat on amphibian communities in high elevation lakes in northeastern Oregon. Northwest Science 76: 240-248

Campbell TS (2005) Eradication of introduced carnivorous lizards from the Cape Coral area. Final report to the Charlotte harbor National Estuary Program, Fort Myers, Florida 1-30

Case TJ, Bolger DT (1991) The role of introduced species in shaping the distribution and abundance of island reptiles, Evolutionary Ecology 5: 272 – 290

Cassey P, Blackburn TM, Jones KE, Lockwood JL. (2004) Mistakes in the analysis of exotic species establishment: source pool designation and correlates of introduction success among parrots (Aves: Psittaciformes) of the world. Journal of Biogeography 31: 277-284

Castillo D, Clarke A (2003) Trap-neuter-release methods ineffective in controlling domestic cat "colonies" on public lands. Natural Areas Journal 23, 247-253

Challies CN (1975) Feral pigs *(Sus scrofa)* on Auckland Island: status, and effects on vegetation and nesting sea birds. New Zealand Journal of Zoology 2: 479-90

Chen T, Lue K (1998) Ecological notes on feral populations of Trachemys scripta elegans in northern Taiwan. Chelonian Conservation and Biology 3: 87-90

Churcher P, Lawton J (1987) Predation by domestic cats in an English village. Journal of Zoology (London) 212: 493-455

Clapperton KD, Day TD (2001) Cost-effectiveness of exclusion fencing for stoat and other pest control compared with conventional control. DOC Science Internal Series 14. Department of Conservation, Wellington, New Zealand

Clout MN, Ericksen K (2000) Anatomy of a disastrous success: the brushtail possum as an invasive species. In Montague T (ed) The Brushtail Possum: Biology, Impact and Management of an Introduced Marsupial. Landcare Research, Lincoln, New Zealand. 1-9

Coblentz, BE, Coblentz BA (1985) Control of the Indian mongoose Herpestes auropunctatus on St. John, US Virgin Islands. Biological Conservation 33: 281-288

Corbet GB, Hill JE (1992) The mammals of the Indomalayan region: a systematic review. Oxford University Press.

Courtenay WR Jr., Stauffer JR Jr. (1990) The introduced fish problem and the aquarium fish industry. Journal of the World Aquaculture Society 21: 145-159

Cowan PE (2001) Advances in New Zealand mammalogy 1999-2000: brustail possum. Journal of the Royal Society of New Zealand 31: 15-29

Darwin C (1872) The origin of species, 6th Edition. John Murray, London, United Kingdom.

Daszak P, Cunningham AA, Hyatt AD (2003) Infectious disease and amphibian population declines. Diversity and Distributions 9: 141-150

De Buffrenil V, Rimblot-Baly F (1999) Female reproductive output in exploited Nile monitor lizard populations in Sahelian Africa. Canadian Journal of Zoology 77: 1530-1539

Doncaster CP, Dickman CR, Macdonald DW (1990) Feeding ecology of red foxes (*Vulpes vulpes*) in the city of Oxford, England. Journal of Mammalogy, 71: 188–194

Eason C, Morgan D, Clapperton B (1992) Toxic bait and baiting strategies for feral cats. Proceedings of the Vertebrate Pest Conference 15: 371-376

Ebbert S (2000) Successful eradication of introduced arctic foxes from large Aleutian Islands. Proceedings of the Vertebrate Pest Conference 19: 127-132

Elton CS (1958) The ecology of invasions by animals and plants. Methuen and Co., Ltd., London, United Kingdom

Enge KM, Krysko KL, Hankins KR, Campbell TS, King FW (2004) Status of Nile monitor in southwestern Florida. Southeastern Naturalist 3: 571-582.

Everard CO, Everard JD (1988) Mongoose rabies. Review of Infectious Diseases 10 (4): S610-614

Fellow DP, Paton PWC (1988) Behavioral response of cattle egrets to population control measures in Hawaii. Proceedings of the Vertebrate Pest Conference 13: 315-318

Figley WK, VanDruff LW (1982) The ecology of urban mallards. Wildlife Monograph No. 82, The Wildlife Society, Bethesda, Maryland, USA.

Fitzgerald BM (1990) House cat. In King C.M (ed) The handbook of New Zealand mammals. Oxford University Press, Auckland, New Zealand 330-348

Fitzwater W (1994) House cats (feral). In Hygnstrom S, Timm R, Larson G (eds) Prevention and control of wildlife damage. University of Nebraska Cooperative Extension Service, Lincoln, Nebraska, United States. C-45 – C-49

Gillespie GR (2001) The role of introduced trout in the decline of the spotted tree frog in south-eastern Australia. Biological Conservation 100: 187-198

Gorman, ML (1975) The diet of feral Herpestes auropunctatus (Carnivora: viveriidae in the Fijiian Islands. Journal of Zoology 175: 273-278.

Griffin CR, King CM, Savidge JA, Cruz F, Cruz JB (1989) Effects of introduced predators on island birds: contemporary case histories from the Pacific. In Ouelle V (ed), Proceedings of the XIX Ornithological Congress. 1: 687-698

Groombridge B, editor (1992) Global biodiversity. Status of the Earth's living resources. Chapman & Hall, London, England.

Grubb TC (1976) Adaptiveness of foraging in the cattle egret. Wilson Bulletin 88: 145-148

Hancock J, Elliot H (1978) The Herons of the World. Harper and Row Publishing, New York, United States

Henry VG, Conley RH (1972) Fall foods of European wild hogs in the Southern Appalachians. Journal of Wildlife Management 36: 854-60

Hess SC, Banko PC, Goltz DM, Danner RM, Brinck KW (2004) Strategies for reducing feral cat threats to endangered Hawaiian birds. Proceedings of the Vertebrate Pest Conference 21: 21-26

Hoffman RL, Larson GL, Samora B (2004) Responses of Ambystoma gracile to removal of introduced nonnative fish from a mountain lake. Journal of Herpetology 38: 578-585

Jackson JA (1978) Alleviating problems of competition, predation, parasitism, and disease in endangered birds. In Temple S (ed) Endangered birds:

management techniques for preserving threatened species. Proceedings of the Symposium on Management Techniques for Preserving Endangered Birds. University of Wisconsin Press, Madison, Wisconsin, United States. 75-84

Kadlec JA (1971) Effects of introducing foxes and raccoons on herring gulls colonies. J Wildl Manage 35: 625-636

Kaiser B, Pitt WC, Burnett K (2006) Economic impact of coqui frogs in Hawaii. Proceedings of Ecology in an era of globalization, Ecological Society of America. Merida, Mexico.

Kats LB, Ferrer RP (2003) Alien predators and amphibians declines: Review of two decades of science and the transition to conservation. Diversity & Distributions 9: 99 110

King C (1989) The natural history of weasels and stoats. Christopher Helm, London.

King WB (1985) Island birds: will the future repeat the past? In Moors PJ (ed). Conservation of island birds. International Council for Bird Preservation Technical Publication No. 3. Cambridge, United Kingdom 3-15

Kinnear JE, Sumner NR, Onus ML (2002) The red fox in Australia—an exotic predator turned biocontrol agent. Biological Conservation 108: 335–35

Kitchell JF, Schindler DE, Ogutu-Ohwayo R, Reinthal PN (1997) The Nile Perch in Lake Victoria: Interactions Between Predation and Fisheries. Ecological Applications 7: 653-664

Koehn JD, Mackenzie RF (2004) Priority management actions for alien freshwater fish species in Australia. New Zealand Journal of Marine and Freshwater Research 38: 457–472

Kraus F, Campbell EW (2002) Human-mediated escalation of a formerly eradicable
problem: the invasion of Caribbean frogs in the Hawaiian Islands. Biological Invasions 4: 327-332

Kraus F, Campbell EW, Allison A, Pratt T (1999) Eleutherodactylus frog introductions to Hawaii. Herpetological Review 30: 21-25

Lockwood JL (1999) Using Taxonomy to Predict Success among Introduced Avifauna: Relative Importance of Transport and Establishment. Conservation Biology 13: 560-567

Long JL (1981) Introduced birds of the world. University Books, New York, United States.

Long JL (2003) Introduced mammals of the world. CSIRO Publishing, Canberra, Australia.

Lowe S, Browne M, Boudjelas S, De Poorter M (2004) 100 of the world's worst invasive alien species. World Conservation Union (IUCN), Gland, Switzerland.

Luiselli L, Akani GC, Capizzi D (1999) Is there any interspecific competition between Dwarf Crocodiles (Osteolaemus tetraspis) and Nile Monitors (Varanus niloticus) in the swamps of central Africa? Journal of Zoology, London 247: 127–131

Luiselli L, Capula M, Capizzi D, Filippi E, Jesus VT, Anibaldi C (1997) Problems for conservation of pond turtles (Emys orbicularis) in central Italy: is the introduced red-eared turtle (Trachemys scripta) a serious threat. Chelonian Conservation and Biology 2: 417-419

Luntz S (1998) Virus can't be used to control cane toads. Australasian Science 19: 10 pp

Maezono Y, Kobayashi R, Kushahara M, Miyashita T (2005) Direct and indirect effects of exotic bass and bluegill on exotic and native organisms in farm ponds. Ecological Applications 15: 638-650

Marti CD (1992) Barn owl. In Poole A, Gill F (eds) The Birds of North America. The Academy of natural Science, Philadelphia 1-16

Martinez-Solano I, Barbadillo LJ, Lapena M (2003) Effect of introduced fish on amphibian species richness and densities at a montane assemblage in the Sieer de Neila, Spain. Herteptological Journal 13: 167-173

McCoid MJ (1995) Non-native reptiles and amphibians. In Laroe, ET, Farris GS, Puckett CE, Doran PD, Mac MJ (eds) Our living resources: a report to the nation on the distribution, abundance, and health of U.S. plants, animals, and ecosystems. U.S. Department of the Interior, National Biological Service, Washington, D.C., United States 433-437

McDonald RA, Larivière S (2001) Review of international literature relevant to stoat control. Science for Conservation 170. Department of Conservation, Wellington, New Zealand

McDonald RA, Murphy EC (2000) A comparison of the management of stoats and weasels in Great Britain and New Zealand. In Griffiths HI. (ed) Mustelids in a modern world. Backhuys Publishers, Leiden, The Netherlands 21–40

McDowall RM (2004) Shoot first, and then ask questions: a look at aquarium fish imports and invasiveness in New Zealand. New Zealand Journal of Marine and Freshwater Research 38: 503–510

McKeown S (1978) Hawaiian reptiles and amphibians. Oriental Publishing Company Honolulu. United States

McNeely JA, Schutyser F (2003) Invasive Species: a global concern bubbling to the surface. International Conference on the Impact of Global Environmental Problems on Continental and Coastal Marine Waters, Geneva, Switzerland

Mooney HA, Hobbs RJ, Editors. 2000. Invasive Species in a Changing World. Island Press, Washington, D.C. United States

Moyle PB (1973) Effects of introduced bullfrogs, (Rana catesbeiana), on the native frogs of the San Joaquin Valley, California. Copeia 1: 18-22

Moyle PB (1986) Fish introductions into North America: patterns and ecological impact. In Billings WD, Golley F (eds) Ecological Studies 58: Ecology of biological invasions of North America and Hawaii. Springer-Verlag, New York 27-43

Moyle PB, Cech JJ (1996) Fishes: an introduction to ichthyology. Prentice Hall, Upper Saddle River, New Jersey, United States

Murdoch WW (1969) Switching in general predators: experiments on predators specificity and stability of prey populations. Ecological Monographs 39: 335-354

Myers JH, Simberloff D, Kuris AM, Carey JR (2000) Eradication revisited: dealing with exotic species, Trends in Ecology & Evolution 15: 316-320

National Invasive Species Council (2001) Meeting the national invasive species challenge: national invasive species management plan. United States Department of Interior, Washington, D.C. United States

Nellis DW, Everard COR (1983) The biology of the mongoose in the Caribbean Islands. Studies on the Fauna of Curacao and other Caribbean Islands 64: 1-162

Nellis DW, Small V (1983) Mongoose predation on sea turtle eggs and nests Biotropica 15: 159-160

Nogales N, Martin A, Tershy BR, Donlan CJ, Veitch D, Puerta N, Wood B, Alonso J. (2004) A review of feral cat eradication on islands. Conservation Biology 18: 310-319

Nowak RM (1991) Walker's mammals of the world II, Fifth edition, Johns Hopkins University Press, Baltimore, Maryland, United States

O'Donnell CFJ, Dilks PJ, Elliott GP (1996) Control of a stoat (Mustela erminea) population irruption to enhance mohua (yellowhead) (Mohoua ochrocephala) breeding success in New Zealand. New Zealand Journal of Zoology 23: 279–286

OTA (Office of Technology Assessment) (1993) Harmful Non-Indigenous Species in the United States. Washington (DC): Office of Technology Assessment, US Congress, United States

Park K (2004) Assessment and management of invasive alien predators. Ecology and Society 9, 12-28

Parkes J, Murphy E (2002) Management of introduced mammals in New Zealand. New Zealand Journal of Zoology 30: 335-359

Pech R (2000) Biological control of vertebrate pests. Proceedings of the Vertebrate Pest Conference 19, 206-211

Pell AS, Tidemann CR (1997). The impact of two exotic hollow-nesting birds on two native parrots in savannah and woodland in eastern Australia. Biological Conservation 79: 145-153

Pimental D, Lech L, Zuniga R, Morrison D (2000) Environmental and economic costs associated with non-indigenous species in the United States. BioScience 50: 53-65

Pitt WC, Sin H (2004a) Dermal toxicity of citric acid based pesticides to introduced Eleutherodactylus frogs in Hawaii. USDA, APHIS, WS, NWRC. Hilo, Hawaii, United States

Pitt WC, Sin H (2004b) Testing citric acid use on plants. Landscape Hawaii July/August 5/12

Pough FH, Andrews RM, Cadle JE, Crump ML, Savitzky AH, Wells KD (1998) Herpetology, Prentice Hall Inc. United States

Reshetnikov AN (2003) The introduced fish, rotan, depresses populations of aquatic animals (macroinvertebrates, amphibians, and a fish). Hydrobiologia 510: 83-90

Rodda GH Fritts TH (1992) The impact of the introduction of the colubrid snake Boiga irregularis on Guam's lizards. Journal of Herpetology 26: 166-174

Rodda GH, Fritts TH, Chiszar D (1997) The disappearance of Guam's

wildlife. BioScience 47: 565–574
Roy S, Jones C, Harris S (2002) An ecological basis for control of the mongoose in Mauritius: is eradication possible? In Veitch C, Clout M (eds) Turning the tide: the eradication of invasive species. World Conservation Union (IUCN), Gland, Switzerland 266-273
Ruesink JL, Parker IM, Groom MJ, Kareiva PM (1995) Reducing the Risks of Nonindigenous Species Introductions BioScience 45: 465-477
Savidge JA (1987) Extinction of an island forest avifauna by an introduced snake. Ecology 68, 660-668
Savidge JA (1988) Food habits of Boiga irregularis, an introduced predator on Guam. Journal of Herpetology 22: 275-282
Seaman G, Randall J (1962) The mongoose as a predator in the Virgin Islands. Journal Mammalogy 43: 544-546
Seidel ME, Franz R (1994) Amphibians and reptiles (exclusive of marine turtles) of the Cayman Islands. In Brunt MA and Davies JE (eds) The Cayman Islands: natural history and biogeography. Kluwer Academic Publishers, The Netherlands 407-434
Seward NW, VerCauteren KC, Witmer GW and Engemann RM (2004) Feral swine impacts on agriculture and the environment. Sheep and Goat Research Journal 19: 34-40
Shine R (1991) Strangers in a strange land: ecology of Australian colubrid snakes. Copeia 1991: 120-131
Short J, Turner B, Risbey D (2002) Control of feral cats for nature conservation. Wildlife Research 29: 475-487
Simberloff D (1995) Why do introduced species appear to devastate island more than mainland areas? Pacific Science 49: 87-97
Simberloff, D (1996) Impacts of introduced species in the United States. Consequences 2: 13-24.
Simberloff D (2003) How much information on population biology is needed to manage introduced species. Conservation Biology 17, 83–92.
Simberloff D, Schmitz DC, Brown TC (1997) Strangers in Paradise: Impact and Management of Nonindigenous Species in Florida, Island Press, Washington D.C., United States
Simberloff D, Von Holle B (1999) Positive interactions of nonindigenous species: invasional meltdown? Biological Invasions 1: 21-32
Smith DG, Polhemus JT, VanderWerf EA (2000). Efficacy of fish-flavored diphacinone bait blocks for controlling Small Indian Mongooses (Herpestes auropunctatus) populations in Hawaii. 'Elepaio 60: 47-51
Speakman JR (1991) The impact of predation by birds on bat populations in the British Isles. Mammal Review 21: 123–142
Stone CP, Anderson SJ (1988) Introduced animals and Hawaii's natural area. Proceedings of the Vertebrate Pest Conference 13: 134-140
Stone CP, Dusek M, Aeder, M (1994) Use of an anticoagulant to control mongooses in Nene breeding habitat. 'Elepaio 54: 73-78
Sweeney JR, Sweeney JM and Sweeney SW (2003) Feral hog. In Feldhamer GA, Thompson BC and Chapman JA (eds). Wild Mammals of North America,

The John Hopkins University Press, Baltimore, Maryland, United States 1164-1179
Sugimura K, Yamada F, Miyamoto A. (2005) Population trend, habitat change and conservation of the unique wildlife species on Amami Island, Japan. Global Environmental Research 6: 79-89
Telfair RC (1994) Cattle egret. In Poole A, Stettenheim P, Gill F (eds) The Birds of North America. The Academy of Natural Sciences of Philadelphia, Philadelphia, United States 1-32
Tomich PQ (1986) Mammals of Hawaii. Second Edition, Bishop Museum Press, Honolulu, Hawaii, United States
Towns, DR, Atkinson IAE, Daugherty CH (1990) Ecological restoration of New Zealand islands. Papers presented at conference on ecological restoration of New Zealand islands 1998. Department of Conservation, Wellington, New Zealand
UH (University of Hawaii at Manoa) (2006) Coqui Frog Control for Homeowners. Miscellaneous Pests MP-5. University of Hawaii at Manoa, College of Tropical Agriculture and Human Resources, Office of Communication Services, Honolulu, Hawaii.
USFWS (United States Fish and Wildlife Service) (1999) Draft Revised Recovery Plan for Hawaiian Waterbirds, Second Revision. United States Fish and Wildlife Service, Portland, Oregon, 107 pp
Van't Woudt BD (1990) Roaming, stray, and feral domestic cats and dogs as wildlife problems. Proceedings of Vertebrate Pest Conference 14: 291-295.
Vargas J, Landaeta C, Simonetti JA (2002) Bats as prey of barn owls in a tropical savanna in Bolivia. Journal of Raptor Research 36: 146-148
Veitch CR (2001) The eradication of feral cats (Felis catus) from Little Barrier Island, New Zealand. New Zealand Journal of Zoology 28: 1-12
Veitch CR, Clout MN (eds) (2002) Turning the tide: the eradication of invasive species. Invasive Species Specialist Group, Species Survival Commission, World Conservation Union, Gland, Switzerland
Vice DS, Pitzler ME (2002) Brown treesnake control: economy of scales. Clark L (ed) Human conflicts with wildlife: economic considerations. National Wildlife Research Center, Fort Collins, Colorado, United States 127- 131
Vice DS, RM Engeman, Vice DL (2005) A comparison of three trap designs for capturing brown treesnakes on Guam. Wildlife Research 32: 355-359
Vilella FJ (1998) Biology of mongoose (Herpestes javanicus) in a rain forest in Puerto Rico. Biotropica 30: 120-125
Vitousek PM, D'Antonio CM, Loope LL, Westbrooks R (1996) Biological invasions as global environmental change. American Scientist 84: 468–47
Warner RD (1984) Occurrence and impact of zoonoses in pet dogs and cats at United States Air Force bases. American Journal of Public Health 74, 1239-1243
Wiewandt TA (1977) Ecology, behavior, and management of the Mona Island ground iguana, Cyclura stejnegeri. Ph.D. Dissertation, Cornell University, Ithaca, New York. United States

Whittaker RJ (1998) Island biogeography: ecology, evolution and conservation. Oxford University Press, Oxford, United Kingdom
Williamson I (1999) Competition between the larvae of the introduced cane toad Bufo
marinus (Anura : Bufonidae) and native anurans from the Darling Downs area of southern Queensland. Australian Journal of Ecology 24: 636-643
Witmer GW, Lewis JC (2001) Introduced wildlife of Oregon and Washington. In Johnson D, O"Neil T (eds) Wildlife-habitat relationships in Oregon and Washington. Oregon State University Press, Corvallis, Oregon, United States 423-443
Witmer GW, Sanders RB and Taft AC (2003) Feral swine---are they a disease threat to livestock in the United States? Proceedings of the Wildlife Damage Management Conference 10: 316-325
Witte F, Goldschmidt T, Goudswaard PC, Ligtvoet W, van Oijen MJP, Wanink, JH (1992) Species extinction and concomitant ecological changes in Lake Victoria. Netherlands Journal of Zoology 42: 214-232
Wood B, Tershy BR, Hermosillo MA, Donlan CJ, Sanchez JA, Keitt BS, Croll DA, Howald GR, Biavaschi N (2002) Removing cats from islands in north-west Mexico. Veitch C, Clout M (eds) Turning the tide: the eradication of invasive species. World Conservation Union (IUCN), Gland, Switzerland 374-380

13 Predator-prey interaction of Brazilian Cretaceous toothed pterosaurs: a case example

André J. Veldmeijer[1], Marco Signore[2] and Enrico Bucci[3]

[1]PalArch Foundation Amsterdam, Natural History Museum Rotterdam, Mezquitalaan 23, 1064 NS Amsterdam, veldmeijer@palarch.nl,
[2]Dipartimento di Scienze della Terra, Università degli Studi di Napoli "Federico II", L.go S.Marcellino 10, 80132 Napoli, Italy, normanno@marcosignore.it, [3]Istituto di Biostrutture e Bioimmagini del CNR, Via Mezzocannone, 16, I-80134 Naples, Italy, bucci@chemistry.unina.it

13.1 Abstract

This chapter presents an overview of evidence of predator-prey relationships in pterosaurs, with a focus on the Cretaceous (Santana Formation) pterosaurs from Chapado do Araripe, northeastern Brazil. The examples from the fossil record of pterosaurs as prey is scanty; the situation of pterosaurs as predators is not much better. However, especially for pterosaurs as predator, secondary evidence provides much insight in the life of these extinct predators. Here, we present a simple geometric model that proves the suggested way of predations of the toothed and crested taxa of the Anhangueridae.

Keywords: Pterosaur, *Anhanguera, Coloborhynchus, Ciorhynchus*, Santana Formation, predator, skimming.

13.2 Introduction

Our knowledge on pterosaurs, the first flying vertebrates, has much increased during the last 30 years. New sites, such as the Brazilian Chapada do Araripe and, more recently, the western part of the Liaoning Province in China, provided scientists with spectacularly preserved, often 3-dimensionally, largely complete specimens of predominantly Cretaceous age. Pterosaurs from earlier eras, especially from the Jura, originate mainly from the well-known Lagerstätte in Sollnhofen.

For years, the focus of research has been on taxonomy and it is only in the last few decades that more research is done on other topics, such as functional morphology (mainly concerning flying and, to a lesser extend, walking; for an overview of scientific research in pterosaurs see Veldmeijer, 2003). True palaeobiological studies however are few (Kellner, 1994 is the only dealing solely with this topic regarding Brazilian pterosaurs) and can be regarded as an area still to be explored in pterosaur palaeontology. Undoubtedly one of the reasons for this slow progress in research is that most pterosaurs, especially from Brazil, come to the scientist through commercial dealers; the lack of stratigraphic and taphonomic data hinders to a certain degree palaeoecological interpretation. Furthermore, mathematical models and experimental palaeontology is still not much used in pterosaur research: a biomechanical approach related to feeding behaviour is not well developed. The research conducted in these fields predominantly explore pterosaurs flight, although various studies are currently being undertaken beyond flight alone.

This chapter presents the current state of knowledge of one of these palaeobiological topics, viz. feeding, on the basis of the fossil record from Brazil (Chapada do Araripe; Lower Cretaceous Santana Formation) and offers a case example of secondary evidence of predator-prey interaction in the pterosaur fossil record in the form of a biomechanical calculation that proves previously proposed fishing techniques.

13.3 Pterosaurs as prey

In general, the evidence of pterosaurs as prey is rare, which might be due to taphonomic reasons: the skeleton of pterosaurs are extremely fragile, with bone wall often less than one mm thick (for instance Wellnhofer, 1985; Veldmeijer, 2003). This makes the chance of fossilization, a rare phenomenon in itself, even more of an extraordinary event. Only few examples of direct evidence are known to date, one of which comes from the site that is the focus of this chapter. An example of the basal pterosaur *Preondactylus buffarinii* from the Upper Triassic (Norian) of Rio Seazza Vally, Italy as prey has been reported by Dalla Vecchia et al (1989). The authors suggest that the remains of this pterosaur, found in a gastric pellet, were spewed up by a predatory fish. Another example comes from the Brazilian Santana Formation and is a series of three cervical vertebrae of an Early Cretaceous pterosaur with a predatory tooth still embedded (Buffetaut et al 2004). The authors have identified the tooth as one from a spinosaurid theropod dinosaur who has bitten the neck of the pterosaur.

According to the authors, the neck must have been fresh and articulated because the three cervical vertebrae are still associated. Furthermore (ibidem: 33) "the vertebrae remain in articulation and lack evidence of etching by gastric juices." However, the neck might have been bitten off after the reported sixth cervical with the spinosaurid tooth, upon which the predator swallowed the seventh (and perhaps subsequent) vertebrae, leaving the series vertebrae reported.

Some fossils show evidence of broken limbs, sometimes with clear evidence of infections (for instance Bennett 2003; Kellner & Tomida 2000; Tischlinger 1993; Wellnhofer 1970, 1991a), but it is not clear at all whether these fractures were caused by attack from a predator or not.

13.4 Pterosaurs as predator

Unequivocal evidence of predator-prey relationships, in which pterosaurs act as the predator, is even more rare than pterosaurs as prey. Most pterosaur fossils are found in deposits associated with water (see Unwin 2005 for an overview). This is true for most of the fossils from Chapada do Araripe. Although it is tempting to conclude from this evidence alone that these animals were piscivorous, it seems too easy to do so just because they were found there: the shallow and calm water bodies are the best places for fossilization due to the lack of currents and rapid burial rate (note that the exact stratigraphic position and taphonomic data from excavations might be of much help here). Also, morphological features might suggest a non-piscivorous habit as well. Research in progress seems to suggest that many, if not all toothless Tapejaridae were not piscivorous but frugivorous or even granivorous. Wellnhofer & Kellner (1991) already suggested for *Tapejara wellnhoferi* a frugivorous diet but Kellner & Campos suggested in their description of *Thallasodromeus sethi* (2002), a much larger member of Tapejaridae, that it skimmed the water for fish, not unlike the black skimmer (*Rynchops niger*), a bird with a sharp-edge lower jaw which "ploughs through the surface of the water during flight" (Wellnhofer 1991a: 160; see also Unwin 2005: 289).

However, other evidence clearly point to piscivorous diet for the toothed and crested taxa. The interlocking teeth at the expanded front part of the jaws, especially if they are as big as seen in *Coloborhynchus*, seem perfectly adapted to catch slippery prey such as fish and the smaller teeth more posteriorly to hold on and transport the fish further towards the throat. The teeth in the various toothed taxa from Brazil differ considerably (see Unwin 2001; but especially Veldmeijer 2006; Veldmeijer et al 2005, 2006).

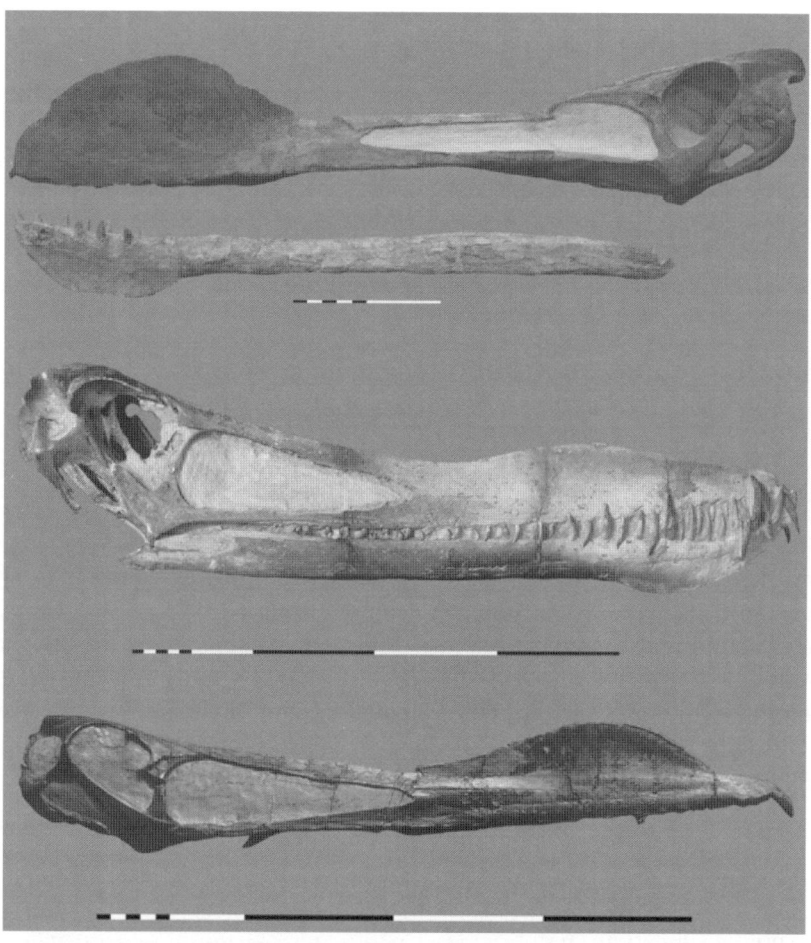

Fig. 1. Crested, tooted taxa from Chapada do Araripe. Top: Cranium and mandible of *Criorhynchus mesembrinus*, BSP 1987 I 46 (courtesy of the Bayerische Staatssammlung für Paläontologie und historische Geologie, Munich). Middle: Cranium and mandible of *Coloborhynchus piscator*, NSM-PV 19892 (courtesy of National Science Museum, Tokyo). Bottom: *Anhanguera blittersdorffi*, MN 4805-V (courtesy of Museu Nacional, Rio de Janeiro). Scalebars in cm. Photographs by E. Endenburg/A.J. Veldmeijer

Note also that the shape of the teeth is basically built in order to spear a slippery prey more than to cut or slash. The teeth in *Coloborhynchus* are fairly large, especially in the expanded part, whereas the teeth in

Anhanguera are smaller and more numerous in general. In both taxa there is a distinct difference in diameter of the various teeth, but in particular between those in the expanded front part and the more posterior part. In *Criorhynchus* the teeth are fairly large too but the difference in diameter is much smaller. Taken this together with the difference in morphology, including the position of the crest (starting at the very front in *Coloborhynchus* and *Criorhynchus* but more posteriorly in *Anhanguera*), the mandibular sagittal groove and the corresponding palatinal sagittal ridge (extremely big in *Criorhynchus* but much smaller or absent in the other taxa), the absence of the expansion of the front part (present in other taxa: robust in *Coloborhynchus* but much more smooth in *Anhanguera*) and the wingspan (about 3.5 m for *Anhanguera* and about 6 m for *Coloborhynchus*) it seems hard to believe they all obtained their food in the same way and moreover, fed on the same food. The described differences in morphology suggests otherwise. Would not it be more likely that the various taxa specialised in their food and/or in the techniques of getting it? This, however, does not necessarily mean there was not a certain overlap.

A feature not unimportant to consider is that the body of these pterosaurs is extremely small (Veldmeijer 2003), which means that there would not have been space for big fish. Furthermore, possibly their stomachs might not have been able to contain a large fish without substantially altering flight capabilities, as in modern bats (Altringham 1996). This would mean either that they hunted small fish and/or pre-digested fish before swallowing, the teeth not being used to chew. The second suggestion would need at least a system like cheeks in order to avoid the fall of the prey from the mouth during pre-digestion, or of gular sacs. Although throat pouches have been reported for some pterosaurs (for instance Wellnhofer 1978, 1991a), it is uncertain whether the Cretaceous toothed taxa from Brazil had pouches.

Another uncertain point is: were the pterosaurs able to correct the sudden jerk to the side of the body if the wing tip touched the water surface, or would they crash? The impact of the wing with water would have caused serious problems of stability as the leading surface is rigid (made of bone) and not plyable as the feathers of a bird, for instance. Of course, they could have soured over the water, but this almost certainly rule out skimming.

The lifestyle proposed for *Coloborhynchus* and all crested pterosaurs such as *Anhanguera*, (and *Criorhynchus* despite the above-mentioned morphological differences), is that of a fish predator, able to catch its pray by extracting it from the water without interrupting the flight proposed first by Wellnhofer (1987, 1991a, b). The theory, although generally accepted, has never been supported by mathematical models. Below, however, we will present a model, supporting Wellnhofer's theory.

Besides the fact that *Anhanguera* and *Coloborhynchus* remains are found in ancient coastal areas and in association with sea fauna (or, rarely in lake deposits, like the Mongolian specimen from the late Lower Cretaceous of Khuren-Dukh) morphological evidence supporting the hypothesis on fishing include the shape and disposition of the teeth, and, as we will see, the exaggerated proportions of the head in conjunction with the ability to bend the neck very deeply downward. Most of these characters are shared with the vast majority of advanced pterosaurs, so that the considerations made in the following paragraphs should apply, at a qualitative level, to all of them.

The specimen of *Anhanguera santanae* AMNH 22555, described by Wellnhofer (1991b) includes the complete neck. It is beyond the scope of the present work to discuss in detail this name (see Veldmeijer 2003 for details) but it is important to note the fact that he reconstructed the mandible with a dentary sagittal crest that does not terminate at the front. This was based on the assumption that, since the premaxillary sagittal crest was situated between the nasoantorbital fenestra and terminated well before the front of the snout, this would also be the case with the mandibular sagittal crest. However, recent finds of *Anhanguera* (referred specimen in Kellner & Tomida 2000; Veldmeijer et al 2006) included more complete mandibles, clearly showing that the dentary sagittal crest terminates at the tip of the mandible. The model presented below will focus on *Coloborhynchus*.

13.5 Fishing technique: the model

To see how the mentioned characteristics came into play when the animal was hunting for fish, let us consider a model situation exemplified in figure 2.

An adult *Coloborhynchus* localized a prey near the sea surface. To catch its prey, it is entering the sea surface with the tip of its snout at point P, with a sea current of speed C, at angle α and at a speed V_0 (Fig. 2). The best fishing strategy would be to evaluate the position of the fish with high accuracy before entering the water (a characteristic which may be related to the 'frontal' position of the eye and to adaptations in the brain, Witmer et al. 2003, note that Witmer et al mix up, in figure 4, c & d, the skulls of *Anhanguera santanae* AMNH 22555 with *Coloborhynchus piscator* [cf.

Fig. 1 of the present work]) after which to enter the water as near to the prey as possible.

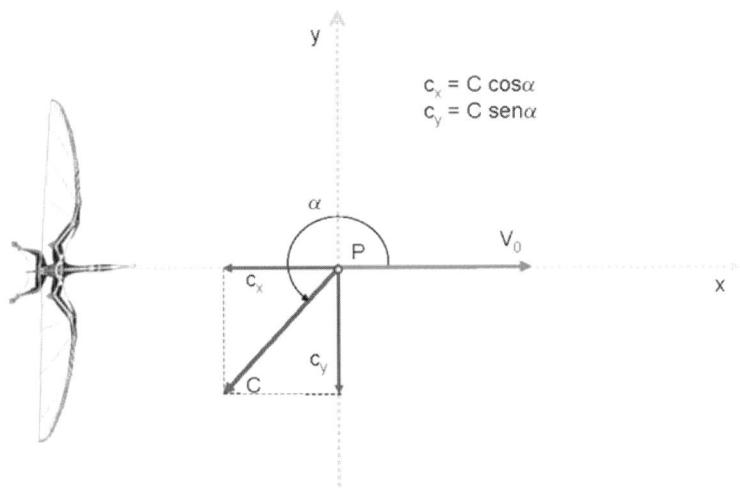

Fig. 2. Geometric model for studying the effect of the sea current on an adult *Coloborhynchus* when entering the water surface at point P to catch a prey. The animal is seen in dorsal view, flying in direction x at speed V_0. The sea current is represented by its speed vector C. C_x and C_y are two vectors resulting from the decomposition of C along the animal flying direction (C_x) and orthogonally to it (C_y), while α is the angle between the animal direction of flight x and the sea current C

This strategy would in fact minimize the prey's awareness of a danger by decreasing the time separation between the noisy entrance in the water and the attack to a minimum. We then assume that P is the position of the prey, or at least very near to it. For the sake of simplicity, we choose a Cartesian reference system having its x-axis aligned with the flight direction of the animal. The current C is not co-linear with the direction of flight, so that it can be decomposed in two orthogonal component vectors, C_x and C_y, along the direction x and y respectively, which have the dimension of a velocity.

Let us first consider the moment the pterosaur sinks the tip of its snout in the water at point P.

When an object is sinking, the pressure exerted by the water, which counteracts the sinking, is related to the shape of the object. The sinking profile of *Coloborhynchus* depends on the angle of sinking of the head δ at point P, as illustrated in figure 3, where the main body of the mouth is represented by a cylinder and the crests are represented by a cyan lamina. The smaller the angle (δ) between the head and the water surface, the bigger is the angle marked as γ (Fig. 3).

In an early National Advisory Committee for Aeronautics (NACA) paper Von Karman (1929) derived a formula for the maximum pressure on a wedge impacting water and depending on the angle γ. The formula was derived to help designers of sea planes and sea plane floats to calculate the stresses during landing. The formula for maximum pressure on a cylindrical wedge with an angle γ between the water and one plane of the wedge is given by:

$$P_{max} = \rho\, s_v^2\, \pi \text{ by cotangent of } \gamma \qquad (13.1)$$

where ρ is the water density, s_v is the sinking velocity, and γ is the angle between the water surface and the inferior surface of the sinking object, as illustrated in figure 3A.

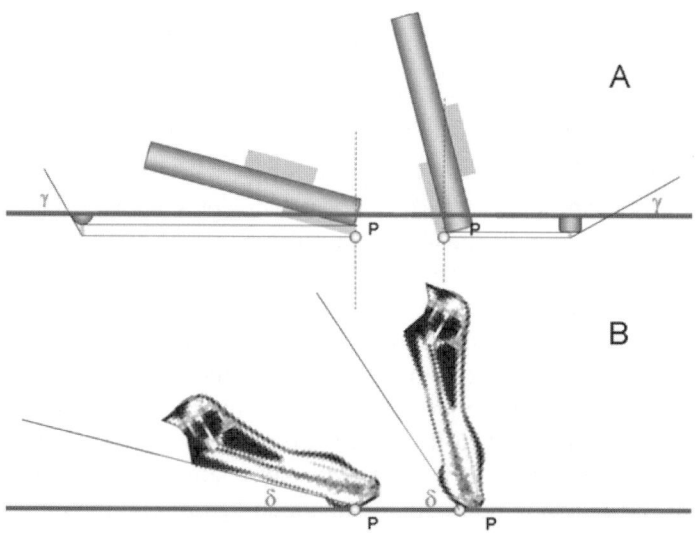

Fig. 3. Sinking of the *Coloborhynchus* head into water. Low δ angles are preferred to minimize water attrition, thanks to the dentary sagittal crest on the tip of the mandible

As evident from figure 3 and equation [1], a low δ, causing an increase in γ, is preferred to optimise the sinking. The fact that the mandible has a dentary sagittal crest terminating to the tip of the jaw supports a way of entering the water engaging the mandible first, consistent with low δ values, as opposing to the "high δ" way. This evidence support the idea that *Coloborhynchus* hunted by soaring over the sea surface, entering it low over the water, with the mandible as flat on the water surface as possible, as opposed to predators which hunt by falling on the prey from high elevations.

Let us consider now the effects of the current component C_x at point P (Fig. 2), once the tip of the skull is in the water. Since this component is co-linear with the pterosaur flight direction, it can effect the position of the skull, displacing it from point P along the x axis. In *Coloborhynchus*, however, this effect is reduced to a minimum by the two crests on the tip of the skull. In case C_x had the same direction of the animal flight, the dentary sagittal crest came again into play by reducing its effect; in the opposite case, the premaxillary sagittal crest, which was very soon submerged after the first impact with water at point P, exerted the same action. This is in accordance with the observations made by Wellnhofer (1991b). The presence of the crests, more than a secondary sexual signal, seems thus related to the hunting behaviour of *Coloborhynchus*, so that fossils missing it should be considered to be different species.

Once the snout sinks into the water, in order to keep the mouth on position P and to allow the grasping of the prey, the head of the pterosaur has to move relatively towards the body of the animal, which is still proceeding at a speed V_0 from point a to point b (Fig. 4A). This is achievable by bending the neck downward, as proposed by Wellnhofer (1991b) and shown in figure 4A.

The maximum time which the animal can spend on point P, i.e. the time to grasp the prey and to manipulate it as to ensure a firm grip ("grasping time"), coincides with the time spent by the animal to cover the distance ab, if ab is the maximal distance which can be covered without moving the tip of the snout from P. This time depends directly on the length of the distance ab and inversely on V_0. To keep V_0 high, i.e. to preserve the speed of the attack on P, and to maintain a long grasping time, ab must be as long as possible. The maximum length of ab depends on the length of the segment bP, i.e. the overall maximal spanning of the segment made by the neck and the head, and on the minimal angle β at which the neck could be bent downward, as exemplified in figure 4B. Thus, the evolutionary trend producing big, long heads and long, downward highly flexible necks in advanced pterosaurs can be explained as an adaptation to maintain a high speed attack, without compressing too much the grasping time.

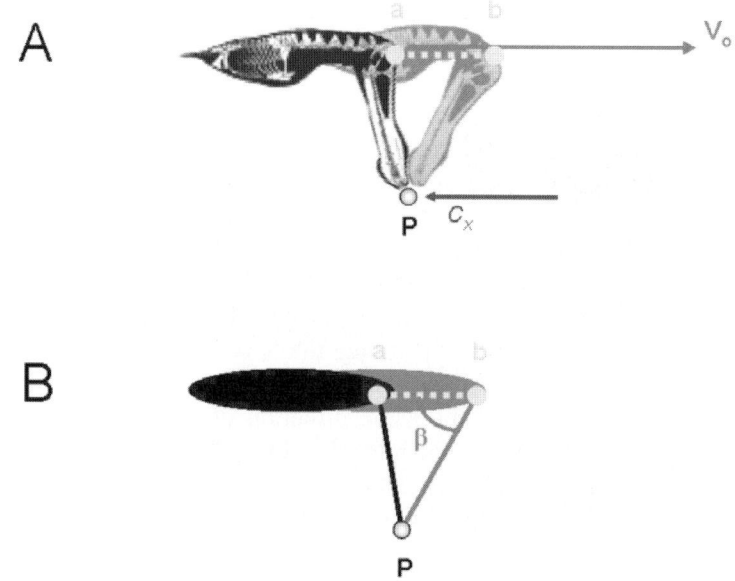

Fig. 4. The dorsoventral flexure of the neck of *Coloborhynchus* maintains the tip of the head on the prey at point P, increasing the "grasping time". This time depends directly on the length of ab and on β, being the maximal distance covered without moving the tip of the snout from P and the minimal angle at which the neck could be bent downward respectively

Let us consider now the effects of the speed and direction of the sea current at point P on the current component C_y (Fig. 2). The effects are reported in figure 5. As it might be expected, the orthogonal velocity C_y experimented by the animal increases as the current speed increases, and reaches its maximum when the current is orthogonal to the direction of the animal.

Since C_y is null before the animal arrives at point P, it is also directly proportional to a force applied on the head of the animal at a direction y, which tends to divert the head of the animal.

An animal able to balance a C_y of 5 m s^{-1} without being diverted would be able to hunt in the light grey region of the graph, i.e. would be able to arrive at point P at any angle with the current if the speed of the current is up to 5 m s^{-1} (18 Km h^{-1}), but would be nearly restricted to co-linearity with the current for speed of 10 m s^{-1} (36 Km h^{-1}).

To oppose C_y an animal can exert a balancing force by the wings, by the muscles of the neck or by the passive resistance of the skeleton.

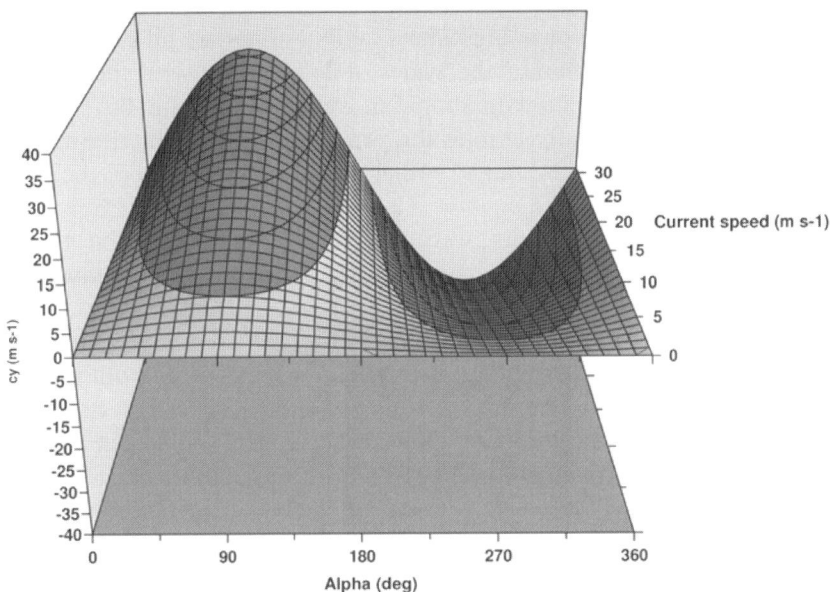

Fig. 5. Relation between the angle alpha (see Fig. 2), the sea current speed and the vector Cy (see Fig. 2). Light grey regions of the graph represent the combinations of current speed and angle alpha for which an animal able to balance a Cy component of 5 m s^{-1} could catch its prey without being diverted by the current; dark grey regions represent the combinations for which it could not

Contrary to the birds, the first solution is not achievable for large pterosaurs, since the space between the animal and the water surface would not be sufficient to manoeuvre the large wings (risk of crashing, see above). The second solution would require to increase the body mass, an obvious disadvantage for any flying animal. The third solution is indeed observed in most of the large pterosaurs, which had a neck skeleton virtually unable to bend laterally. Every cervical vertebra, in fact, is elongated, and tightly connected to the adjacent ones, which give rise to a very rigid neck backbone. Such a strong, rigid structure is ideal to support the force arising from the C_y orthogonal component of the sea current, increasing the "light grey region" represented in the graph of figure 5.

13.6 Conclusions

The idea of Wellnhofer of fishing techniques in crested Anhanguerid pterosaurs is supported by simple geometric considerations. Furthermore, it shows that the rigid, elongated neck, the elongated head structure, the presence of dentary and/or premaxillary sagittal crests are all characters in favour of an animal entering the water at low angles with the surface, penetrating the water strata with a rapid movement, snatching the prey and manipulating it, and finally exiting the water, all without interrupting the flight. As a consequence, we suggest that the crests on the tip of the snout of *Anhanguera, Coloborhynchus, Criorhynchus*, and all the other, comparable crested pterosaurs were used as useful adaptation to this hunting strategy, and, therefore, sexual display did not seem to have been the primary function.

In conclusion, we have demonstrated that the crested Anhanguerid pterosaurs *Coloborhynchus* and *Criorhynchus* were perfectly built to catch their prey in just one way: gliding or flying with their lower jaw very close to the water surface, and lowering their necks suddenly to snatch the fish. Although this way of fishing might be assumed for the other crested taxon, *Anhanguera*, as well because of the presence of crests, the exact consequences of the more posterior position of the premaxillary sagittal crest has yet to be determined.

13.7 Acknowledgements

We like to thank the two reviewers for valuable comments and suggestions, which improved the manuscript considerably.

The following persons are acknowledged for kindly allowing access to the collections under their care or for helping with visiting it (in alphabetical order): T. Bürgin, D.A. Campos, C. Collins, F. Dalla Vecchia, M. Dorling, E. Frey, J. Gamble, A. Keefer, A.W.A. Kellner, G. Mauricio, H. Mayr, A.C. Milner, S. Nabana, M.A. Norell, U. & S. Oberli, Y. Okazaki, M. Oshima, I. Rutzky, Y. Takakuwa, Y. Tomida, J. de Vos, P. Wellnhofer, R. Wild. Unfortunately, material in Berlin remained inaccessible.

The study of various collections by AJV (Germany, Rio de Janeiro and New York) has been made possible due to the financial support by the Jan Joost ter Pelkwijkfonds, Stichting Molengraaff Fonds, Mej. A.M Buitendijkfonds and Mr. & Mrs. Endenburg. The study of the material in various collections in Japan was made possibly by the Netherlands Organization for Scientific Research (NWO). Due to the grant of the Egypt Exploration Society for studying archaeological material in Cambridge, AJV was able to study some of the type specimens from the Cambridge

Greensands. The Natural History Museum Rotterdam is kindly thanked for their support.

References

Altringham JD (1996) Bats. Biology and behaviour. Oxford, Oxford University Press
Bennet SC (2003) A survey of pathologies of large pterodactyloid pterosaurs. Palaeontology 46: 185–198
Buffetaut E, Martill DM, Escuillié F (2004) Pterosaurs as part of a spinosaur diet. Nature 430: 33
Dalla Vecchia FM, Muscia G, Wild R (1989) Pterosaur remains in a gastric pellet from the Upper Triassic (Norian) of Rio Seazza Valley (Udine, Italy). Gortania 10: 121-132
Karman Von T (1929) The impact of seaplane floats during landing. NACA TN 321, Washington, DC
Kellner AWA (1994) Remarks on pterosaur taphonomy and paleoecology. Acta Geologica Leopoldensia 39, 1: 175-189
Kellner AWA, Campos DA (2002) The function of the cranial crest and jaws of a unique pterosaur from the Early Cretaceous of Brazil. Science 297: 389–392
Kellner AWA, Tomida Y (2000) Description of a new species of Anhangueridae (Pterodactyloidea) with comments on the pterosaur fauna from the Santana Formation (Aptian–Albian), northeastern Brazil. Tokyo, National Science Museum (National Science Museum Monographs 17)
Tischlinger, H (1993) Überlegungen zur Lebensweise der Pterosaurier anhand eines verheilten Obrschenkelbruches bei Pterodactylus kochi (Wagner). Archaeopteryx 11: 63-71
Unwin, DM (2001) An overview of the pterosaur assemblage from the Cambridge Geensand (Cretaceous) of eastern England. Mitteilungen Museum für Naturkunde Berlin, Geowissenchaftliche Reihe 4: 189–221
Unwin DM (2005) The pterosaurs from deep time. New York, PIPress
Veldmeijer AJ (2003) Description of *Coloborhynchus spielbergi* sp. nov. (Pterodactyloidea) from the Albian (Lower Cretaceous) of Brazil. Scripta Geologica 125: 35–139
Veldmeijer AJ (2006) Toothed pterosaurs from the Santana Formation (Cretaceous; Aptian-Albian) of northeastern Brazil. A reappraisal on the basis of newly described material. PhD Thesis University of Utrecht, The Netherlands (http://igitur-archive.library.uu.nl/dissertations/2006-0201-200610/UUindex.html)
Veldmeijer AJ, Signore M, Meijer HJM (2005) *Brasileodactylus* (Pterosauria, Pterodactyloidea, Anhangueridae); an update. Cranium 22, 1: 45–56
Veldmeijer AJ, Signore M, Meijer HJM. (2006). Description of two pterosaur (Pterodactyloidea) mandibles from the Lower Cretaceous Santana Formation, Brazil. Deinsea 11: 67-86.

Veldmeijer AJ, Meijer HJM & Signore M (2006) *Coloborhynchus* from the Lower Cretaceous Santana Formation, Brazil (Pterosauria, Pterodactyloidea, Anhangueridae); an update. PalArch's Journal of Vertebrate Palaeontology 3, 2: 15-29

Wellnhofer P (1970) Die Pterodactyloidea (Pterosauria) der Oberjura–Plattenkalke Süddeutschlands. Munich, Verlag der Bayerischen Akademie der Wissenschaften (Bayerischen Akademie der Wissenschaften, Mathematisch–Naturwissenschaftliche Klasse, Abhandlungen, Neue Folge 141)

Wellnhofer P (1978) Handbuch der Paläoherpetologie. Teil 19. Pterosauria. Stuttgart/New York, Gustav Fischer Verlag.

Wellnhofer P (1985) Neue Pterosaurier aus der Santana–Formation (Apt.) der Chapada do Araripe, Brasilien. Palaeontographica A 187: 105–182

Wellnhofer P (1987) New crested pterosaurs from the Lower Cretaceous of Brazil. Mitteilungen der Bayerische Staatsammlung für Paläontologie und historische Geologie 27: 175–186

Wellnhofer P (1991a) The illustrated encyclopedia of pterosaurs. New York, Crescent Books

Wellnhofer P (1991b) Weitere Pterosaurierfunde aus der Santana–Formation (Apt) der Chapada do Araripe, Brasilien. Palaeontographica A 215: 43–101

Wellnhofer P, Kellner AWA (1991) The skull of Tapejara wellnhoferi Kellner (Reptilia, Pterosauria) from the Lower Cretaceous Santana Formation of the Araripe Basin, northeastern Brazil. Mitteilungen der Bayerische Staatsammlung für Paläontologie und historische Geologie 31: 89–106

Witmer LM, Chatterjee S, Franzosa J, Rowe T. (2003) Neuroanatomy of flying reptiles and implications for flight, posture and behavious. Nature 425: 950-953

Index

"Obligatory Predation", 21
"Selective Predation", 21

Aeglidae, 59, 60, 63, 66, 79
Africa, 7, 8, 9, 10, 11, 14, 18, 19, 21
amphibian, 278, 282, 285
Amphibian, 266, 279
Anhangueridae, 295
Antagonism, 1
antipredator, 224
antipredator trait, 177, 180, 181, 182, 190
Antipredator trait, 185
aposematism, 179, 186
Aposematism, 185
attack deterrence, 182, 185, 190, 194

bird, 266, 267, 268, 269, 270, 271, 272, 273, 274, 276, 277, 278, 279, 280
Bird, 275
Bradoriida, 50
Brazil, 45, 295, 296, 297, 299
brown treesnakes, 265, 267
Brown treesnakes, 277, 280
bullfrog, 268
Bullfrog, 279

Cambrian, 50
Candonidae, 41
cannibalism, 4, 151, 152, 153, 154, 156, 158, 159, 162, 164, 165, 166
Cannibalism, 151, 152, 153
capture deterrence, 182, 185, 187, 190, 193, 199
carrying capacity, 221, 223, 251
cats, 265, 267, 268, 269, 270, 271
Chapada do Araripe, 295, 296, 297, 298
chemoreceptors, 49
Circadian rhythms, 72
consumption deterrence, 182, 185, 189, 190
coyote, 229, 230, 231, 232, 234, 235, 237, 239, 240, 243, 244, 245, 247
Coyote, 249
Cretaceous, 45, 151, 153, 155, 157, 158, 159, 160, 161, 163, 166, 295, 296, 299, 300

crows, 234, 235
Crustacea, 80

Decapoda, 60, 61, 63, 67, 69, 76
deer, 237, 238, 239, 241, 242, 243, 244, 245
dentary sagittal crest, 300, 302

echinoids, 27
egret, 268, 275, 276
Egypt, 27, 28, 36, 37
elk, 224, 225, 241
England, 48

fish, 265, 266, 268, 273, 274, 275, 278, 282, 285
Fish, 281
fox, 233, 234, 235, 240, 241, 246, 247, 248, 249, 250, 273

Gastropod drillholes, 28
grouse, 232, 234, 237, 241, 246, 247

insects, 87, 88, 89, 91, 94, 95, 96, 97, 98, 99, 104, 105, 107, 108, 109, 110, 111, 112, 113, 114, 115, 116

Jurassic, 39

less patient predators, 28
lizard, 272, 274, 275, 277, 278
Lizard, 276

marine fungi, 27
Mediterranean, 124
Mersa Matruh, 27, 28, 33
Middle East, 7, 8, 9, 10, 18, 19, 21
Miocene, 27, 28, 29, 33, 34
mongoose, 265, 268, 269, 272, 280
Mongoose, 271
mosquito, 123, 124

Naticid drillholes, 28
naticid gastropods, 151, 152, 157, 160, 161, 162, 164, 165, 166

Ordovician, 40
Ostracoda, 27, 50, 52, 53
owl, 268, 276, 278

Paleocene-Eocene, 7, 9, 20, 21
parasitism, 4, 39, 43, 48
Parasitism, 40, 43, 44
perch, 265, 275, 281, 282
pest control, 123, 139
PETM, 7, 9, 11, 16, 17, 18, 20, 21
pheasant, 234, 237, 238, 247
pig, 268, 273
Pleistocene, 151, 152, 154, 155, 158, 161, 163, 165, 166
predation, 3, 28, 31, 33, 34, 39, 40, 41, 44, 87, 88
Predation, 1, 2, 28, 40, 41, 53, 87, 89
predator avoidance, 184, 192, 194
Predator avoidance, 182
predator-prey interaction, 171, 173, 175
provinciality, 7
pterosaur, 295, 296, 299, 301, 303, 305
Pterosaur, 295
python, 268, 277

rat, 265, 267, 269, 271, 276, 280
raven, 233, 234, 235, 237, 246, 247, 248, 249
Raven, 241
Russia, 46, 47, 52

Santana Formation, 295, 296
Scavenging, 4
sheep, 225, 226, 228, 231, 232, 237, 238, 241, 245, 246
Silurian, 49, 51, 53

skunk, 234, 235, 240, 247, 248, 250
Solution holes, 28
South Africa, 50, 51, 53
Spitzbergen, 46, 53
stoat, 272
Stoat, 272
survival, 28
swarming, 41, 45, 51
Swarming, 47

Tapejaridae, 297
toad, 268, 275, 279

trophic, 59, 60, 62, 63, 64, 65, 66, 67, 70, 73, 74, 75, 77, 85
Trophic, 59, 60, 76, 85
turkey, 234, 235

United States, 151

waterfowl, 222, 226, 235, 236
Waterfowl, 235, 247
Western Desert, 27, 28, 36
wolf, 225
Wolf, 224

Printing: Krips bv, Meppel
Binding: Stürtz, Würzburg